Step-by-Step Business Math and Statistics

Jin W. Choi
DePaul University

First published in the United States of America in 2011 by University Readers, Inc.

15 14 13 12 11   1 2 3 4 5

Printed in the United States of America

ISBN: 978-1-60927-872-4

www.cognella.com 800.200.3908

Contents

Acknowledgments

I would like to thank many professors who had used this book in their classes. Especially, Professors Bala Batavia, Burhan Biner, Seth Epstein, Teresa Klier, Jin Man Lee, Norman Rosenstein, and Cemel Selcuk had used previous editions of this book in teaching GSB420 Applied Quantitative Analysis at DePaul University. Their comments and feedbacks were very useful in making this edition more user-friendly.

Also, I would like to thank many current and past DePaul University's Kellstadt Graduate School of Business MBA students who studied business mathematics and statistics using the framework laid out in this book. Their comments and feedbacks were equally important and useful in making this book an excellent guide into the often-challenging fields of mathematics and statistics. I hope and wish that the knowledge gained via this book would help them succeed in their business endeavors.

As is often the case with equations and numbers, I am sure this book still has some errors. If you find some, please let me know at jchoi@depaul.edu.

Best wishes to those who use this book.

Jin W. Choi, Ph.D.
Kellstadt Graduate School of Business
DePaul University
Chicago, IL 60604
jchoi@depaul.edu

Part 1. Business Mathematics

There are 4 chapters in this part of business mathematics: Algebra review, calculus review, optimization techniques, and economic applications of algebra and calculus.

Chapter 1. Algebra Review

A. The Number System

The number system is comprised of real numbers and imaginary numbers. Real numbers are, in turn, grouped into natural numbers, integers, rational numbers, and irrational numbers.

1. Real Numbers = numbers that we encounter everyday during a normal course of life → the numbers that are **real** to us.

   i. Natural numbers = the numbers that we often use to count items → counting trees, apples, bananas, etc.: 1, 2, 3, 4, …

      a. odd numbers: 1, 3, 5, …

      b. even numbers: 2, 4, 6, …

   ii. Integers = whole numbers without a decimal point: 0, ± 1, ± 2, ± 3, ± 4, ….

      a. positive integers: 1, 2, 3, 4, …

      b. negative integers: –1, –2, –3, –4, …

   iii. Rational numbers = numbers that can be expressed as a fraction of integers such as a/b (= a÷b) where both a and b are integers

      a. finite decimal fractions: 1/2, 2/5, etc.

      b. (recurring or periodic) infinite decimal fractions: 1/3, 2/9, etc.

   iv. Irrational Numbers = numbers that can NOT be expressed as a fraction of integers = nonrecurring infinite decimal fractions:

      a. n-th roots such as $\sqrt{2}, \sqrt[3]{5}, 7^{\sqrt{3}}$, etc.

      b. special values such as π (=pi), or e (=exponential), etc.

v. Undefined fractions:

a. any number that is divided by a zero such as k/0 where k is any number

b. a zero divided by a zero = 0/0

c. an infinity divided by an infinity = $\frac{\infty}{\infty}$

d. a zero divided by an infinity = $\frac{0}{\infty}$

vi. Defined fractions:

a. a one divided by a very small number →
$\frac{1}{0.0000000001} = \frac{1}{10^{-10}} = 10^{10} = 10{,}000{,}000{,}000 \approx$ a very large number such as a number that can approach ∞

b. a one divided by a very large number →
1/(a large number) = a small number → $\frac{1}{\infty} \approx 0$

c. a scientific notion → the use of exponent

$2.345E{+}2 = 2.345 \times 10^2 = 234.5$

$2.345E{+}6 = 2.345 \times 10^6 = 2{,}345{,}000$

$2.345E{-}2 = 2.345 \times 10^{-2} = 2.345 \cdot \frac{1}{10^2} = 2.345 \cdot \frac{1}{100} = 0.02345$

$2.345E{-}6 = 2.345 \times 10^{-6} =$

$2.345 \cdot \frac{1}{10^6} = 2.345 \cdot \frac{1}{1{,}000{,}000} = 0.000002345$

Similarly, a caret (^) can be used as a sign for an exponent:

X^n = X^n → X^{10} = X^10

Note: **For example, E+6 means move the decimal point 6 digits to the right of the original decimal point whereas E-6 means move the decimal point 6 digits to the left of the original decimal point.**

2. Imaginary Numbers = numbers that are not easily encountered and recognized on a normal course of life and thus, not real enough (or imaginary) to an individual.
→ Often exists as a mathematical conception.

$$i = \sqrt{-1} \qquad \sqrt{-2} = \sqrt{2}i = i\sqrt{2}$$

$$\sqrt{-4} = 2i \qquad (5i)^2 = -25$$

B. Rules of Algebra

1.	$a + b = b + a$	→	$2 + 3 = 3 + 2$	→	5
2.	$ab = ba$	→	$2 \text{ x } 3 = 3 \text{ x } 2$	→	6
3.	$aa^{-1} = 1$ for $a \neq 0$	→	$2 \text{ x } 2^{-1} = 2^0 = 1$		
4.	$a(b + c) = ab + ac$	→	$2 \text{ x } (3 + 4) = 2 \text{ x } 3 + 2 \text{ x } 4$	→	14
5.	$a + (-a) = a - (+a) = 0$	→	$2 + (-2) = 2 - (+2) = 2 - 2 = 0$		
6.	$(-a)b = a(-b) = -ab$	→	$(-2) \text{ x } 3 = 2 \text{ x } (-3)$	→	–6
7.	$(-a)(-b) = ab$	→	$(-2) \text{ x } (-3) = 2 \text{ x } 3$	→	6
8.	$(a + b)^2 = a^2 + 2ab + b^2$	→	$(2 + 3)^2 = 2^2 + 2(2)(3) + 3^2$	→	25
9.	$(a - b)^2 = a^2 - 2ab + b^2$	→	$(2 - 3)^2 = 2^2 - 2(2)(3) + 3^2$	→	1
10.	$(a + b)(a - b) = a^2 - b^2$	→	$(2 + 3)(2 - 3) = 2^2 - 3^2$	→	–5
11.	$\frac{-a}{-b} = (-a)/(-b) = a/b$	→	$\frac{-2}{-3} = (-2)/(-3) = 2/3$	→	$\frac{2}{3}$
12.	$\frac{-a}{b} = \frac{a}{-b} = -1 \cdot \frac{a}{b} = -\frac{a}{b}$	→	$\frac{-2}{3} = (2)/(-3) = \frac{2}{-3}$	→	$-\frac{2}{3}$
13.	$a + \frac{b}{c} = \frac{ac + b}{c}$	→	$2 + \frac{3}{4} = \frac{2 \cdot 4 + 3}{4}$	→	$\frac{11}{4}$
14.	$\frac{a}{b} + \frac{c}{d} = \frac{ad + bc}{bd}$	→	$\frac{2}{3} + \frac{4}{5} = \frac{2 \cdot 5 + 3 \cdot 4}{3 \cdot 5}$	→	$\frac{22}{15}$

15. $a \times \frac{b}{c} = a \cdot \frac{b}{c} = \frac{ab}{c}$ → $2 \times \frac{3}{4} = 2 \cdot \frac{3}{4} = \frac{2 \cdot 3}{4}$ → $\frac{6}{4}$

16. $\frac{\frac{a}{b}}{\frac{c}{d}} = \frac{a}{b} \div \frac{c}{d} = \frac{a}{b} \times \frac{d}{c} = \frac{ad}{bc}$ → $\frac{\frac{2}{3}}{\frac{4}{5}} = \frac{2}{3} \div \frac{4}{5} = \frac{2}{3} \times \frac{5}{4} = \frac{2 \cdot 5}{3 \cdot 4}$ → $\frac{10}{12}$

17. $a^{1/2} = a^{0.5} = \sqrt{a}$ where $a \geq 0$ → $2^{1/2} = 2^{0.5} = \sqrt{2}$ → 1.4142

18. $a^{1/n} = \sqrt[n]{a}$ where $a \geq 0$ → $2^{1/3} = \sqrt[3]{2}$ → 1.2599

19. $\sqrt{ab} = \sqrt{a} * \sqrt{b}$ → $\sqrt{2 \cdot 3} = \sqrt{2} * \sqrt{3} = \sqrt{6}$ → 2.4495

20. $\sqrt{\frac{a}{b}} = \frac{\sqrt{a}}{\sqrt{b}}$ → $\sqrt{\frac{2}{3}} = \frac{\sqrt{2}}{\sqrt{3}}$ → 0.8165

21. $\sqrt{a+b} \neq \sqrt{a} + \sqrt{b}$ → $\sqrt{2+3} \neq \sqrt{2} + \sqrt{3}$

22. $\frac{\sqrt{a}}{b} \neq \sqrt{\frac{a}{b}}$ → $\frac{\sqrt{2}}{3} \neq \sqrt{\frac{2}{3}}$

C. Properties of Exponents → Pay attention to equivalent notations

It is very important that we know the following properties of exponents:

1. $X^0 = 1$ → Note that 0^0 = undefined

2. $\frac{1}{X^b} = X^{-b} = X\wedge(-b)$ → $\frac{1}{X^{10}} = X^{-10} = X\wedge(-10)$

3. $X^a * X^b = X^a \cdot X^b = X^a X^b = X^{a+b}$ → $X\wedge(a+b)$

→ $X^2 * X^3 = X^2 \cdot X^3 = X^2 X^3 = X^{2+3} = X^5$

→ $2^3 \cdot 2^4 = 2^{3+4} = 2^7$ = 128

4. $(X^a)^b = X^{a*b} = X^{a \cdot b} = X^{ab}$ → $X\wedge ab$

→ $(X^2)^3 = X^{2*3} = X^{2 \cdot 3} = X^6$

$\rightarrow (2^3)^4 = 2^{3 \cdot 4} = 2^{12} = 4096$

5. $\dfrac{X^a}{X^b} = X^a \cdot X^{-b} = X^{a-b}$

$\rightarrow \dfrac{X^2}{X^3} = X^2 \cdot X^{-3} = X^{2-3} = X^{-1} = \dfrac{1}{X}$

6. $(XY)^a = X^a * Y^a = X^a \cdot Y^a = X^a Y^a$

$\rightarrow (XY)^2 = X^2 * Y^2 = X^2 \cdot Y^2 = X^2 Y^2$

7. $\sqrt[n]{X} = X^{\frac{1}{n}} = X^{1/n}$

$\rightarrow \sqrt{4} = 4^{\frac{1}{2}} = 4^{1/2} = 4^{0.5} = (2^2)^{0.5} = 2^{2 \cdot 0.5} = 2^1 = 2$

8. $X^{p/q} = (X^{1/q})^p = (X^p)^{1/q} = \sqrt[q]{X^p}$

$\rightarrow 2^{10/5} = 2^2 = (2^{1/5})^{10} = (2^{10})^{1/5} = \sqrt[5]{2^{10}} = 2^2 = 4$

$\rightarrow 8^{2/3} = (2^3)^{2/3} = 2^2 = \sqrt[3]{8^2} = \sqrt[3]{64} = 4$

D. Linear and Nonlinear Functions

1. Linear Functions

Linear Functions have the general form of:

$$Y = a + b X$$

where Y and X are variables and a and b are constants. More specifically, a is called an intercept and b, a slope coefficient. The most visually distinguishable character of a linear function is that it is a straight line. Note that +b means a positive slope and –b means a negative slope.

2. Nonlinear Functions

There are many different types of nonlinear functions such as polynomial, exponential, logarithmic, trigonometric functions, etc. Only polynomial, exponential and logarithmic functions will be briefly explained below.

i) The n-th degree polynomial functions have the following general form:

$$Y = a + bX + cX^2 + dX^3 + ... + pX^{n-1} + qX^n$$

Or alternatively expressed as:

$$Y = qX^n + pX^{n-1} + ... + dX^3 + cX^2 + bX + a$$

where a, b, c, d, …, p and q are all constant numbers called coefficients and n is the largest exponent value.

Note that the n-th degree polynomial function is named after the highest value of n. For example, when n = 2, it is most often called a quadratic function, instead of a second-degree polynomial function, and has the following form:

$$Y = a + bX + cX^2$$

When n = 3, it is called a third-degree polynomial function or a cubic function and has the following form:

$$Y = a + bX + cX^2 + dX^3$$

ii) Finding the Roots of a Polynomial Function

Often, it is important and necessary to find roots of a polynomial function, which can be a challenging task. An n-th degree polynomial function will have n roots. Thus, a third degree polynomial function will have 3 roots and a quadratic function, two roots. These roots need not be always different and in fact, can have the same value. Even though finding roots to higher-degree polynomial functions is difficult, the task of finding the roots of a quadratic equation is manageable if one relies on either the factoring method or the quadratic formula.

If we are to find the roots to a quadratic function of:

$$aX^2 + bX + c = 0$$

we can find their two roots by using the following quadratic formula:

$$X_1, X_2 = \frac{-b \pm \sqrt{b^2 - 4ac}}{2a}$$

iii) Examples:

Find the roots, X_1 and X_2, of the following quadratic equations:

(a) $X^2 - 3X + 2 = 0$

Factoring Method[1]:

$$X^2 - 3X + 2 = (X-1)\cdot(X-2) = 0$$

Therefore, we find two roots as: $X_1 = 1$ and $X_2 = 2$.

Quadratic Formula[2]:

Note: $a = 1$, $b = -3$, and $c = 2$

$$X_1, X_2 = \frac{-b \pm \sqrt{b^2 - 4ac}}{2a} = \frac{-(-3) \pm \sqrt{(-3)^2 - 4(1)(2)}}{2 \cdot 1}$$

$$= \frac{3 \pm \sqrt{9-8}}{2} = \frac{3 \pm 1}{2} = 1, 2$$

(b) $4X^2 + 24X + 36 = 0$

Factoring Method:

$$4X^2 + 24X + 36 = (2X+6)\cdot(2X+6) = (2X+6)^2 = 4(X+3)^2 = 0$$

Therefore, we find two identical roots (or double roots) as:

$$X_1 = X_2 = -3$$

Quadratic Formula:

Note: $a = 4$, $b = 24$, and $c = 36$

$$X_1, X_2 = \frac{-b \pm \sqrt{b^2 - 4ac}}{2a} = \frac{-(24) \pm \sqrt{(24)^2 - 4(4)(36)}}{2 \cdot 4}$$

---

[1] The factoring method often seems more convenient for people with great experience with algebra. That is, the easiness comes with experience. Those who lack algebraic skill may be better off using the quadratic formula.

[2] In order to use the quadratic formula successfully, one must match up the values for a, b, and c correctly.

$$= \frac{-24 \pm \sqrt{576-576}}{8} = \frac{-24 \pm 0}{8} = -\frac{24}{8} = -3$$

(c) $4X^2 - 9Y^2 = 0$

Factoring Method:

$$4X^2 - 9Y^2 = (2X - 3Y) \cdot (2X + 3Y) = 0$$

Therefore, we find two roots as:

$$X_1 = \frac{3Y}{2} = 1.5Y \text{ and } X_2 = -\frac{3Y}{2} = -1.5Y$$

Quadratic Formula[3]:

Note: $a = 4$, $b = 0$, and $c = -9Y^2$

$$X_1, X_2 = \frac{-b \pm \sqrt{b^2 - 4ac}}{2a} = \frac{-(0) \pm \sqrt{(0)^2 - 4(4)(-9Y^2)}}{2 \cdot 4}$$

$$= \frac{0 \pm \sqrt{0 + 144Y^2}}{8} = \frac{\pm 12Y}{8} = \pm\frac{3}{2}Y = 1.5Y, -1.5Y$$

E. Exponential and Logarithmic Functions

1. Exponential Functions

An exponential function has the form of $Y = a \cdot b^X$ where a and b are constant numbers. The simplest form of an exponential function is $Y = b^X$ where b is called the base and X is called an exponent or a growth factor.

A unique case of an exponential function is observed when the base of e is used. That is, $Y = e^X$ where $e \approx 2.718281828$. Because this value of e is often identified with natural phenomena, it is called the "natural" base[4].

---

[3] One must be very cognizant of the construct of this quadratic equation. Because we are to find the roots associated with X, $-9Y^2$ should be considered as a constant term, like c in the quadratic equation.

[4] Technically, the expression $\left(1 + \frac{1}{n}\right)^n$ approaches e as n increases. That is, as n approaches $+\infty$, $e \approx 2.718281828$.

Examples>

In order to be familiar with how exponential functions work, please verify the following equalities by using a calculator.

a. $5e^2 \cdot e^4 = 5e^{2+4} = 5e^6 = 5 \cdot 403.4287935 = 2017.143967$

b. $(5e^3) \cdot (3e^4) = 15e^7 = 15 \cdot 1096.633158 = 16449.49738$

c. $10e^3 \div 2e^4 = \frac{10}{2} \cdot e^{3-4} = 5e^{-1} = \frac{5}{e} = \frac{5}{2.718281828} = 1.839397206$

2. Logarithmic Functions

The logarithm of Y with base b is denoted as "$\log_b Y$" and is defined as:

$\log_b Y = X$ if and only if $b^X = Y$

provided that b and Y are positive numbers with $b \neq 1$. The logarithm enables one to find the value of X given $2^X = 4$ or $5^X = 25$. In both of these cases, we can easily find X=2 due to the simple squaring process involved. However, finding X in $2^X = 5$ is not easy. This is when knowing a logarithm comes in handy.

Examples>

Convert the following logarithmic functions into exponential functions:

$\log_2 8 = X$ → $2^X = 8$ → $X = 3$

$\log_5 1 = 0$ → $5^0 = 1$

$\log_4 4 = 1$ → $4^1 = 4$

$\log_{1/2} 4 = -2$ → $\left(\frac{1}{2}\right)^{-2} = \left(2^{-1}\right)^{-2} = 2^{+2} = 2^2 = 4$

a. Special Logarithms: A common logarithm and a natural logarithm.

i) A Common Logarithm = a logarithm with base 10 and often denoted without the base value.

That is, $\log_{10} X = \log X$ → read as "a (common) logarithm of X."

ii) A Natural Logarithm = a logarithm with base e and often denoted as 'ln".

That is, $\log_e X = \ln X$ → read as "a natural logarithm of X."

b. Properties of Logarithms

i) Product Property: $\log_b mn = \log_b m + \log_b n$

ii) Quotient Property: $\log_b \frac{m}{n} = \log_b m - \log_b n$

iii) Power Property: $\log_b m^n = n \cdot \log_b m$

Example 1> Using the above 3 properties of logarithm, verify the following equality or inequality by using a calculator.

i) $\ln 30 = \ln(5 \cdot 6) = \ln 5 + \ln 6$ → 3.401197

ii) $\ln \frac{20}{40} = \ln 20 - \ln 40 = \ln 0.5$ → −0.693147

iii) $\frac{\ln 20}{\ln 40} \neq \ln \frac{20}{40}$

iv) $\ln 10^3 = 3 \cdot \ln 10 = \ln 1000$ → 6.907755

Example 2> Find X in $2^X = 5$. (This solution method is a bit advanced.)

In order to find X,

(1) we can take a natural (or common) logarithm of both sides as:
$\ln 2^X = \ln 5$

(2) rewrite the above as: $X \cdot \ln 2 = \ln 5$ by using the Power Property

(3) solve for X as: $X = \frac{\ln 5}{\ln 2}$

(4) use the calculator to find the value of X as:

$$X = \frac{\ln 5}{\ln 2} = \frac{1.6094379}{0.6931471} = 2.321928095$$

Additional topics of exponential and logarithmic functions are complicated and require many additional hours of study. Because it is beyond our realm, no additional attempt to explore this topic is made herein[5].

F. Useful Mathematical Operators

1. Summation Operator = Sigma = Σ → $\sum_{i=1}^{n} X_i = \sum_{i=1}^{n} X_i = \sum X_i$

$$\sum_{i=1}^{n} X_i = X_1 + X_2 + \ldots.. + X_{n-1} + X_n = \text{Sum } X_i\text{'s where i goes from 1 to n.}$$

Examples: Given the following X data, verify the summation operation.

i =	1	2	3	4	5
X_i =	25	19	6	27	23

a. $\sum_{i=1}^{3} X_i = X_1 + X_2 + X_3 = 25 + 19 + 6 = 50$

b. $\sum_{i=1}^{5} X_i = X_1 + X_2 + X_3 + X_4 + X_5 = 25 + 19 + 6 + 27 + 23 = 100$

c. $\sum_{i=3}^{5} X_i = X_3 + X_4 + X_5 = 6 + 27 + 23 = 56$

d. $\sum_{i=1}^{3} X_i + \sum_{i=3}^{5} X_i = (X_1 + X_2 + X_3) + (X_3 + X_4 + X_5)$
$= (25 + 19 + 6) + (6 + 27 + 23) = 50 + 56 = 106$

e. $\sum_{i=1}^{3} X_i - \sum_{i=3}^{5} X_i = (X_1 + X_2 + X_3) - (X_3 + X_4 + X_5)$
$= (25 + 19 + 6) - (6 + 27 + 23) = 50 - 56 = -6$

2. Multiplication Operator = pi = Π → $\prod_{i=1}^{n} X_i = \prod X_i$

$$\prod_{i=1}^{n} X_i = X_1 \cdot X_2 \cdot \ldots. \cdot X_{n-1} \cdot X_n = \text{Multiply } X_i\text{'s where i goes from 1 to n.}$$

[5] For detailed discussions and examples on this topic, please consult high school algebra books such as Algebra 2, by Larson, Boswell, Kanold, and Stiff. ISBN=13:978-0-618-59541-9.

Examples: Given the following X data, verify the multiplication operation.

i =	1	2	3	4	5
X_i =	3	5	6	2	4

a.

$$\prod_{i=1}^{3} X_i = X_1 \cdot X_2 \cdot X_3 = 3 \cdot 5 \cdot 6 = 90$$

b. $$\prod_{i=1}^{5} X_i = X_1 \cdot X_2 \cdot X_3 \cdot X_4 \cdot X_5 = 3 \cdot 5 \cdot 6 \cdot 2 \cdot 4 = 720$$

c. $$\prod_{i=3}^{5} X_i = X_3 \cdot X_4 \cdot X_5 = 6 \cdot 2 \cdot 4 = 48$$

d. $$\prod_{i=1}^{3} X_i + \prod_{i=3}^{5} X_i = (X_1 \cdot X_2 \cdot X_3) + (X_3 \cdot X_4 \cdot X_5)$$
$$= (3 \cdot 5 \cdot 6) + (6 \cdot 2 \cdot 4) = 90 + 48 = 138$$

e. $$\prod_{i=1}^{3} X_i - \prod_{i=4}^{5} X_i = (X_1 \cdot X_2 \cdot X_3) - (X_4 \cdot X_5)$$
$$= (3 \cdot 5 \cdot 6) - (2 \cdot 4) = 90 - 8 = 72$$

f. $$\sum_{i=1}^{2} X_i - \prod_{3}^{5} X_i = (X_1 + X_2) - (X_3 \cdot X_4 \cdot X_5)$$
$$= (3 + 5) - (6 \cdot 2 \cdot 4) = 8 - 48 = 40$$

G. Multiple-Choice Problems for Exponents, Logarithms, and Mathematical Operators:

Identify all equivalent mathematical expressions as correct answers.

1. $(X + Y)^2 =$

a. $X^2 + 2XY + Y^2$ b. $X^2 - 2XY + Y^2$

c. $X^2 + XY + Y^2$ d. $X^2 + 2XY + 2Y^2$

e. none of the above

2. $(X - Y)^2 =$

a. $X^2 + 2XY + Y^2$

b. $(X - Y)(X - Y)$

c. $X^2 - 2XY + Y^2$

d. $X^2 - XY + Y^2$

e. only (b) and (c) of the above

3. $(2X + 3Y)^2 =$

a. $4X^2 + 6YX + 9Y^2$

b. $4X^2 + 12XY + 9Y^2$

c. $2X^2 + 6XY + 3Y^2$

d. $4X^2 + 9Y^2$

e. none of the above

4. $(2X - 3Y)^2 =$

a. $4X^2 - 9Y^2$

b. $2X^2 + 6XY + 3Y^2$

c. $4X^2 - 12XY + 9Y^2$

d. $4X^2 + 9Y^2$

e. none of the above

5. $(2X^3)(6X^{10}) =$

a. $2X^{3+10}$

b. $12X^{30}$

c. $48X^{3/10}$

d. $12X^{13}$

e. none of the above

6. $(12X^6Y^2)(2Y^3X^2)(3X^3Y^4) =$

a. $72X^{11}Y^9$

b. $72X^{12}Y^8$

c. $17X^{10}Y^{10}$

d. $72Y^8 X^{12}$

e. only (b) and (d) of the above

7. $X^2(X + Y)^2 =$

a. $X^2(X^2 + 2XY + Y^2)$

b. $X^{2+2} + 2X^{1+2}Y + X^2Y^2$

c. $X^4 + 2X^3Y + X^2Y^2$

d. all of the above

e. none of the above

8. $\frac{X^3}{2} \cdot \frac{6}{X^2} =$

a. $3X^5$

b. $3X$

c. $3X^{-1}$

d. $12X$

e. none of the above

9. $(2X^3)/(6X^{10}) =$

a. $0.33333333X^{-7}$

b. $\frac{1}{3X^7}$

c. $\frac{1}{3}X^{-7}$

d. only (a) and (c) of the above

e. all of the above

10. $\frac{10}{X^9Y^5} \cdot X^5Y^3 =$

a. $10X^{-4}Y^{-2}$

b. $\frac{10}{X^4Y^2}$

c. $10X^9Y^5X^{-5}Y^{-3}$

d. only (a) and (b) of the above

e. all of the above

11. $24X^{0.5}Y^{1.5} \div 12X^{1.5}Y^{0.5} =$

a. $\frac{2Y}{X}$

b. $\frac{2X}{Y}$

c. $\frac{Y}{2X}$

d. $\frac{Y}{X}$

e. $\frac{X}{Y}$

12. $(64X)^{\frac{1}{2}}(8Y)^{\frac{1}{3}} \div 8X^{2.5}Y^{\frac{4}{3}} =$

a. $\frac{2}{X^2Y}$ b. $\frac{2}{XY}$ c. $\frac{2}{Y^2X}$

d. $\frac{2Y}{X^2}$ e. none of the above

13. $(2X^2)^3 =$

a. $2X^6$ b. $8X^6$ c. $8X^5$

d. $16X^6$ e. $(2X^3)^2$

14. $[(3X^4Y^3)^2]^2 =$ $(9x^8y^6)^2 = 81x^{16}y^{12}$

a. $9X^8Y^7$ b. $9X^{16}Y^{12}$ c. $81X^{16}Y^{10}$

d. $81X^{16}Y^{12}$ e. $81X^8Y^7$

15. $(4X^4Y^3)^2/(2X^2Y^2)^4 =$

a. Y^{-3} b. Y^2 c. $1/Y^2$

d. X^2 e. $1/X^2$

Using the following data, answer Problems 16 – 20.

i =	1	2	3	4	5	6	7
X_i =	30	52	67	22	16	42	34

16. $\sum_{i=1}^{3} X_i =$

a. 6 b. 140 c. 149

d. 104520 e. none of the above

17. $\prod_{i=4}^{6} X_i =$

a.	6	b.	80	c.	120
d.	14784	e.	none of the above		

18. $\sum_{i=1}^{2} X_i - \prod_{3}^{5} X_i =$

a.	− 23	b.	0	c.	2352
d.	2350	e.	none of the above		

19. $\sum_{i=5}^{7} X_i + \prod_{3}^{5} X_i =$

a.	92	b.	23584	c.	23676
d.	46432	e.	none of the above		

20. $\prod_{i=6}^{7} X_i + \sum_{i=3}^{4} X_i - \prod_{5}^{6} X_i =$

a.	0	b.	1428	c.	89
d.	672	e.	845		

21. Find the value of X in $3^X = 59049$.

a.	20	b.	15	c.	10
d.	5	e.	none of the above		

22. Identify the correct relationship(s) shown below:

a. $X \log 20 = \log_{20} X$

b. $15^X \neq 3^X \cdot 5^X$

c. $5 \ln \frac{2}{5} = \ln 2$

d. $\frac{\ln X}{\ln Y} = \ln X - \ln Y$

e. none of the above is correct.

Answers to Exercise Problems for Exponents and Mathematical Operators

1. $(X + Y)^2 =$

a.* $X^2 + 2XY + Y^2$ because
$(X + Y)(X + Y) = X^2 + XY + YX + Y^2 = X^2 + 2XY + Y^2$

2. $(X - Y)^2 =$

e.* only (b) and (c) of the above because
$(X - Y)(X - Y) = X^2 - XY - YX + Y^2 = X^2 - 2XY + Y^2$

3. $(2X + 3Y)^2 =$

b.* $4X^2 + 12XY + 9Y^2$ because
$(2X + 3Y)(2X + 3Y) = 4X^2 + 6XY + 6YX + 9Y^2$
$= 4X^2 + 12XY + 9Y^2$

4. $(2X - 3Y)^2 =$

c.* $4X^2 - 12XY + 9Y^2$ because
$(2X - 3Y)^2 = (2X - 3Y)(2X - 3Y) = 4X^2 - 6XY - 6YX + 9Y^2$
$= 4X^2 - 12XY + 9Y^2$

5. $(2X^3)(6X^{10}) =$

d.* $12X^{13}$ because $(2)(6)X^{3+10} = 12X^{3+10} = 12X^{13}$

6. $(12X^6Y^2)(2Y^3X^2)(3X^3Y^4) =$

a.* $72X^{11}Y^9$ because $(12)(2)(3)X^{6+2+3}Y^{2+3+4} = 72X^{11}Y^9$

7. $X^2(X + Y)^2 =$

d.* all of the above because
$X^2(X^2 + 2XY + Y^2) = X^{2+2} + 2X^{1+2}Y + X^2Y^2 = X^4 + 2X^3Y + X^2Y^2$

8. $\frac{X^3}{2} \cdot \frac{6}{X^2} =$

b.* $3X$ because $\frac{6X^3}{2X^2} = 3X^{3-2} = 3X$

9. $(2X^3)/(6X^{10}) =$

e.* all of the above because $(2/6)X^{3-10} = (1/3)X^{-7} = 0.3333X^{-7} = \frac{1}{3X^7}$

10. $\frac{10}{X^9Y^5} \cdot X^5Y^3 =$

d.* only (a) and (b) of the above because

$$10X^{-9}Y^{-5}X^5Y^3 = 10X^{-9+5}Y^{-5+3} = 10X^{-4}Y^{-2} = \frac{10}{X^4Y^2}$$

11. $24X^{0.5}Y^{1.5} \div 12X^{1.5}Y^{0.5} =$

a.* $\frac{2Y}{X}$ because $\frac{24X^{0.5}Y^{1.5}}{12X^{1.5}Y^{0.5}} = 2X^{0.5-1.5}Y^{1.5-0.5} = 2X^{-1}Y^1 = \frac{2Y}{X}$

12. $(64X)^{\frac{1}{2}}(8Y)^{\frac{1}{3}} \div 8X^{2.5}Y^{\frac{4}{3}} =$

a.* $\frac{2}{X^2Y}$ because $\frac{(64X)^{\frac{1}{2}}(8Y)^{\frac{1}{3}}}{8X^{2.5}Y^{\frac{4}{3}}} = \frac{(64)^{\frac{1}{2}}X^{\frac{1}{2}}(8)^{\frac{1}{3}}Y^{\frac{1}{3}}}{8X^{2.5}Y^{\frac{4}{3}}}$

$$= \frac{(8)X^{\frac{1}{2}}(2)Y^{\frac{1}{3}}}{8X^{2.5}Y^{\frac{4}{3}}} = X^{\frac{1}{2}-2.5}(2)Y^{\frac{1}{3}-\frac{4}{3}} = 2X^{-2}Y^{-1} = \frac{2}{X^2Y}$$

13. $(2X^2)^3 =$

b.* $8X^6$ because $(2X^2)(2X^2)(2X^2) = (2)^3X^{2+2+2} = 2^3X^{2x3} = 8X^6$

14. $[(3X^4Y^3)^2]^2 =$

d.* $81X^{16}Y^{12}$ because $[3^2X^{4x2}Y^{3x2}]^2 = 3^{2x2}X^{8x2}Y^{6x2} = 81X^{16}Y^{12}$

15. $(4X^4Y^3)^2/(2X^2Y^2)^4 =$

c.* $1/Y^2$ because

$$(4X^4Y^3)^2(2X^2Y^2)^{-4} = [(2^2)^2X^8Y^6][(2)^{-4}X^{-8}Y^{-8}] = Y^{-2} = 1/Y^2$$

16. $\sum_{i=1}^{3} X_i =$

c.* 149 because $\sum_{i=1}^{3} X_i = X_1 + X_2 + X_3 = 30 + 52 + 67 = 149$

17. $\prod_{i=4}^{6} X_i =$

d.* 14784 because $\prod_{i=4}^{6} X_i = X_4 \cdot X_5 \cdot X_6 = 22 \cdot 16 \cdot 42 = 14784$

18. $\sum_{i=1}^{2} X_i - \prod_{i=3}^{5} X_i =$

e.* none of the above

because $\sum_{i=1}^{2} X_i - \prod_{i=3}^{5} X_i = (X_1 + X_2) - (X_3 \cdot X_4 \cdot X_5)$

$= (30 + 52) - (67 \cdot 22 \cdot 16) = 82 - 23584 = 23502$

19. $\sum_{i=5}^{7} X_i + \prod_{i=3}^{5} X_i =$

c.* 23676

because $\sum_{i=5}^{7} X_i + \prod_{i=3}^{5} X_i = (X_5 + X_6 + X_7) + (X_3 \cdot X_4 \cdot X_5)$

$= (16 + 42 + 34) + (67 \cdot 22 \cdot 16) = 92 + 23584 = 23676$

20 $\prod_{i=6}^{7} X_i + \sum_{i=3}^{4} X_i - \prod_{i=5}^{6} X_i =$ 20.

e.* 845

because $\prod_{i=6}^{7} X_i + \sum_{i=3}^{4} X_i - \prod_{i=5}^{6} X_i = (X_6 + X_7) + (X_3 + X_4) - (X_5 \cdot X_6)$

$= (42 \cdot 34) + (67 + 22) - (16 \cdot 42) = 1428 + 89 - 672 = 845$

21. Find the value of X in $3^X = 59049$.

c.* 10

In order to find X,
(1) we can take a natural (or common) logarithm of both sides as:

$$\ln 3^X = \ln 59049$$

(2) rewrite the above as: $X \cdot \ln 3 = \ln 59049$ by using the Power Property
(3) solve for X as: $X = \dfrac{\ln 59049}{\ln 3}$
(4) use the calculator to find the value of X as:

$$X = \frac{\ln 59049}{\ln 3} = \frac{10.9861}{1.09861} = 10$$

22. Identify the correct relationship(s) shown below:

e.* none of the above is correct.

Note that

a. $X\log 20 = \log 20^X \neq \log_{20} X$ b. $15^X = (3\cdot 5)^X = 3^X \cdot 5^X$

c. $5\ln\frac{2}{5} = \ln(\frac{2}{5})^5 \neq \ln 2$ d. $\frac{\ln X}{\ln Y} \neq \ln\frac{X}{Y} = \ln X - \ln Y$

H. Graphs

In economics and other business disciplines, graphs and tables are often used to describe a relationship between two variables – X and Y. X is often represented on a horizontal axis and Y, a vertical axis.

1. A Positive-Sloping Line and a Negative-Sloping Line

For example, a function of Y = 2 + 0.5X, as plotted below, has an intercept of 2 and a positive slope of +0.5. Therefore, it rises to the right (and declines to the left) and thus, is characterized as a positive sloping or upward sloping line. It shows a pattern where as X increases (decreases), Y increases (decreases). This relationship is also known as a direct relationship.

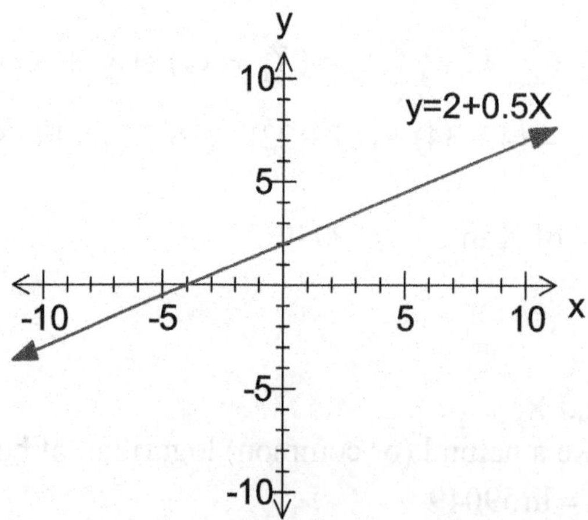

On the other hand, a function of Y = 2 – 0.5X as plotted below, has an intercept of 2 and a negative slope of –0.5. Therefore, it declines to the right (and rises to the left) and thus, is characterized as a negative sloping or downward sloping line. It shows a pattern where as X increases (decreases), Y decreases (increases). That is, because X and Y move in an opposite direction, it is also known as an indirect or inverse relationship.

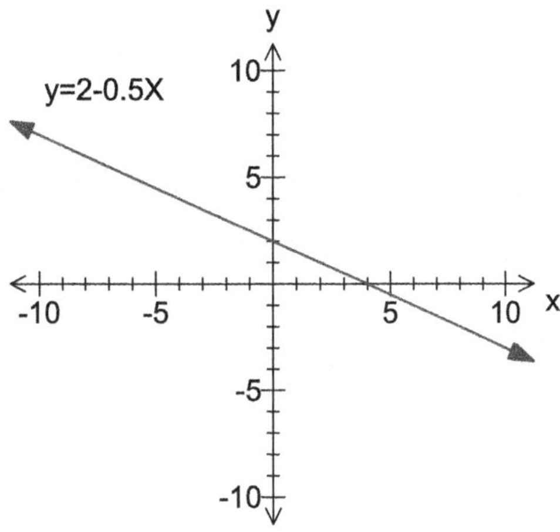

2. Shifts in the Lines

Often, the line can move up or down as the value of the intercept changes, while maintaining the same slope value. When the following two equations are plotted in addition to the original one we plotted above, we can see how the two lines differ from the original one by their respective intercept values:

Original Line: Y = 2 + 0.5X ← The middle line

New Line #1: Y = 6 + 0.5X ← The top line
New Line #2: Y = –2 + 0.5X ← The bottom line

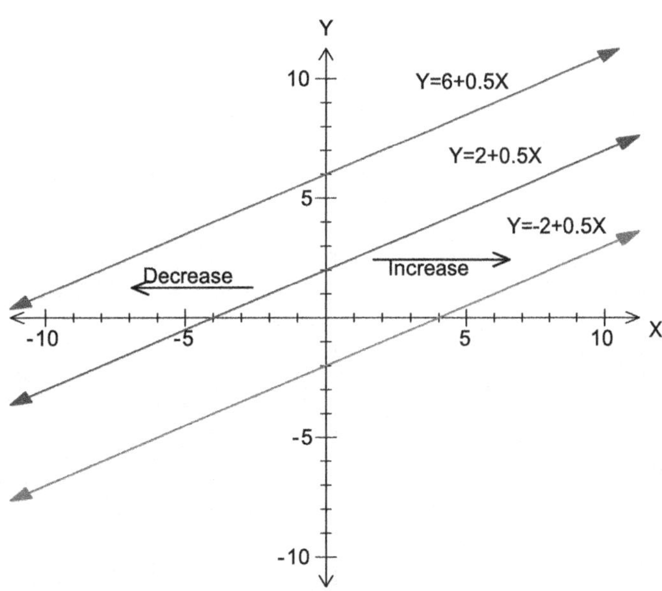

Note 1: As the intercept term increases from 2 to 6, the middle line moves up to become the top line. This upward shift in the line indicates that the value of X has decreased while the Y value was held constant (or unchanged). Thus, the upward shift is the same as a shift to the left and indicates a decrease in X given the unchanged (or same) value of Y.

Note 2: As the intercept term decreases from 2 to –2, the middle line moves down to become the bottom line. This downward shift in the line indicates that the value of X has increased while the Y value was held constant (or unchanged). Thus, the downward shift is the same as a shift to the right and indicates an increase in X given the unchanged (or same) value of Y.

Note 3: This observation is often utilized in the demand and supply analysis of economics as a shift in the curve. A leftward shift is a "decrease" and a rightward shift is an "increase."

3. Changes in the Slope

When the value of a slope changes, holding the intercept unchanged, we will note that the line will rotate around the intercept as the center. Let's plot two new lines in addition to the original line as follows:

Original Line: $Y = 2 + 0.5X$ ← The original (=middle) line

New Line #1: $Y = 2 + 2X$ ← The top line
New Line #2: $Y = 2 + 0X = 2$ ← The flat line
New Line #3: $Y = 2 - 0.5X$ ← The bottom line

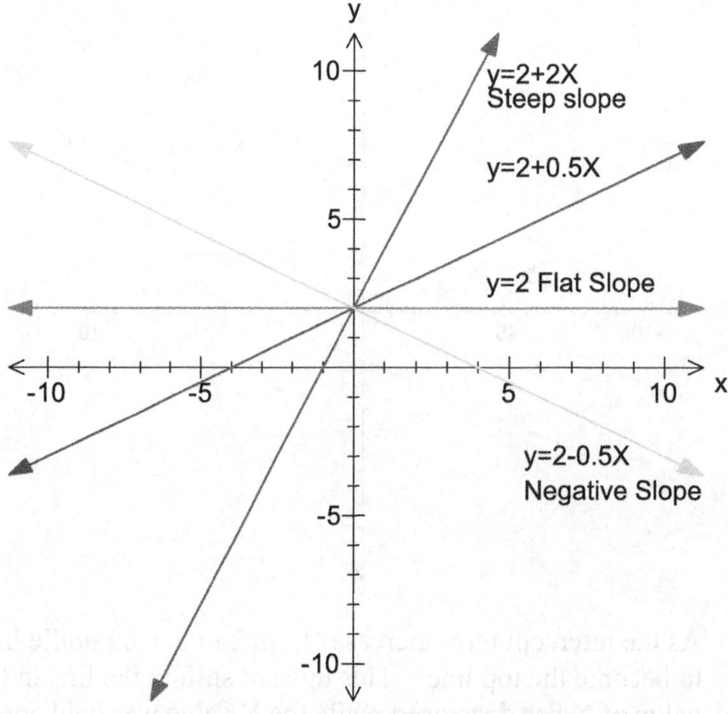

Note that the steepness (or flatness) of the slope as the value of the slope changes. Likewise, note the relationship among a flat, a positive, and a negative slope.

I. Applications: Compound Interest

1. The Concept of Periodic Interest Rates

Assume that the annual percentage rate (APR) is $(r \cdot 100)$%. That is, if an APR is 10%, then r = 0.1. Also, define FV = future value, PV = present value, and t = number of years to a maturity.

i) Annual compounding for t years → $FV = PV \cdot (1+r)^t$

ii) Semiannual compounding for t years → $FV = PV \cdot (1+\frac{r}{2})^{2t}$

iii) Quarterly compounding for t years → $FV = PV \cdot (1+\frac{r}{4})^{4t}$

iv) Monthly compounding for t years → $FV = PV \cdot (1+\frac{r}{12})^{12t}$

v) Weekly compounding for t years → $FV = PV \cdot (1+\frac{r}{52})^{52t}$

vi) Daily compounding for t years → $FV = PV \cdot (1+\frac{r}{365})^{365t}$

vii) Continuous compounding for t years[6] → $FV = PV \cdot e^{rt}$

Examples>

Assume that \$100 is deposited at an annual percentage rate (APR) of 12% for 1 year.

i) Annual compounding → one 1-year deposit → 1 interest calculation

$$FV = PV \cdot (1+r)^t = \$100 \cdot (1+0.12)^1 = \$112.00$$

ii) Semiannual compounding →two ½-year deposits → 2 interest calculations in 1 year

$$FV = PV \cdot (1+\frac{r}{2})^{2t} = \$100 \cdot (1+\frac{0.12}{2})^{2\cdot 1} = \$100 \cdot (1+0.06)^2 = \$112.36$$

iii) Quarterly compounding →four ¼-year deposits → 4 interest calculations in 1 year

$$FV = PV \cdot (1+\frac{r}{4})^{4t} = \$100 \cdot (1+\frac{0.12}{4})^{4\cdot 1} = \$100 \cdot (1+0.03)^4 = \$112.55$$

[6] Do you remember that this is an exponential function with a natural base of e?

iv) Monthly compounding → twelve 1/12-year deposits → 12 interest calculations in 1 year

$$FV = PV \cdot (1+\frac{r}{12})^{12t} = \$100 \cdot (1+\frac{0.12}{12})^{12\cdot 1} = \$100 \cdot (1+0.01)^{12} = \$112.68$$

v) Weekly compounding →fifty-two 1/52-year deposits → 52 interest calculations in 1 year

$$FV = PV \cdot (1+\frac{r}{52})^{52t} = \$100 \cdot (1+\frac{0.12}{52})^{52\cdot 1} = \$100 \cdot (1+0.0023077)^{52} = \$112.73$$

vi) Daily compounding →365 1/365-year deposits → 365 interest calculations in 1 year

$$FV = PV \cdot (1+\frac{r}{365})^{365t} = \$100 \cdot (1+\frac{0.12}{365})^{365\cdot 1} = \$100 \cdot (1+0.000328767)^{365} = \$112.74$$

vii) Continuous compounding for 1 year → continuous interest calculations

$$FV = PV \cdot e^{rt} = \$100 \cdot e^{0.12\cdot 1} = \$100 \cdot e^{0.12} = \$112.75$$

Examples>

Calculate the annual rate of return (ROR) based on the various compounding schemes shown above.

i) For annual compounding,

$$\text{ROR} = \frac{P_1 - P_0}{P_0} = \frac{112-100}{100} = 0.12 \rightarrow 12\%$$

ii) For semi-annual compounding,

$$\text{ROR} = \frac{P_1 - P_0}{P_0} = \frac{112.36-100}{100} = 0.1236 \rightarrow 12.36\%$$

iii) For monthly compounding,

$$\text{ROR} = \frac{P_1 - P_0}{P_0} = \frac{112.68 - 100}{100} = 0.1268 \rightarrow 12.68\%$$

Note: The rate of return on an annual basis is known as the Annual Percentage Yield (APY). Even though APR may be the same, APY will increase as the frequency of compounding increases → because an interest is earned on an interest more frequently.

2. Annuity Calculation

Annuity Formulas:

$$FV = \frac{A \cdot [(1+i)^n - 1]}{i}$$

$$PV = \frac{A \cdot [(1+i)^n - 1]}{i \cdot (1+i)^n}$$

where A= the fixed annuity amount; n = the number of periods; and i = a periodic interest rate. Of course, FV = the future (or final or terminal) value and PV = the present (or current) value.

Examples

i) If you obtain a 30 year mortgage loan of $100,000 at an annual percentage rate (APR) of 6%, what would be your monthly payment?

Answer:

$$100{,}000 = \frac{A \cdot [(1 + \frac{0.06}{12})^{12x30} - 1]}{\frac{0.06}{12} \cdot (1 + \frac{0.06}{12})^{12x30}}$$

Therefore, A=$599.55

ii) If you invest $1,000 a month in an account that is guaranteed to yield a 10% rate of return per year for 30 years (with a monthly compounding), what will be the balance at the end of the 30-year period?

Answer:

$$FV = \frac{\$1{,}000 \cdot [(1+\frac{0.1}{12})^{360} - 1]}{\frac{0.1}{12}} = \$2{,}260{,}487.92$$

iii) If you are guaranteed of a 10% rate of return for 30 years, how much should you save and invest each month to accumulate $1 million at the end of the 30-year period?

Answer:

$$\$1{,}000{,}000 = \frac{A \cdot [(1+\frac{0.1}{12})^{360} - 1]}{\frac{0.1}{12}}$$

Therefore, A=$442.38

iv) Suppose that you have saved up $100,000 for your retirement. You expect that you can continuously earn 10% each year for your $100,000. If you know that you are going to live for 15 additional years from the date of your retirement and that the balance of your retirement fund will be zero at the end of the 15-year period, how much can you withdraw to spend each month?

Answer:

$$\$100{,}000 = \frac{A \cdot [(1+\frac{0.1}{12})^{180} - 1]}{\frac{0.1}{12} \cdot (1+\frac{0.1}{12})^{180}}$$

Therefore, A=$1,074.61

v) Assume the same situation as Problem 4 above, except that now you have to incorporate an annual inflation rate of 3%. What will be the possible monthly withdrawal, net of inflation?

Answer:

$$\$100{,}000 = \frac{A \cdot [(1+\frac{0.1-0.03}{12})^{180} - 1]}{\frac{0.1-0.03}{12} \cdot (1+\frac{0.1-0.03}{12})^{180}}$$

Therefore, A=$898.83

Note: Combining Answers to Problems (iv) and (v) above, it means that you will be actually withdrawing $1,074.61 per month but its purchasing power will be equivalent to $898.83. This is because inflation only erodes the purchasing power; it does not reduce the actual amount received. If one goes through a professional financial planning, the financial planner will expand on this simple assumption to a more complex and realistic scenario.

vi) Assuming only annual compounding, how long will it take to double your investment if you earn 10% per year?

Answer[7]**:**

$$A \cdot (1+0.1)^x = 2A$$

$$\therefore 1.1^x = 2$$

Now, take the natural logarithm of both sides as follows:

$$\ln 1.1^x = \ln 2$$

$$X \cdot \ln 1.1 = \ln 2$$

$$\therefore X = \frac{\ln 2}{\ln 1.1} = 7.2725 years$$

vii) Assuming monthly compounding, how long will it take to double your investment if you earn 10% per year?

Answer:

$$A \cdot (1+\frac{0.1}{12})^x = 2A$$

$$1.0083333^x = 2$$

$$\ln 1.0083333^x = \ln 2$$

$$X \cdot \ln 1.0083333 = \ln 2$$

$$\therefore\ X = \frac{\ln 2}{\ln 1.0083333} = 83.5 months = 6.96 years$$

[7] When either the natural logarithm or the common logarithm is taken, the exponent, X, as in this case, will become a coefficient as shown herein. Then, use the calculator with a "ln" function to complete the calculation.

viii) Assume that you have a 30-year, $100,000 mortgage loan at an annual percentage rate (APR) of 6%. How long will it take you to pay off this loan if you pay off $1,000 a month?

Answer: Use the information on Answers to Problem 1 as follows:

$$100{,}000 = \frac{1{,}000 \cdot [(1+\frac{0.06}{12})^X - 1]}{\frac{0.06}{12} \cdot (1+\frac{0.06}{12})^X}$$

Therefore,

$$100{,}000 \cdot \frac{0.06}{12} \cdot (1+\frac{0.06}{12})^X = 1{,}000 \cdot [(1+\frac{0.06}{12})^X - 1]$$

$$500 \cdot (1+\frac{0.06}{12})^X = 1000 \cdot (1+\frac{0.06}{12})^X - 1000$$

$$500 \cdot (1.005)^X = 1000$$

$$X \cdot \ln 1.005 = \ln 2$$

$$X = \frac{\ln 2}{\ln 1.005} = 138.975 months = 11.58 years$$

J. Inequalities

1. If a > 0 and b > 0, then (a+b) > 0 and ab > 0

   If a=7 and b=5, then (7+5) > 0 and (7)(5) > 0

2. If a > b, then (a–b) > 0

   If a=7 and b=5, then (7–5) > 0

3. If a > b, then (a+c) > (b+c) for all c

   If a=7 and b=5, then (7+c) > (5+c) → 7 > 5

4. If a > b and c > 0, then ac > bc

   If a=7 and b=5 and c=3, then (7)(3) > (5)(3) → 21 > 15

5. If $a > b$ and $c < 0$, then $ac < bc$

If a=7 and b=5 and c= –3, then $(7)(-3) < (5)(-3) \rightarrow -21 < -15$

K. Absolute Values and Intervals

1. $|X| = X$ if $X \geq 0$ and $|X| = -X$ if $X \leq 0$

Examples>

a. $|+5| = +5 = 5$ and $|-5| = -(-5) = +5 = 5$

b. $|+10| = 10$ and $|-10| = -(-10) = 10$

2. If $|X| \leq n$, then $-n \leq X \leq n$

Examples>

a. If $|X| \leq 5$, then $-5 \leq X \leq 5$

b. If $|X-2| \leq 5$, then $-5 \leq X-2 \leq 5 \rightarrow -5+2 \leq X \leq 5+2$

$\rightarrow -3 \leq X \leq 7$

c. If $|2X+4| < 10$, then $-10 < 2X+4 < 10 \rightarrow -14 < 2X < 6$

$\rightarrow -7 < X < 3$

3. If $|X| > n$, then $X > n$ if $X > 0$ or $-X > n$ if $X < 0$

Note that when a negative number is multiplied to both sides of the inequality sign, the direction of the inequality sign reverses.

Examples>

a. If $|X| > 5$,
then, $X > 5$ or $-X > 5 \rightarrow X < -5$

b. If $|X-3| > 5$,
then, $(X-3) > 5 \rightarrow X > 8$

or $-(X-3) > 5 \rightarrow (X-3) < -5 \rightarrow X < -2$

c. If $|6 - 3X| > 12$,

then, $(6 - 3X) > 12 \rightarrow -3X > 6 \rightarrow X < -2$

or

$-(6 - 3X) > 12 \rightarrow (6 - 3X) < -12 \rightarrow -3X < -18 \rightarrow X > 6$

L. A System of Linear Equations in Two Unknowns

Given the following system of linear equations, solve for X and Y.

$$3X + 2Y = 13$$
$$4Y - 2X = 2$$

1. Solution Method 1: The Substitution Method

(1) Rearrange the bottom equation for X as follows:

$$2X = 4Y - 2 \quad \rightarrow \quad X = 2Y - 1$$

(2) Substitute this X into the top equation as follows:

$$3(2Y - 1) + 2Y = 13 \quad \rightarrow \quad 8Y = 16 \quad \rightarrow \quad Y = 2$$

(3) Substitute this Y into any of the above equation for X value:

$$X = 2Y - 1 = 2\times 2 - 1 = 3$$

(4) Verify if the values of X and Y satisfy the system of equations:

$$3X + 2Y = 3\times 3 + 2\times 2 = 13$$
$$4Y - 2X = 4\times 2 - 2\times 3 = 2$$

(5) Verification completed and solutions found.

2. Solution Method 2: The Elimination Method

(1) Match up the variables as follows:

$$3X + 2Y = 13$$
$$-2X + 4Y = 2$$

(2) Multiply either of the two equations to find a common coefficient. (Y is chosen to be eliminated and thus, the top equation is multiplied by 2 as follows:)

$$2 \times 3X + 2 \times 2Y = 2 \times 13 \quad \rightarrow \quad 6X + 4Y = 26$$

(3) Subtract the bottom equation from the adjusted top equation in (2) above and obtain:

$$\begin{array}{l} 6X + 4Y = 26 \\ -(-2X + 4Y = 2) \\ \hline \end{array} \quad \text{or} \quad \begin{array}{l} 6X + 4Y = 26 \\ 2X - 4Y = -2 \\ \hline \end{array}$$

$$6X - (-2X) + 4Y - 4Y = 26 - 2 \quad \text{or} \quad 6X + 2X + 4Y - 4Y = 26 - 2$$
$$8X = 24 \qquad\qquad 8X = 24$$
$$X = 3 \qquad\qquad X = 3$$

(4) Substitute this X into any of the above equation for Y value:

$$6X + 4Y = 26$$
$$6 \times 3 + 4Y = 26$$
$$4Y = 26 - 18$$
$$Y = 2$$

(5) Verify if the values of X and Y satisfy the system of equations:

$$3X + 2Y = 3 \times 3 + 2 \times 2 = 13$$
$$4Y - 2X = 4 \times 2 - 2 \times 3 = 2$$

(6) Verification completed and solutions found.

3. An Example

Suppose that you have \$10 with which you can buy apples (A) and oranges (R). Also, assume that your bag can hold only 12 items – such as 12 apples, or 12 oranges, or some combination of apples and oranges. If the apple price is \$1 and the orange price is \$0.50, how many apples and oranges can you buy with your \$10 and carry them home in your bag?

Answer:

The Substitution Method:

(1) identify relevant information:

Budget Condition: $A + 0.5R = 10$
Bag-Size Condition: $A + R = 12$

(2) convert the Bag-Size Condition as:

$$A = 12 - R$$

(3) substitute $A = 12 - R$ into the Budget Condition as:

$$(12 - R) + 0.5R = 10$$

$$-0.5R = -2 \rightarrow R=4$$

(4) substitute R=4 into (2) above and find:

$$A = 12 - 4 = 8$$

(5) verify the answer of A=8 and R=4 by plugging them into the above two conditions as:

Budget condition: $8 + 0.5(4) = 10$
Bag-Size condition: $8 + 4 = 12$

Because both conditions are met, the answer is A=8 and R=4.

The Elimination Method:

(1) identify relevant information:

Budget Condition: $A + 0.5R = 10$
Bag-Size Condition: $A + R = 12$

(2) subtract the bottom equation from the top:

$$-0.5R = -2 \rightarrow R=4$$

(3) plug this R=4 into either one of the two conditions above:

$$A + 0.5(4) = 10 \rightarrow A=8$$

Or $A + (4) = 12$ → $A=8$

(4) verify the answer of A=8 and R=4 by plugging them into the above two conditions as:

Budget condition: $8 + 0.5(4) = 10$
Bag-Size condition: $8 + 4 = 12$

Because both conditions are met, the answer is A=8 and R=4.

4. Solve the following simultaneous equations by using both the substitution and elimination methods:

a. $20X + 4Y = 280$
$10Y - 9X = 110$

Answer: X=10 and Y=20

b. $2X + 7 = 5Y$
$3Y + 7 = 4X$

Answer: X=4 and Y=3

c. $(1/3)X - (1/4)Y = -37.5$
$3Y - 5X = 330$

Answer: X=120 and Y=310

Note that there is no way of telling which solution method – the substitution or the elimination – is superior to the other. Even though the elimination method is often preferred, it is the experience and preference of the solver that will decide which method would be used.

M. Examples of Algebra Problems

1. For your charity organization, you had served 300 customers who bought either one hot dog at \$1.50 or one hamburger at \$2.50, but never the two together. If your total sales of hot dogs (HD) and hamburgers (HB) were \$650 for the day, how many hot dogs and hamburgers did you sell?

2. You are offered an identical sales manager job. However, Company A offers you a base salary of \$30,000 plus a year-end bonus of 1% of the gross sales you make for that year. Company B, on the other hand, offers a base salary of \$24,000 plus a year-end bonus of 2% of the gross sales you make for that year.

a. Which company would you work for?

b. If you can achieve a total sale of $1,000,000 for either A or B, which company would you work for?

3. A fitness club offers two aerobics classes. In Class A, 30 people are currently attending and attendance is growing 3 people per month. In Class B, 20 people are regularly attending and growing at a rate of 5 people per month. Predict when the number of people in each class will be the same.

4. Everybody knows that Dr. Choi is the best instructor at DePaul. When a student in GSB 420 asked about the midterm exam, he said the following:

   a. "The midterm exam will have a total of 100 points and contain 35 problems. Each problem is worth either 2 points or 5 points. Now, you have to figure out how many problems of each value there are in the midterm exam."

   b. "The midterm exam will have a total of 108 points and there are twice as many 5-point problems than 2-point problems. Each problem is worth either 2 points or 5 points. Now, you have to figure out how many problems of each value there are in the midterm exam."

5. Your boss asked you to prepare a company party for 20 employees with a budget of $500. You have a choice of ordering a steak dinner at $30 per person or a chicken dinner at $25 per person. (All tips are included in the price of the meal.)

   a. How many steak dinners and chicken dinners can you order for the party by using up the budget?

   b. How many steak dinners and chicken dinners can you order for the party if the budget increases to $550?

6. Your father just received a notice from the Social Security Administration saying the following:

   "If you retire at age 62, your monthly social security payment will be $1500. If you retire at age 66, your monthly social security payment will be $2100."

   a. Your father is asking you to help decide which option to take. What would you tell him? Do not consider the time value of money. (Hint:

Calculate the age at which the social security income received will be the same.)

b. The Social Security Administration has given your father one more option as: "If you retire at age 70, your monthly social security payment will be $2800." What would you now tell him? Do not consider the time value of money. (Hint: Calculate the age at which the social security income received will be the same.)

Answers to Above Examples of Algebra Problems

1. Quantity Condition: HD + HB = 300
Sales Condition: 1.50 HD + 2.50 HB = 650

Solving these two equations simultaneously, we find

HD* = 100 and HB* = 200

2.a. We have to identify the break-even sales (S) for both companies as follows:

Compensation from A = $30,000 + 0.01 S
Compensation from B = $24,000 + 0.02 S

Therefore,

Compensation from A = Compensation from B

$30,000 + 0.01 S = $24,000 + 0.02 S
S* = $600,000

Conclusion: If you think you can sell more than $600,000, you had better work for B. Otherwise, work for A.

2.b. Since you can sell more than $600,000, such as $1 million, work for B and possibly realize a total compensation of $44,000 (=$24,000 + 0.02 x ($1 million)). If you work for A, you would receive $40,000 (=$30,000 + 0.01 x ($1 million)).

3. Attendance in A = 30 + 3(Months)
Attendance in B = 20 + 5(Months)

Attendance in A = Attendance in B

Therefore, 30 + 3(Months) = 20 + 5(Months)

Months* = 5

4.a. Total Points: 2X + 5Y = 100
Number of Problems: X + Y = 35

where X = the number of 2 point problems and Y = the number of 5 point problems.

Therefore, X*=25 and Y*=10

4.b. Total Points: 2X + 5Y = 108
Number of Problems: 2X = Y

where X = the number of 2 point problems and Y = the number of 5 point problems.

Therefore, X*=9 and Y*=18

5.a. Total Number of Employees: S + C = 20
Budget: 30S + 25C = 500

where S = number of steak dinner and C = chicken dinner.

Therefore, C*=20 and S*=0

5.b. Total Number of Employees: S + C = 20
Budget: 30S + 25C = 550

where S = number of steak dinner and C = chicken dinner.

Therefore, C*=10 and S*=10

6.a. Total Receipt between 62 and X = (X – 62)*1500*12
Total Receipt between 66 and X = (X – 66)*2100*12
Total Receipt between 62 and X = Total Receipt between 66 and X

That is, (X – 62)*1500*12 = (X – 66)*2100*12
Therefore, X* = 76

That is, if your father can live longer than 76 of age, he should start receiving the social security payment at 66 of age. Otherwise, he should retire at 62.

6.b. If the retirement decision is between 62 vs. 70:

$$(X - 62)*2100*12 = (X - 70)*2800*12$$

Therefore, $X^* = 79.23$

That is, if your father can live longer than 79.23 of age, he should retire at 70 of age. Otherwise, he should retire at 62.

If the retirement decision is between 66 vs. 70:

$$(X - 66)*2100*12 = (X - 70)*2800*12$$

Therefore, $X^* = 82$

That is, if your father can live longer than 82 of age, he should retire at 70 of age. Otherwise, he should retire at 66.

Exercise Problems on Chapter 1. Algebra Review

This exercise problem set has 21 problems worth 21 points.

Expand the following equations:

1. $(2X - 3Y)(6X + 5Y) =$

a.	$8X + 2Y$	b.	$12X^2 - 8XY - 15Y^2$
c.	$12X^2 + 8XY - 15Y^2$	d.	$2X^2 - 18XY + Y^2$
e.	only (b) and (c) of the above		

2. $(X + 1)^3 =$

a.	$X^3 + 3X^2 + 3X + 1$	b.	$X^3 - 3X^2 - 3X + 1$
c.	$X^3 + 1$	d.	$3X + 3$
e.	none of the above		

3. $(21\text{-}6)/5 \div (18\text{-}9)/(9\text{-}8) =$

a.	1/3	b.	3
c.	(15/5)(9/1)	d.	only (a) and (c) of the above
e.	none of the above		

Factor the following equations:

4. $a^2 - 9Y^2 =$

a.	$(a - 3Y)(a + 3Y)$	b.	$(a + 3Y)(a + 3Y)$
c.	$(a + 3Y)(-a + 3Y)$	d.	$(a - 3Y)(a - 3Y)$
e.	none of the above		

5. $X^2Y - XY^2 =$

a.	$Y^*(X - XY)$	b.	$X^*(XY - Y)$
c.	$X^2{}^*(Y - XY)$	d.	$Y^2{}^*(X - XY)$
e.	$XY^*(X - Y)$		

Reduce the following equations:

6. $$\frac{\frac{1}{2}+\frac{2}{3}}{4\frac{2}{3}-\frac{5}{6}} =$$

a. $\frac{7}{9}$ b. $\frac{7}{23}$ c. $\frac{23}{7}$

d. only (a) and (b) of the above e. only (a) and (c) of the above

7. $\frac{2x}{x-y} - \frac{x-1}{x+y} =$

a. $\frac{2x(x-y)(x+y)-(x-1)(x-y)(x+y)}{(x-y)(x+y)}$

b. $\frac{2x(x+y)-(x-1)(x-y)}{(x-y)(x+y)}$ c. $\frac{x^2+3xy+x-y}{x^2+y^2}$

d. only (b) and (c) of the above e. none of the above

8. $64^{-1.5} =$

a. $-\frac{1}{64^{3/2}}$ b. $\frac{1}{8^{-3}}$ c. $\frac{1}{512}$

d. all of the above e. none of the above

Identify the value of X:

9. $(1+X)^{12} = 1.12683$

a. $X \approx 0.1$ b. $X \approx 0.01$ c. $X \approx 0.001$
d. $X \approx 1$ e. none of the above

Identify the range(s) of X values that satisfy the following absolute values:

10. $|X+5| \leq 7$

a. $X \leq 2$ b. $X \leq 12$
c. $-12 \geq X \geq 2$ d. $-(X+5) \leq 7 \leq (X+5)$
e. $-12 \leq X \leq 2$

11. $|2X| < X+5$

a. $(X+5) < 2X < -(X+5)$ b. $-5/3 < X < 5$

c. $-2X < X+5 < 2X$ d. only (a) and (b) of the above
e. none of the above

12. Given $\left|\frac{x-\mu}{s}\right| \le t$, solve for μ if t=3, x= –5, and s=2.

a. $-11 \le \mu \le 1$ b. $-1 \le \mu \le 11$
c. $-11 \ge \mu$ or $1 \le \mu$ d. $11 \ge \mu$ or $-1 \le \mu$
e. none of the above

The following is called the quadratic formula that is used to solve the roots of a quadratic equation:

Given $aX^2 + bX + c = 0$

$$X_1, X_2 = \frac{-b \pm \sqrt{b^2 - 4ac}}{2a}$$ where X_1 and X_2 are the roots.

Solve for the roots of the following equations by using the above quadratic equation:

13. $\frac{1}{9}X^2 - 9Y^2 = 0$

a. $X = -9Y, 9Y$ b. $X = -9Y, 3Y$
c. $X = -9, 9$ d. $X = -3, 3$
e. none of the above

14. $-1.7X^2 + 5.1X = 3.4$

a. $X = 2, 3$ b. $X = -1, -2$ c. $X = 1, 2$
d. $X = -2, 4$ e. none of the above

Solve the following systems of simultaneous equations.

15. 3Y + 15 = 15X
13X – 19 = 2Y

a. X = 1; Y = 2 b. X = 2; Y = 9 c. X = 3; Y = 9
d. X = 3; Y = 10 e. None of the above

16. 2.5X + 3Y – 28 = 0
6X – 4.5Y + 3 = 0

a. X = 2; Y = 8 b. X = 4; Y = 6 c. X = 6; Y = 4
d. X = 8; Y = 2 e. None of the above

Solve the following annuity problems:

17. If you deposit $100 in a bank account that yields an annual percentage rate (APR) of 6%, what will be the balance at the end of a one-year period if there is a semiannual compounding?

a. $106.00 b. $106.09 c. $103.00
d. $112.00 e. none of the above

18. If you are to deposit $1000 into your 401(k) every year for 30 years until your retirement, how much will you retire with if you are to earn 5% per year on your 401(k)?

a. $66,438.85 b. $60,000.00 c. $30,000.00
d. $1,000,000 e. none of the above

19. If you are to deposit $1000 into your 401(k) every year for 30 years until your retirement, how much will you retire with if you are to earn 10% per year on your 401(k)?

a. $33.000.00 b. $160,000.05 c. $164,494.02
d. $1,000,000 e. none of the above

20. If you just received a 15-year $300,000 mortgage, what will be your monthly payment if the interest rate is 8%?

a. $2,866.96 b. $2,860.00 c. $2,800.00
d. $2,900.01 e. none of the above

21. Which of the following is correct?

a. $\frac{\infty}{\infty} = 1$ b. $\frac{0}{8} = 0$ c. $X^0 = 0$

d. all of the above e. only (a) and (b) of the above

Chapter 2. Calculus Review

A. An Introduction to Calculus Review

The following 4 questions are to be answered in this chapter on Calculus Review:

1. What is calculus?

   There are two types of calculus, addressing different issues:

   a. Differential calculus → measures a slope of a function
   b. Integral calculus → measures an area underneath a function

   Note that differential calculus is the primary topic of this course and is most often used in economics, finance, and management.

2. Why do we need differential calculus in business?

   To solve optimization[8] problems such as profit maximization, cost minimization, revenue maximization, customer satisfaction maximization, etc.

3. How do we do it?

   Know the 6 rules of differentiation that will be discussed herein below[9].

4. How do we use it?

   Differential calculus is often used as a tool (or language) to understand various economic, financial, and business phenomena better as will be soon elaborated herein with examples and applications.

B. Derivatives: A Definition

A function is a mathematical expression of how one variable is related to another variable(s). If a variable Y is related to a variable X, it can be expressed as: $Y = f(X)$. This means that given a value of X, the corresponding value of Y can be

---

[8] Optimization refers to the process of finding either a maximum or a minimum of a given function or relation.

[9] Often, due to simplicity of problems, one can get by with the basic 3 rules of differentiation if one utilizes effectively the rules of exponents in solving optimization problems.

determined. This initial or original function is called the "primary" function and is read as "**Y is** a function of **X**," which means that Y is determined by X.

If Y is related to many variables such as X, W, and Z, its primary functional form is expressed as: $Y = f(X, W, Z)$ and read as: "Y is a function of X, W, and Z."

In economics, when an equation is written as above, Y is called a dependent or endogenous variable and X (W and Z also) is called an independent or exogenous variable. This designation implies that X (W and Z) causes or determines Y.

A derivative (function) = a function obtained by differentiating a primary function with respect to a particular variable of one's interest = the derivative of Y with respect to X can be denoted in many different ways as shown below:

$\frac{dY}{dX}$ or $\frac{\partial Y}{\partial X}$ → Leibniz notation

Y' or f' → Lagrangean notation

$\dot{Y}$ → Newtonian notation

Among these and other notations for a first derivative (function), the Leibniz and Lagrangean notations are most often used in business whereas the Newtonian notation is most often used in physics.

Any derivative function of a primary function can be thought of as a slope function and generically expressed as $\frac{\Delta Y}{\Delta X}$ where:

$$\frac{\Delta Y}{\Delta X} = slope = \frac{\text{the change in } Y}{\text{the change in } X} = \frac{\text{the vertical displacement}}{\text{the horizontal displacement}} = \frac{\text{the rise}}{\text{the run}}.$$

Therefore, it is critically important to note that the following expressions for the first derivative function are treated be equivalent[10]:

$$Slope = \frac{\text{the rise}}{\text{the run}} = \frac{\Delta Y}{\Delta X} = \frac{dY}{dX} = \frac{\partial Y}{\partial X} = Y'$$

Technically, the correct meaning of a derivative can be better defined as:

[10] This statement is not true because partial derivatives are not equivalent to total derivatives. However, for simplicity, convenience, and practicality, we will not make the distinction unless it becomes absolutely necessary.

$$\frac{dY}{dX} = Y' = \lim_{\Delta X \to 0} \frac{\Delta Y}{\Delta X} = \lim_{\Delta X \to 0} \frac{f(X + \Delta X) - f(X)}{\Delta X}$$

That is, the derivative means finding the limit of $\frac{\Delta Y}{\Delta X}$ as ΔX approaches zero → This process algebraically identifies a slope of a line that is tangent to the primary function at a specific value of X and it can have a different value as X takes a different value[11].

For example, the slope of a linear line such as "Y=a+bX" is a constant or fixed number, b. That is, at every point on the linear line, the slope is the same as shown in the graph below.

That is, Slope = $\frac{rise}{run} = \frac{\Delta Y}{\Delta X} = \text{b}$

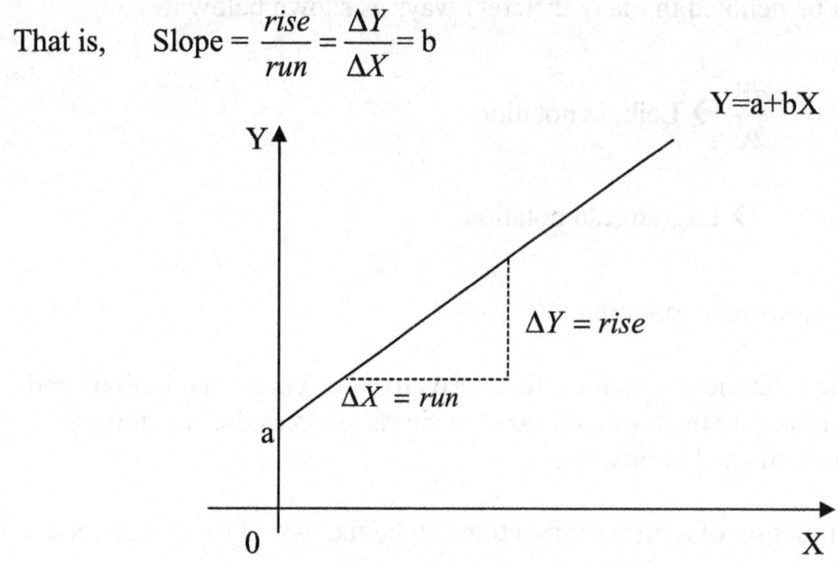

How about the slope of a constant function such as "Y = a"? Regardless of the magnitude of the constant, a, because it is a perfectly flat line, parallel to the horizontal axis, its rise is zero. That is, because $rise = \Delta Y = 0$, its slope is zero as shown in the graph below.

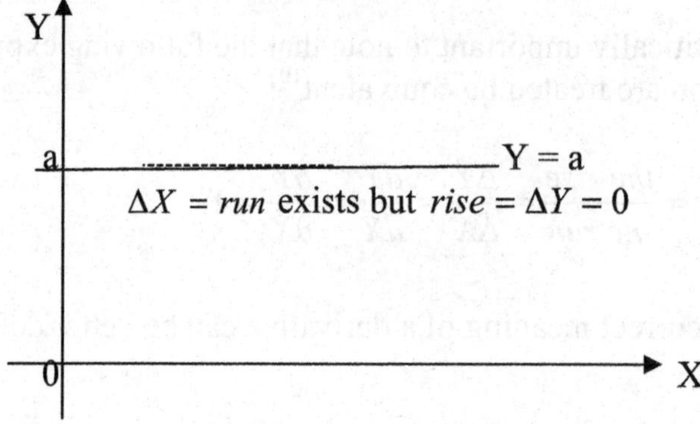

[11] For a more detailed description on this topic, visit http://en.wikipedia.org/wiki/Derivative.

However, the slope of a nonlinear function will vary as the shape of the primary function varies at different values of X chosen[12]. This is self-evident as shown in the following graph where the slope (value) is different at a different value of X such as X_1 and X_2.

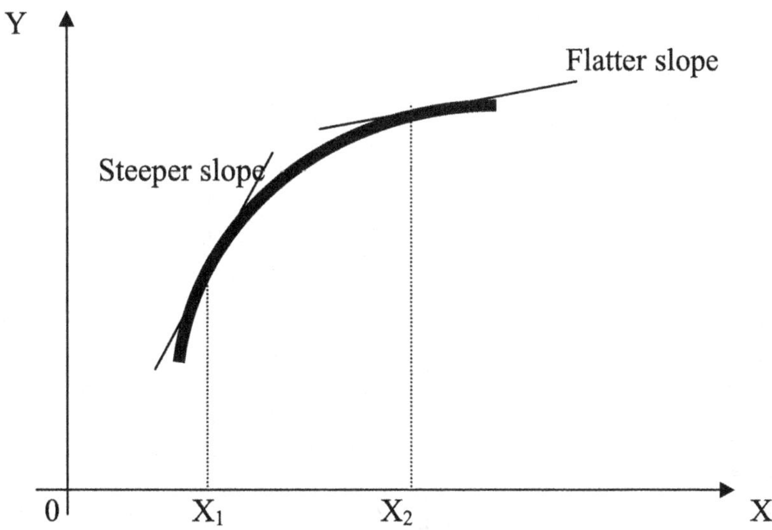

C. **Rules for Differentiation for One Variable:**

The following rules of differentiation generalize and formularize what we intuitively understand about the nature of a slope (function) given a primary function as discussed above. That is, the following differentiation rules provide the process of finding the slope function methodically, not intuitively[13].

1. The Constant Function Rule:

Given a primary function of $Y = a$ where a is any constant,

then, its derivative (or slope function) is: $\frac{dY}{dX} = Y' = 0$

Example 1:

If $Y = 10$, then $\frac{dY}{dX} = Y' = 0$

---

[12] To minimize the possible error in measuring the slope of a primary function at a chosen value of X, the change in X is allowed to be very very small, approaching almost a zero. We call this an infinitesimal change in X.

[13] This means that one must memorize these differentiation rules to find a derivative (or slope) function.

Example 2:

If Y = 10 million, then $\frac{dY}{dX} = Y' = 0$

2. The Power Function Rule:

Given a primary function of $Y = aX^b$

then, its derivative (with respect to X) is: $\frac{dY}{dX} = Y' = baX^{b-1}$

Example 1:

If $Y = 10X^3$, then $\frac{dY}{dX} = Y' = 3 \cdot 10X^{3-1} = 30X^2$

Example 2:

If Y = X, then $\frac{dY}{dX} = Y' = 1 \cdot X^{1-1} = 1 \cdot X^0 = 1$

Example 3:

If $Y = bX$, then $\frac{dY}{dX} = Y' = 1 \cdot bX^{1-1} = 1 \cdot bX^0 = b$

3. The Sums and Differences Rule[14]:

Given a primary function of $Y = U \pm W$

where $U = U(X)$ and $W = W(X)$

then, its derivative (with respect to X) is: $\frac{dY}{dX} = Y' = \frac{dU}{dX} \pm \frac{dW}{dX}$

Example 1:

If $Y = 5X^4 + 7X + 3$,

[14] Note that this is not much of a rule. It simply states that when one term is added (=summed) to or subtracted (=differenced) from another term in a primary function, you must keep the same mathematical operation of addition or subtraction when finding the derivative. If this generic formula is hard to understand, focus on the examples by paying attention to how the plus (+) and minus (-) signs are kept unchanged from a primary function to a derivative function.

then $\frac{dY}{dX} = Y' = 4 \cdot 5X^{4-1} + 1 \cdot 7X^{1-1} + 0 = 20X^3 + 1 \cdot 7 \cdot 1 = 20X^3 + 7$

Example 2:

If $Y = 5X^2 - 8X - 4$,

then $\frac{dY}{dX} = Y' = 2 \cdot 5X^{2-1} - 1 \cdot 8X^{1-1} - 0 = 10X - 8$

Example 3:

If $Y = a + bX$, then $\frac{dY}{dX} = Y' = 0 + 1 \cdot bX^{1-1} = 1 \cdot bX^0 = b$

4. The Product Rule:

Given a primary function of $Y = U \cdot W$

where $U = U(X)$ and $W = W(X)$

then, its derivative (with respect to X) is:

$$\frac{dY}{dX} = Y' = \frac{dU}{dX} \cdot W + \frac{dW}{dX} \cdot U$$

Example 1:

If $Y = 2X(X^2 + 1)$, then

$$\frac{dY}{dX} = Y' = 2 \cdot (X^2 + 1) + (2X) \cdot 2X = 2X^2 + 2 + 4X^2 = 6X^2 + 2$$

Example 2:

If $Y = (2X^3 - X)(X + 1)$, then

$$\frac{dY}{dX} = Y' = (3 \cdot 2X^{3-1} - 1)(X + 1) + (1) \cdot (2X^3 - X) = (6X^2 - 1)(X + 1) + (2X^3 - X)$$

5. The Quotient Rule:

Given a primary function of $Y = \frac{U}{W}$

where $U = U(X)$ and $W = W(X)$

then, its derivative (with respect to X) is:

$$\frac{dY}{dX} = Y' = \frac{\frac{dU}{dX} \cdot W - \frac{dW}{dX} \cdot U}{W^2}$$

Example 1:

If $Y = \frac{1}{X}$, then $\frac{dY}{dX} = Y' = \frac{0 \cdot X - (1 \cdot X^{1-1}) \cdot 1}{X^2} = -\frac{1}{X^2}$

Example 2:

If $Y = \frac{X+1}{2X^2}$, then

$$\frac{dY}{dX} = Y' = \frac{1 \cdot (2X^2) - (X+1) \cdot 4X}{(2X^2)^2} = \frac{2X^2 - 4X^2 - 4X}{4X^4}$$

$$= \frac{-2X^2 - 4X}{4X^4} = \frac{-2X - 4}{4X^3} = \frac{-X - 2}{2X^3}$$

6. The Chain Rule (=The Function-of-Functions Rule)

Given a primary function of $Y = f(U)$ where $U = g(X)$

then, its derivative (with respect to X) is: $\frac{dY}{dX} = \frac{dY}{dU} \cdot \frac{dU}{dX}$

Example 1:

If we are to take a derivative of:

$Y = (3X^2 + 4)^2$

then, we can alternatively express it as a function of functions as follows:

Let $Y = U^2$ and $U = 3X^2 + 4$

We can now apply the chain rule and find:

$$\frac{dY}{dU} = 2U \text{ and } \frac{dU}{dX} = 6X$$

Therefore,

$$\frac{dY}{dX} = Y' = \frac{dY}{dU} \cdot \frac{dU}{dX} = 2U \cdot 6X = 12(3X^2 + 4) \cdot X = 36X^3 + 48X$$

Example 2:

If $Y = 5 \cdot (3X^2 + 4)^{10}$

We can let $Y = 5U^{10}$ and $U = 3X^2 + 4$

We now note that $\frac{dY}{dU} = 50U^9$ and $\frac{dU}{dX} = 6X$

Therefore,

$$\frac{dY}{dX} = Y' = \frac{dY}{dU} \cdot \frac{dU}{dX} = 50U^9 \cdot 6X = 300X \cdot U^9 = 300X(3X^2 + 4)^9$$

D. Algebraic Short-cuts for the Product Rule and the Quotient Rule

Out of these 6 rules, one may only need the constant function rule, the power function rule, and the sums-and-differences rule to solve many economics problems. That is, one can often bypass the product rule, the quotient rule, and the chain rule by simplifying or reducing the primary function. In order to simplify equations, it is essential that one must memorize the following exponent rules:

Exponent Rule Refresher:

i) $X^0 = 1$ ii) $X^{-a} = \frac{1}{X^a}$ iii) $X^{\frac{b}{a}} = \sqrt[a]{X^b}$

iv) $\frac{X^a}{X^b} = X^{a-b}$ v) $(X^a) \cdot (X^b) = X^{a+b}$ vi) $(X^a)^b = X^{a*b}$

The following examples show how manageable differentiation – that is, taking a derivative of Y with respect X – is if you memorize the exponent rules and apply them to identify the primary function before you apply the differentiation rules.

1. $Y = \frac{1}{X} = X^{-1}$ $\rightarrow \frac{dY}{dX} = Y' = (-1) \cdot X^{-1-1} = -X^{-2} = -\frac{1}{X^2}$

2. $Y = \frac{3}{4X^2} = \frac{3}{4} X^{-2}$ $\rightarrow \frac{dY}{dX} = Y' = (-2) \cdot \frac{3}{4} \cdot X^{-2-1} = -\frac{3}{2} X^{-3} = -1.5X^{-3} = -\frac{3}{2X^3}$

3. $Y = 3 \cdot \sqrt[4]{X^3}$ $\rightarrow Y = 3X^{\frac{3}{4}}$ $\rightarrow \frac{dY}{dX} = Y' = \frac{3}{4} \cdot 3 \cdot X^{\frac{3}{4}-1} = \frac{9}{4} X^{-\frac{1}{4}} = \frac{9}{4 \cdot \sqrt[4]{X}}$

4. $Y = (X+1)(X^2-1)$ $\rightarrow Y = X^3 + X^2 - X - 1$ $\rightarrow \frac{dY}{dX} = Y' = 3X^2 + 2X - 1$

5. $Y = \frac{X+1}{2X^2}$ $\rightarrow Y = 0.5X^{-1} + 0.5X^{-2}$

Therefore, $\frac{dY}{dX} = Y' = -0.5X^{-2} - X^{-3} = \frac{-0.5X - 1}{X^3} = \frac{-X-2}{2X^3}$

6. $Y = \frac{(2X-1)^2}{(2X)^3}$ $\rightarrow Y = \frac{4X^2 - 4X + 1}{8X^3} = 0.5X^{-1} - 0.5X^{-2} + 0.125X^{-3}$

Therefore, $\frac{dY}{dX} = Y' = -0.5X^{-2} + X^{-3} - 0.375X^{-4}$

THEREFORE, WHENEVER POSSIBLE, ALWAYS EXPAND AND/OR SIMPLIFY FIRST BEFORE TAKING A DERIVATIVE! Then, in many cases, you can get by with the basic 3 rules of the Constant Function rule, the Power Function rule, and the Sums-and-Differences rule.

E. The First Derivative and the Second Derivative

The function obtained by taking a derivative of a primary function for the first time is called a derivative. However, its official and real name is a "first" derivative because the derivative is taken for the "first" time. Therefore, what has been described up to this point is the process of taking the first derivative of Y with respect to X. That is, we found the first derivative of Y with respect to X.

If one takes a derivative of the first derivative with respect to X – that is, take a derivative of the derivative function or take a derivative of a primary function "second" time – it is called the "second" derivative of a primary function. The rules for taking a second derivative is exactly the same as the rules applied for taking the first derivative.

The same procedure and concept applies to the third derivative, the fourth derivative, etc. That is, a third derivative is found by taking a derivative of the second derivative and a fourth derivative is found by taking a derivative of the third derivative, etc.

However, there are following common notations for the second derivative:

$\frac{d^2Y}{dX^2}$ → Leibniz Notation or Y'' → Lagrangean Notation

Examples: Find a second derivative of the following functions:

1) Given $Y = 2X^4 - 3X^{-2} + 4X - 10$, then $\frac{dY}{dX} = Y' = 8X^3 + 6X^{-3} + 4$.

Therefore, $\frac{d^2Y}{dX^2} = Y'' = 24X^2 - 18X^{-4}$

2) Given $Y = 2X^{0.5} + 10X^2 + 4X^{-1} + 110$, then $\frac{dY}{dX} = Y' = X^{-0.5} + 20X - 4X^{-2}$.

Therefore, $\frac{d^2Y}{dX^2} = Y'' = -0.5X^{-1.5} + 20 + 8X^{-3}$

Note 1: In order to find the second derivative, one must find the first derivative. There is no short-cut or by-pass to this process.

Note 2: The first derivative is the slope function of the primary function. The second derivative is, then, the slope of the slope function. The meaning of the third derivative and beyond is, however, hard to interpret and thus, not used much in economics or finance.

F. Total Derivatives and Partial Derivatives

For simplicity, it is defined herein that total differentiation is for the case of one independent variable whereas partial differentiation is for the case of more-than-one independent variables.[15] **The partial differentiation is similar to total differentiation in all aspects except that only one independent variable is allowed to change while all other variables are assumed to be held as unchanged or constant[16].**

To distinguish partial differentiation from total differentiation, it utilizes a lower-case Greek letter, delta(∂), instead of a lower-case English letter, d, and is denoted as follows:

$$\text{Given } Y = f(X,Z,W) = 3X^2 - 2XZ^3 + 5XW - XZ^2 + ZW^2 - 10$$

A partial derivative with respect to X:

$$\left.\frac{dY}{dX}\right|_{Z,W=\text{constant}} = \frac{\partial Y}{\partial X} = 6X - 2Z^3 + 5W - Z^2$$

A partial derivative with respect to Z:

$$\left.\frac{dY}{dZ}\right|_{X,W=\text{constant}} = \frac{\partial Y}{\partial Z} = -6XZ^2 - 2XZ + W^2$$

A partial derivative with respect to W:

$$\left.\frac{dY}{dW}\right|_{X,Z=\text{constant}} = \frac{\partial Y}{\partial W} = 5X + 2ZW$$

Note: Due to simplicity and convenience, the lower-case Greek letter, delta(∂), is used for partial derivatives, instead of the lower-case English letter, d, and the bar.

Example:

Find partial derivatives of the following function with respect to X, Z, and W:

$$Y = f(X,Z,W) = X^2Z^2W^2 - 3XZ^3W^4 + 6Z^4W^5 - 10XZW - 100X$$

Solution:

$$\frac{\partial Y}{\partial X} = Y'(X) = Y_X' = 2XZ^2W^2 - 3Z^3W^4 - 10ZW - 100$$

---

[15] This is a simplified and convenient description of total vs. partial differentiation. In differential calculus, however, there exists total differentiation for the case of more-than-one independent variables.

[16] The key to partial differentiation is treating all other variables except the one being differentiated by as constants such as a fixed number of say, 2.

$$\frac{\partial Y}{\partial Z} = Y'(Z) = Y_Z' = 2X^2ZW^2 - 9XZ^2W^4 + 24Z^3W^5 - 10XW$$

$$\frac{\partial Y}{\partial W} = Y'(W) = Y_W' = 2X^2Z^2W - 12XZ^3W^3 + 30Z^4W^4 - 10XZ$$

G. Practice Problems for Calculus Review

In differentiating the following Y functions with respect to X, **identify and select all correct mathematical expressions as the right answer.**

1. $Y = 200X + 3000 + 20X^2$

   a. $\frac{dY}{dX} = Y' = 200 + 3000 + 20X^2$

   b. $\frac{dy}{dX} = Y' = 200 + 40X^2$

   c. $\frac{dY}{dX} = Y' = 200 + 40X$

   d. $\frac{dY}{dX} = Y' = 40X^2 + 200$

   e. only (b) and (d) of the above

2. $Y = \frac{1}{2}X^2 - \frac{2}{3}X^{-3} + 10X - 5$

   a. $Y' = \frac{1}{2}X - \frac{2}{3}X^{-2} + 10X - 5$

   b. $Y' = \frac{1}{2}X - \frac{2}{3}X^{-4} + 10X - 5$

   c. $Y' = X + \frac{2}{3}X^{-4} + 10$

   d. $Y' = X + 2X^{-4} + 10$

   e. $Y' = X - 2X^{-4} + 10$

3. $Y = 25 - 5X + 3X^{1.5}$

   a. $Y' = -5 + 4.5X^{0.5}$

   b. $Y' = -5 + 3X^{0.5}$

   c. $Y' = 4.5X^{\frac{1}{2}} - 5$

   d. $Y' = -5X + 4.5X^{0.5}$

   e. only (a) and (c) of the above

4. $Y = aX^{\frac{1}{a}} + 2bX^{\frac{2}{b}} - c$

a. $Y' = -a^2 X^{\frac{1}{a}-1} + 2b^2 X^{\frac{2}{b}-1}$ b. $Y' = -X^{\frac{1}{a}-1} - 4X^{\frac{2}{b}-1}$

c. $Y' = X^{\frac{1}{a}-1} + 4X^{\frac{1}{b}}$ d. $Y' = X^{\frac{1}{a}-1} + 4X^{\frac{2}{b}-1}$

e. only (c) and (d) of the above

5. $Y = (X+1)(2X-3)$

a. $Y' = 2X - 2$ b. $Y' = 4X - 1$

c. $Y' = 4X - 2$ d. $Y' = 4X + 5$

e. $Y' = 2X - 1$

6. $Y = (2X+1)^2 X^2$

a. $Y' = 16X^3 + 12X^2 + 2X$ b. $Y' = 8X^3 + 4X^2$

c. $Y' = X^2(8X+4) + (4X^2 + 4X + 1) \cdot 2$

d. $Y' = 6X^2 + 2X$ e. none of the above is correct.

7. $Y = (X+1)^2(X^2+1)$

a. $Y' = 2(X+1)(2X)$ b.
$Y' = 2(X+1)(X^2+1) + (X^2+1)(2X)$

c. $Y' = 2X$ d. $Y' = 4X^3 + 6X^2 + 4X + 2$

e. only (b) and (d) of the above

8. $Y = \dfrac{2X-1}{3X}$

a. $Y' = \frac{2}{3}$ b. $Y' = \frac{3 \cdot (2X - 1)}{(3X)^2}$ c. $Y' = \frac{4X + 1}{3X^2}$

d. $Y' = \frac{1}{3}X^2$ e. $Y' = \frac{1}{3X^2}$

9. $Y = \frac{10X + 1}{X^2}$

a. $Y' = 10X^{-2} + 2X^{-3}$ b. $Y' = -10X^{-2} - 2X^{-3}$

c. $Y' = -\frac{10}{X^2} - \frac{2}{X^3}$

d. only (a) and (b) of the above e. only (b) and (c) of the above

10. $Y = \frac{10X^2 + 2}{2 + 3X}$

a. $Y' = 5X^2 + \frac{2}{3}X$ b. $Y' = \frac{30X^2 + 40X - 6}{(2 + 3X)^2}$

c. $Y' = \frac{90X^2 + 40X + 6}{(2 + 3X)^2}$ d. $Y' = \frac{-30X^2 - 40X + 6}{(2 + 3X)^2}$

e. none of the above

Answers to Practice Problems for Calculus Review

1. $Y = 200X + 3000 + 20X^2$

Answer: c.* $\frac{dY}{dX} = Y' = 200 + 40X$

By applying the constant function rule, the sums-and-differences rule and the power function rule, we find:

$$\frac{dY}{dX} = Y' = 200X^{1-1} + 0 + 2 \cdot 20X^{2-1} = 200 + 40X$$

2. $Y = \frac{1}{2}X^2 - \frac{2}{3}X^{-3} + 10X - 5$

Answer: d.* $Y' = X + 2X^{-4} + 10$

By applying the constant function rule, the sums-and-differences rule and the power function rule, we find:

$$\frac{dY}{dX} = Y' = 2 \cdot \frac{1}{2}X^{2-1} - (-3) \cdot \frac{2}{3}X^{-3-1} + 1 \cdot 10X^{1-1} - 0$$

$$= \cancel{2} \cdot \frac{1}{\cancel{2}}X^{2-1} - (-\cancel{3}) \cdot \frac{2}{\cancel{3}}X^{-3-1} + 1 \cdot 10X^{1-1} - 0$$

$$= X + 2X^{-4} + 10$$

3. $Y = 25 - 5X + 3X^{1.5}$

Answer: e.* only (a) and (c) of the above

By applying the constant function rule, the sums-and-differences rule and the power function rule, we find:

$$\frac{dY}{dX} = Y' = 0 - 1 \cdot 5X^{1-1} + 1.5 \cdot 3X^{1.5-1}$$

$$= 0 - 5X^0 + 4.5X^{0.5} = -5 + 4.5X^{0.5} = -5 + 4.5X^{\frac{1}{2}} = 4.5X^{\frac{1}{2}} - 5$$

Therefore, both a. $Y' = -5 + 4.5X^{0.5}$

and c. $Y' = 4.5X^{\frac{1}{2}} - 5$ are correct.

4. $Y = aX^{\frac{1}{a}} + 2bX^{\frac{2}{b}} - c$

Answer: d.* $Y' = X^{\frac{1}{a}-1} + 4X^{\frac{2}{b}-1}$

By applying the constant function rule, the sums-and-differences rule and the power function rule, we find:

$$\frac{dY}{dX} = Y' = \frac{1}{a} \cdot aX^{\frac{1}{a}-1} + \frac{2}{b} \cdot 2bX^{\frac{2}{b}-1} - 0 = X^{\frac{1}{a}-1} + 4X^{\frac{2}{b}-1}$$

5. $Y = (X+1)(2X-3)$

Answer: b.* $Y' = 4X - 1$

First, applying the simplification method, we find

$$Y = (X+1)(2X-3) = (2X^2 + 2X - 3X - 3) = 2X^2 - X - 3$$

Next, by applying the constant function rule, the sums-and-differences rule and the power function rule, we find:

$$\frac{dY}{dX} = Y' = 2 \cdot 2X^{2-1} - 1 \cdot X^{1-1} - 0 = 4X - 1$$

Of course, we could have used the product rule as follows:

$$\frac{dY}{dX} = Y' = (X+1) \cdot (2) + (2X-3) \cdot (1) = 4X - 1$$

6. $Y = (2X+1)^2 X^2$

Answer: a.* $Y' = 16X^3 + 12X^2 + 2X$

First, applying the simplification method, we find

$$Y = (2X+1)^2 X^2 = (4X^2 + 4X + 1) \cdot X^2 = 4X^4 + 4X^3 + X^2$$

Next, by applying the constant function rule, the sums-and-differences rule and the power function rule, we find:

$$\frac{dY}{dX} = Y' = 4 \cdot 4X^{4-1} + 3 \cdot 4X^{3-1} + 2 \cdot X^{2-1} = 16X^3 + 12X^2 + 2X$$

Of course, we could have used the product rule and the chain rule as follows:

Defining $U = (2X+1)$ and $W = X^2$, we can express the original function as

$$Y = U^2 \cdot W$$

Furthermore, we have to define $V = U^2$ and thus, $Y = V \cdot W$

Therefore,

$$\frac{dY}{dX} = Y' = \frac{dV}{dX} \cdot W + \frac{dW}{dX} \cdot V = \left[\frac{dV}{dU} \cdot \frac{dU}{dX}\right] \cdot W + \frac{dW}{dX} \cdot V$$

$$= 2U \cdot 2 \cdot W + 2X \cdot U^2 = 4 \cdot (2X+1) \cdot X^2 + 2X \cdot (2X+1)^2$$
$$= 16X^3 + 12X^2 + 2X$$

Note: Using the product rule and the chain rule can have some disadvantages due to their complexity. For our purpose in GSB 420, an emphasis should be given to the simplification method whenever possible.

7. $Y = (X+1)^2(X^2+1)$

Answer: d.* $Y' = 4X^3 + 6X^2 + 4X + 2$

First, applying the simplification method, we find

$$Y = (X+1)^2(X^2+1) = (X^2+2X+1) \cdot (X^2+1) = X^4 + 2X^3 + 2X^2 + 2X + 1$$

Next, by applying the constant function rule, the sums-and-differences rule and the power function rule, we find:

$$\frac{dY}{dX} = Y' = 4 \cdot X^{4-1} + 3 \cdot 2X^{3-1} + 2 \cdot 2X^{2-1} + 2X^{1-1} + 0 = 4X^3 + 6X^2 + 4X + 2$$

Of course, we could have used the product rule and the chain rule as follows:

Defining $U = (X+1)$ and $W = X^2 + 1$, we can express the original function as

$$Y = U^2 \cdot W$$

Furthermore, we have to define $V = U^2$ and thus, $Y = V \cdot W$

Therefore,

$$\frac{dY}{dX} = Y' = \frac{dV}{dX} \cdot W + \frac{dW}{dX} \cdot V = \left[\frac{dV}{dU} \cdot \frac{dU}{dX}\right] \cdot W + \frac{dW}{dX} \cdot V$$

$$= 2U \cdot 1 \cdot W + 2X \cdot U^2 = 2 \cdot (X+1) \cdot (X^2+1) + 2X \cdot (X+1)^2$$
$$= (2X^3 + 2X + 2X^2 + 2) + 2X \cdot (X^2 + 2X + 1)$$
$$= 4X^3 + 6X^2 + 4X + 2$$

Note: Once again, using the product rule and the chain rule can have some disadvantages due to their complexity. For our purpose in GSB 420, an emphasis should be given to the simplification method whenever possible.

8. $Y = \dfrac{2X-1}{3X}$

Answer: e.* $Y' = \dfrac{1}{3X^2}$

First, applying the simplification method, we find

$$Y = \frac{2X}{3X} - \frac{1}{3X} = \frac{2}{3} - \frac{1}{3} \cdot X^{-1}$$

Next, by applying the constant function rule, the sums-and-differences rule and the power function rule, we find:

$$\frac{dY}{dX} = Y' = (-1) \cdot (-\frac{1}{3}) \cdot X^{-1-1} = \frac{1}{3} X^{-2} = \frac{1}{3X^2}$$

Of course, we could have used the quotient rule as follows:

$$\frac{dY}{dX} = Y' = \frac{3X \cdot (2) - (2X-1) \cdot 3}{(3X)^2} = \frac{6X - 6X + 3}{9X^2} = \frac{1}{3X^2}$$

9. $Y = \dfrac{10X+1}{X^2}$

Answer: e.* only (b) and (c) of the above

Because $Y' = -10X^{-2} - 2X^{-3} = -\dfrac{10}{X^2} - \dfrac{2}{X^3}$

First, applying the simplification method, we find

$$Y = \frac{10X+1}{X^2} = \frac{10X}{X^2} + \frac{1}{X^2} = 10X^{-1} + X^{-2}$$

Next, by applying the constant function rule, the sums-and-differences rule and the power function rule, we find:

$$\frac{dY}{dX} = Y' = -10X^{-2} - 2X^{-3} = -\frac{10}{X^2} - \frac{2}{X^3}$$

Of course, we could have used the quotient rule as follows:

$$\frac{dY}{dX} = Y' = \frac{X^2 \cdot (10) - (10X+1) \cdot 2X}{(X^2)^2} = \frac{-10X^2 - 2X}{X^4} = -\frac{10}{X^2} - \frac{2}{X^3} = -10X^{-2} - 2X^{-3}$$

10. $Y = \dfrac{10X^2 + 2}{2 + 3X}$

Answer: b.* $Y' = \dfrac{30X^2 + 40X - 6}{(2+3X)^2}$

In this situation where simplification is not possible, we must resort to the quotient rule as follows:

$$\frac{dY}{dX} = Y' = \frac{(2+3X)(20X) - (10X^2+2)(3)}{(2+3X)^2}$$

$$= \frac{40X + 60X^2 - 30X^2 - 6}{(2+3X)^2} = \frac{30X^2 + 40X - 6}{(2+3X)^2}$$

Exercise Problems on
Chapter 2. Calculus Review

This set of exercise problems has 20 problems, worth a total of 20 points.

In differentiating the following functions with respect to X, **select all algebraically correct expressions as answers.**

1. $Y = 2X + 4$

   a. $\dfrac{dY}{dX} = Y' = 2$ b. $\dfrac{dY}{dX} = Y' = 2X^2$

   c. $\dfrac{dY}{dX} = Y' = 2X^{-1}$ d. only (a) and (b) of the above

   e. none of the above

2. $Y = 25X^2 - 25X + 25$

   a. $Y' = 2 \cdot 25X - 25X$ b. $Y' = 50X - 25$

   c. $Y' = 50X - 25X^{-1}$

   d. all of the above e. none of the above

3. $Y = aX^n + bX^m - cX + d$

   a. $Y' = anX^n + mbX^m - cX$ b. $Y' = anX^{n-1} + mbX^{m-1} - 1$

   c. $Y' = anX^{n-1} + mbX^{m-1} - c$ d. $Y' = aX^{n-1} + bX^{m-1} - c$

   e. $Y' = nX^{n-1} + mX^{m-1} - 1 \cdot c - d$

4. $Y = \dfrac{2}{3}X^3 + \dfrac{5}{4}X^2 - \dfrac{2}{5}X^0$

   a. $Y' = 2X^2 + 2.5X$ b. $Y' = \dfrac{2}{3}X^2 + \dfrac{5}{2}X - \dfrac{2}{5}$

c. $Y' = \frac{2}{3}X^2 + \frac{5}{2}X$

d. $Y' = \frac{2}{3 \cdot 3}X^2 + \frac{5}{2}X$

e. $Y' = 2X^2 - 2.5X$

5. $Y = 20X^5X^3X^2$

a. $Y' = 5 \cdot 20X^4X^3X^2$

b. $Y' = 3 \cdot 5 \cdot 20X^4X^2X^2$

c. $Y' = 2 \cdot 3 \cdot 5 \cdot 20X^4X^2X$

d. $Y' = 200X^{10}$

e. $Y' = 200X^9$

6. $Y = 5X^3(2X + 3)$

a. $Y' = 10X^4 + 15X^3$

b. $Y' = 40X^3 + 45X^2$

c. $Y' = 15X^2 \cdot (2X + 3)$

d. only (a) and (b) of the above

e. none of the above

7. $Y = 2(X^2 - 3X - 1)(X + 2)$

a. $Y' = (2X^2 - 6X - 2)(X + 2)$

b. $Y' = 2X^3 - 2X^2 - 14X$

c. $Y' = 6X^2 - 4X - 14$

d. only (a) and (b) of the above

e. none of the above

8. $Y = \frac{25}{X - 1}$ (Hint: Use the quotient rule)

a. $Y' = -\frac{25}{(X-1)^2}$

b. $Y' = \frac{25}{(X-1)^2}$

c. $Y' = -\frac{25}{(X-1)}$

d. $Y' = \frac{25}{(X-1)}$

e. $Y' = 0$

9. $Y = \dfrac{25X + 2}{X^2 + 2}$ (Hint: Use the quotient rule)

a. $Y' = \dfrac{25X^2 + 4X - 50}{(X^2 + 2)}$

b. $Y' = \dfrac{-25X^2 - 4X + 50}{(X^2 + 2)^2}$

c. $Y' = \dfrac{-25X^2 - 4X + 50}{(2X)^2}$

d. $Y' = \dfrac{25X^2 + 4X - 50}{(X^2 + 2)^2}$

e. none of the above

10. $Y = \dfrac{5X^2 - 2X + 1}{X}$ (Hint: Use the reduction/simplification method)

a. $Y' = \dfrac{5X^2 - 2}{X^2}$

b. $Y' = 5 - X^{-2}$

c. $Y' = \dfrac{10X - 2}{X^2}$

d. only (a) and (b) of the above

e. none of the above

Find **the SECOND derivative** of the following functions with respect to X:

11. $Y = 3X^2 - 2X + 5$

a. $Y'' = 6X - 2$

b. $Y'' = 6$

c. $Y'' = 2 \cdot 3X^2 - 1 \cdot 2X$

d. $Y'' = 6X$

e. none of the above

12. $Y = aX^n - bX^{-m} + c$

a. $Y'' = naX^{n-1} + mbX^{-m-1}$

b. $Y'' = (n-1)naX^{n-2} + (-m-1)mbX^{-m-2}$

c. $Y'' = naX^{n-2} + mbX^{-m-2}$

d. only (a) and (c) of the above e. none of the above

4 • 2(x−1) (2x−2)•4

13. $Y = 4(X-1)^2$ (Hint: Use the reduction/simplification method)

a. $Y'' = 8X - 8$ b. $Y'' = 8X$

c. $Y'' = 8X + 8$ d. $Y'' = 8$

e. none of the above

14. $Y = 2(X+1)X^{-0.5}$ (Hint: Use the reduction/simplification method)

a. $Y'' = -0.5X^{-1.5} + 1.5X^{-2.5}$ b. $Y'' = 2X^{-0.5} - (X+1)X^{-1.5}$

c. $Y'' = X^{-0.5} - X^{-1.5}$ d. only (b) and (c) of the above

e. none of the above

15. $Y = 2\sqrt{X}$ 2X½ = ½ • 2x^(½−1) = −½x^(−1½) 1x^(−½)

a. $Y'' = X^{-0.5}$ b. $Y'' = \frac{1}{\sqrt{X}}$

c. $Y'' = -0.5X^{-1.5}$ d. only (a) and (b) of the above

e. none of the above

16. $Y = 6\sqrt[3]{X}$ 6x^(1/3) = 1/3 • 6x^(1/3−1)

a. $Y'' = 2X^{-2/3}$ b. $Y'' = -\frac{4}{3}X^{-5/3}$

c. $Y'' = -\frac{4}{3}X^{1/3}$ d. only (b) and (c) of the above

e. none of the above

17. $Y = \frac{1}{X}$

a. $Y'' = -X^{-2}$

b. $Y'' = 2X^{-3}$

c. $Y'' = \frac{2}{X^3}$

d. only (b) and (c) of the above

e. all of the above

18. $Y = \frac{X^2 - 1}{2X}$

a. $Y'' = \frac{1}{2} \cdot X - \frac{1}{2} X^{-1}$

b. $Y'' = 0.5 + 0.5X$

c. $Y'' = \frac{2X^2 + 2}{4X^2}$

d. $Y'' = -\frac{1}{X^3} = -X^{-3}$

e. all of the above

19. $Y = \frac{X+1}{X-1}$

a. $Y'' = -\frac{2}{(X-1)^2}$

b. $Y'' = \frac{4X-2}{(X-1)^4}$

c. $Y'' = \frac{4}{(X-1)^3}$

d. only (b) and (c) of the above

e. all of the above

20. $Y = \frac{2X - 3}{3X^2}$

a. $Y'' = \frac{-6X^2 + 18X}{9X^4} = \frac{-2X + 6}{3X^3}$

b. $Y'' = \frac{6X^5 - 18X^4}{9X^8} = \frac{2X - 9}{3X^4}$

c. $Y'' = \frac{12X^5 - 54X^4}{9X^8} = \frac{4X - 18}{3X^4}$

d. only (b) and (c) of the above

e. none of the above

Chapter 3. Optimization Methods

Optimization is a process of finding a value(s) of X that maximizes or minimizes a primary function.[17] The uniqueness of solving an optimization problem is that wherever a slope value is equal to zero at a given X value, then that X value identifies an optimum point – either a maximum of a minimum[18] - in the primary function of Y.

We can visualize this concept with the aid of the following graph:

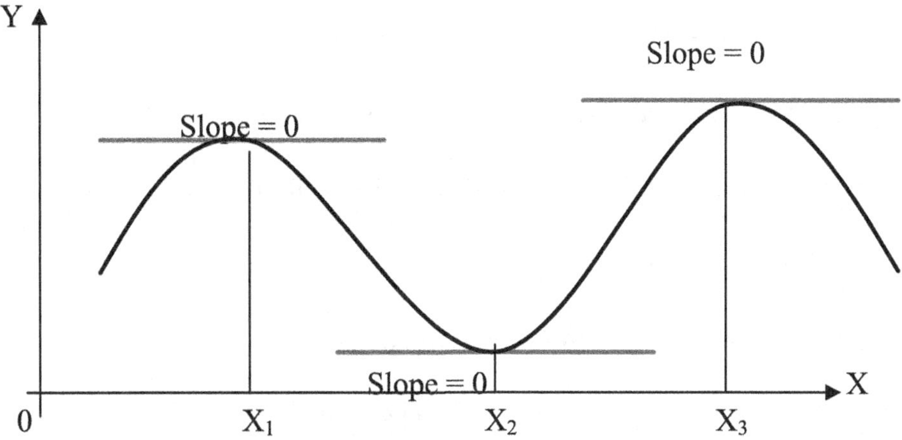

Because economics and businesses ask such questions as:

- What level of output will **maximize** a company's sales and profits?
- What level of output will **minimize** a company's total costs and average costs?
- How much should a company spend on advertisement to **maximize** sales?
- What price should a company charge to **maximize** profits?
- How many workers should a company hire to **maximize** output?
- How many units of X and Y to consume to **maximize** one's happiness?

it is essential that we know HOW to apply the concept of derivatives to economic and business problems. **Note that identifying the value of X where the slope of Y is equal to zero is the first key to solving an optimization problem.** Solving optimization problems requires a two-stage process, however.

The first stage involves the task of finding an optimum point – be it a maximum or a minimum – of a primary function by identifying the value of X that will make the slope function equal to zero. We can accomplish this task by identifying the first derivative function and set it equal to zero. This process is known as finding the **First Order**

[17] The third case of an inflexion point will not be discussed herein.

[18] Given multiple optimal points, a X value that is associated with a slope of zero identifies any "local" optimum. The largest or the smallest Y value associated with X is called a "global" optimum.

Condition (FOC). The second stage involves identifying the sign of the second derivative, known as the Second Ordr Condition (SOC).

These two stages to solving an optimization problem can be categorized into following steps:

Step 1. Identify the primary (or objective) function.

Step 2. Find its first derivative (or slope) function and set it equal to zero → this is called the FOC.

Step 3. Solve the FOC and identify the value of X.

Step 4. Find the second derivative function and check for its sign → the SOC.

For example, if we are to find the value of X, X*, that will maximize the following function:

$$Y = -X^2 + 4X - 3$$

we must faithfully follow the procedures described above:

Step 1: Identify the primary function← The primary function is known in this case. If not, known, you must identify it.

Step 2. Find its first derivative and set it equal to zero as follows:

$$\text{Slope} = \frac{rise}{run} = \frac{\Delta Y}{\Delta X} = \frac{dY}{dX} = -2X + 4 = 0 \quad \leftarrow \text{FOC}$$

Step 3: Solving this FOC, we get X*=2.

However, our job is not complete because we are not sure if this X value of 2 is associated with a maximum or a minimum point of the primary function[19]. Thus, additional work is needed to identify whether it is associated with a maximum or a minimum. This additional task is accomplished by taking the derivative of the first derivative, which is called "taking a second derivative" of the primary function. If the second derivative yields a negative value at the X value where the slope is zero, we conclude that the X value found is associated with a maximum. On the other hand, if the second derivative yields a positive value at the X value where the slope is zero, we conclude that the X value found is associated with a minimum. This is called the **Second Order Condition (SOC)** and can be summarized as follows: (Pay attention to the notation for the second derivative.)

SOC:

[19] Remember that when a slope is equal to zero, it means an optimum exists but it does not identify if it is a maximum or a minimum.

If $\frac{d^2Y}{dX^2} = Y'' < 0$,

then, the primary function has a Maximum at X* found via solving $\frac{dY}{dX} = 0$.

If $\frac{d^2Y}{dX^2} = Y'' > 0$,

then, the primary function has a Minimum at X* found via solving $\frac{dY}{dX} = 0$.

To complete the solution to the above example of $Y = -X^2 + 4X - 3$, its SOC is found as follows:

$\frac{d^2Y}{dX^2} = Y'' = -2 < 0$ → Therefore, the primary function has a Maximum at X*=2 as found above. In fact, we note that the maximum Y value of 1 ($= Y = -2^2 + 4 \cdot 2 - 3$) is found at X=2.

Examples of Solving Optimization Problems

1. Given the following relationship between sales (S) in millions of dollars and advertisement expenditures (A) in millions of dollars:

$$S = 20 + 10A - 2.5A^2$$

What level of advertising expenditures will MAXIMIZE sales revenue?

Solution Process:

Step 1. The primary (=objective) function is given in this case.
Step 2. Take the first derivative and set it equal to zero:

$$\frac{dS}{dA} = 10 - 5A = 0$$

Step 3. Solve for A.

$10 - 5A = 0$ → A=2

Step 4. In order to make sure that A=2 is associated with a maximum, we must check the SOC as follows:

$$\frac{d^2S}{dA^2} = S'' = -5 \rightarrow$$ Because this second derivative yields a negative value, one can conclude that a maximum sales is associated with A=2.

2. Given the following profit function (π) in millions of dollars and output (Q) in units of thousands:

$$\pi = -4 - 5Q + 3Q^2 - \frac{Q^3}{3}$$

Identify the level of output, Q*, that will MAXIMIZE the profit.

Solution Process:

Step 1. The primary (=objective) function is given in this case.

Step 2 Take the first derivative and set it equal to zero. [Recall that this is known as the First Order Condition (FOC)].

$$\frac{d\pi}{dQ} = \pi' = -5 + 6Q - Q^2 = 0$$

Step 3. Solve for Q. Because the above FOC is a quadratic equation, we can solve for Q by using the following quadratic formula:

Given $aQ^2 + bQ + c = 0$

$$Q = \frac{-b \pm \sqrt{b^2 - 4ac}}{2a}$$

That is, given the FOC as:

$$\frac{d\pi}{dQ} = \pi' = -5 + 6Q - Q^2 = 0$$

We can identify the values of coefficients as:

$a = -1$, $b = 6$, and $c = -5$

Therefore,

$$Q = \frac{-6 \pm \sqrt{6^2 - 4 \cdot (-1) \cdot (-5)}}{2 \cdot (-1)} = \frac{-6 \pm \sqrt{16}}{-2} = 3 \pm 2 = 1, 5$$

Because we have two roots for Q (of 1 and 5), we need to check which one of the two is associated with a maximum and which one, with a minimum, by checking the Second Order Condition (SOC). In order to do this, we must plug in each root value into the second derivative equation and check its sign → note that its magnitude does not matter but its sign does.

Step 4. Check the SOC.

That is,

$$\text{Given } \frac{d^2\pi}{dQ^2} = \pi'' = 6 - 2Q$$

When Q=1, we get

$$\left.\frac{d^2\pi}{dQ^2}\right|_{Q=1} = \pi''\big|_{Q=1} = \pi''@(Q=1) = 6 - 2(1) = 4 > 0.$$

Therefore, a MINIMUM profit exists at Q=1. In fact, by plugging in this Q=1 into the profit function, we find:

$$\pi = -4 - 5Q + 3Q^2 - \frac{Q^3}{3} = -4 - 5(1) + 3(1)^2 - \frac{(1)^3}{3} = -6.33$$

On the other hand,

When Q=5, we get

$$\left.\frac{d^2\pi}{dQ^2}\right|_{Q=5} = \pi''\big|_{Q=5} = \pi''@(Q=5) = 6 - 2(5) = -4 < 0.$$

Therefore, a MAXIMUM profit exists at Q=5. Once again, by plugging in this Q=5 into the profit function, we find:

$$\pi = -4 - 5Q + 3Q^2 - \frac{Q^3}{3} = -4 - 5(5) + 3(5)^2 - \frac{(5)^3}{3} = 4.33$$

The conclusion is:

If the firm produces Q=1, it will experience a maximum loss of $6.33 million whereas if it produces Q=5, it will earn a maximum profit of $4.33 million.

3. An Example of a Multivariate Profit Function:

Assume that a firm sells the same product in two different markets. Q_1 is the quantity sold in Market 1 and Q_2 is the quantity sold in Market 2. The firm's profit (π) function is given below:

$$\pi = -5 - 8Q_1 + 10Q_2 - Q_1^2 - 3Q_2^2 + 4Q_1Q_2$$

Identify the levels of output, Q_1 and Q_2, that will MAXIMIZE the profit.

Solution Process:

Step 1. The primary (=objective) function is given in this case.

Step 2. Find the FOCs and set them equal to zero:

(Note that because there are two independent variables of Q_1 and Q_2, there must be two FOCs, resulting from partial differentiation with respect to Q_1 and Q_2.)

That is,

Partial derivative with respect to Q_1 is:

$$\frac{\partial \pi}{\partial Q_1} = -8 - 2Q_1 + 4Q_2 = 0 \qquad (1)$$

Partial derivative with respect to Q_2 is:

$$\frac{\partial \pi}{\partial Q_2} = 10 - 6Q_2 + 4Q_1 = 0 \qquad (2)$$

Step 3. Solve them – FOC (1) and FOC (2) above – simultaneously:

The elimination method is used in this case as follows:

i) Multiply (1) by 2 to eliminate Q_1.

$$\frac{\partial \pi}{\partial Q_1} \times 2 = -8 \times 2 - 2 \times 2Q_1 + 2 \times 4Q_2 = -16 - 4Q_1 + 8Q_2 = 0 \quad (3)$$

ii) Add (3) and (2).

$$-6+2Q_2=0 \quad \rightarrow \quad \text{Therefore, } Q_2=3$$

iii) Plug this value of Q_2=3 into either Equation (1) or (2) and get

$$-8-2Q_1+4Q_2=-8-2Q_1+4(3)=0 \quad \rightarrow \quad \text{Therefore, } Q_1=2$$

Or

$$10-6Q_2+4Q_1=10-6(3)+4Q_1=0 \quad \rightarrow \quad \text{Therefore, } Q_1=2.$$

iv) The profit at $Q_1 = 2$ and $Q_2 = 3$ is:

$$\pi=-5-8Q_1+10Q_2-Q_1^2-3Q_2^2+4Q_1Q_2$$

$$=-5-8(2)+10(3)-(2)^2-3(3)^2+4(2)(3)=2$$

Step 4. In order to check if the above Q_1 and Q_2 values of 2 and 3 are associated with a maximum profit, instead of a minimum profit, we must check their Second Order Condition as is the routine for the one-independent variable case. However, the mathematics needed to carry out this task is beyond our scope and thus, WE DO NOT CHECK THE SECOND ORDER CONDITION FOR THE CASE OF MULTI-INDEPENDENT VARIABLES. **We assume that the SOC is fulfilled when we work with a multi-variable case.**

Advanced Topic: Constrained Optimization

All of the above examples showed the application of an optimization technique to solving economic problems that had no constraint. However, in economics and other business disciplines such as finance, marketing, and management, there are situations where one may wish to know a solution to a problem that comes with one or more constraints.

For example, one may wish to know the optimum levels of TV or radio advertisements to run in order to achieve maximum sales or profits, given that the advertisement budget is fixed. Or in some cases, one may wish to know what quantities to sell in Markets 1 and 2 under the condition that the total quantity to sell is fixed.

In situations like this where there are equality constraints, we use a technique known as the Lagrangean (Multiplier) Method. The process for using this method is as follows:

1) Identify the primary objective function such as: $S = f(A_1, A_2)$

2) Identify the constraint and convert it as a zero-equality function. (Always put the numerical value first and subtract from it the variables!)[20] → call this the Converted Constraint.

3) Construct the NEW function called a Lagrangean (Multiplier) function as:

L = Primary Objective Function + λ(Converted Constraint)

Note that the sign associated with the lambda is always positive! This lambda is called the Lagrangean Multiplier and indicates an opportunity cost (or shadow price) of the constrained (or limited) resource.

4) Find a set of first derivative functions with respect to each of the variables in the Lagrangean function and set them equal to zero → That is, find a set of FOCs[21].

5) Solve them simultaneously → It is almost always the easiest to form a new equation by eliminating the Lagrangean multiplier, λ, from the first two FOCs; and then, use this newly formed equation with the equality constraint to solve for the optimum values of independent variables such as A_1 and A_2.

6) Conclude the values of the variables found as optimum solution.

7) That is, skip (or omit) the process of verifying a maximum or a minimum by using the SOC → This is because the verification procedure requires advanced matrix algebra.

This process of using a Lagrangean (Multiplier) Method for solving a constrained optimization problem[22] can be best described by an example as follows:

An Example of a Lagrangean (Multiplier) Method:

Your company sells DVDs in two regions: 1 and 2. Your sales staff came up with the following relationships between sales and advertising expenses in each of these regions:

$$S_1 = 50 + 10A_1 - A_1^2$$
$$S_2 = 20 + 5A_2 - 0.5A_2^2$$

where:

---

[20] If the arrangement is reversed such that the constraint is expressed as variables minus the numerical value, then the sign associated with the lambda in Step 3 should be negative.
[21] The number of FOCs will be equal to the number of independent variables including λ.
[22] This type of a Lagrangean function construction deals with the equality-constraints. For non-equality constraints, other advanced methods are used.

S_i = sales revenue in Region "i" (millions of \$)
A_i = advertising expenditures in Region "i" (millions of \$)

Your boss (or you) provides a total advertising budget of \$7 million. Identify the advertising expenses for each region that will maximize your company's total sales (from these two regions).

The following steps can be taken:

1) Identify the primary (or objective) function as

$$\text{Maximize} \quad S = S_1 + S_2$$
$$= (50 + 10A_1 - A_1^2) + (20 + 5A_2 - 0.5A_2^2)$$

2) Identify the constraint and convert it as a zero-equality function as follows: (Remember to put the numerical value first and subtract from it the variables!)

$$A_1 + A_2 = 7 \quad \rightarrow \quad 7 - A_1 - A_2 = 0$$

Note: These two pieces of information are usually combined and stated as follows:

$$\text{Maximize} \quad S = S_1 + S_2 = (50 + 10A_1 - A_1^2) + (20 + 5A_2 - 0.5A_2^2)$$
$$\text{Subject to:} \quad A_1 + A_2 = 7$$

3) Construct the NEW function called a Lagrangean function, L, as follows: (Remember that the sign associated with the lambda is positive because of the rearranged constraint in Step 2 above!) → Pay attention to how the Converted Constraint is incorporated into the Lagrangean function.

$$L = \text{Primary Objective Function} + \lambda(\text{Converted Constraint})$$
$$= S + \lambda(7 - A_1 - A_2)$$
$$= S_1 + S_2 + \lambda(7 - A_1 - A_2)$$
$$= (50 + 10A_1 - A_1^2) + (20 + 5A_2 - 0.5A_2^2) + \lambda(7 - A_1 - A_2)$$
$$= 70 + 10A_1 - A_1^2 + 5A_2 - 0.5A_2^2 + 7\lambda - \lambda A_1 - \lambda A_2$$

4) Find a set of first derivative functions with respect to each of the variables in the Lagrangean function and set them equal to zero → That is, find a set of FOCs as:

$$\frac{\partial L}{\partial A_1} = 10 - 2A_1 - \lambda = 0 \quad \rightarrow \quad (1)$$

$$\frac{\partial L}{\partial A_2} = 5 - A_2 - \lambda = 0 \qquad \rightarrow \qquad (2)$$

$$\frac{\partial L}{\partial \lambda} = 7 - A_1 - A_2 = 0 \qquad \rightarrow \qquad (3)$$

5) Solve these FOCs simultaneously → **Always get rid of λ first by using the first two equations of (1) and (2)** → Then, solve simultaneously the resulting equation with the third equation of (3) for either X or Y. That is,

Subtracting Equation (2) from (1)[23], we eliminate λ and find

$$5 - 2A_1 + A_2 = 0 \qquad \rightarrow \qquad (4)$$

Adding Equations (3) and (4), we find

$$12 - 3A_1 = 0 \qquad \rightarrow \qquad \text{Therefore, } A_1 = 4$$

Substituting this value of $A_1 = 4$ into either Equation (3) or (4) above, we find

$A_2 = 3$

Substituting $A_1 = 4$ and $A_2 = 3$ into either Equation (1) or (2) above, we find

$\lambda = 2$

6) Conclude the values of the variables found as optimum.

The optimum values are:

$A_1=4$, $A_2 = 3$, and $\lambda = 2$

Substituting these values into the primary objective function of S, we find

$$\begin{aligned} S &= S_1 + S_2 \\ &= (50 + 10A_1 - A_1^2) + (20 + 5A_2 - 0.5A_2^2) \\ &= (50 + 10 \cdot 4 - 4^2) + (20 + 5 \cdot 3 - 0.5 \cdot 3^2) \\ &= 74 + 30.5 = 104.5 \end{aligned}$$

23 This process allows one to form (or identify) a new equation without λ, which can be used with the equality constraint to solve for the independent variables. That is, eliminating λ via the first two FOCs is an easier way to solve this type of 3 variable simultaneous equation systems.

Therefore, by spending an advertising budget of \$4 million and \$3 million on Regions 1 and 2 respectively, your company can realize a maximum sales revenue of \$104.5 million.

The lambda value represents the opportunity cost of not having one additional unit of the constrained resource. In this case, the lambda value is the marginal, incremental, additional, or extra sales that can be generated if there is one additional unit of advertising expenditure available.

Therefore, the interpretation of the lambda value of 2 found in this case is as follows:

If there is one additional dollar that can be spent on advertisement (beyond the current level of \$7 million), it can bring in an additional \$2 in sales.

Another Example of a Lagrangean (Multiplier) Method:

Everything is the same as above. However, the advertising budget has a new constraint of \$7.1 million. That is, solve the following equation for A_1, A_2, λ, S_1, S_2, and S.

$$\text{Maximize} \quad S = S_1 + S_2 = (50 + 10A_1 - A_1^2) + (20 + 5A_2 - 0.5A_2^2)$$
$$\text{Subject to:} \quad A_1 + A_2 = 7.1$$

You should find the following answer:

A_1=4.0333; A_2 = 3.0667;
λ = 1.9333;
S_1 = 74.0655; S_2 = 30.6312; and
S = 104.6967 $\approx$ 104.70.

As you can see via this example, if the advertising budget increases by \$0.1 million, the sales would increase by \$0.2 million (=\$104.70 – \$104.50). Therefore, it is proven that lambda of 2 in the previous example indicates the opportunity cost of not having one additional unit of the constrained resources.

Exercise Problems on Chapter 3. Optimization Methods

This set of exercise problems has 17 problems worth 20 points.

A. **Find the value of X that optimizes the following functions. Identify whether X value found is associated with a maximum or a minimum of the primary function, Y. Find the value of Y at the optimal value of X**

1. $Y = 2X^2 - 4X + 20$

   a. X=1; Minimum; Y=18
   b. X=1; Maximum; Y=18
   c. X=2; Minimum; Y=20
   d. X=2; Maximum; Y=20
   e. none of the above

2. $Y = -4X^2 + 16X + 1$

   a. X=1; Minimum; Y=13
   b. X=1; Maximum; Y=13
   c. X=2; Minimum; Y=17
   d. X=2; Maximum; Y=17
   e. none of the above

3. $Y = -2(X-5)^2$

   a. X=5; Minimum; Y=0
   b. X=5; Maximum; Y=0
   c. X=-5; Minimum; Y=200
   d. X=-5; Maximum; Y=200
   e. none of the above

4. $Y = \frac{1}{3}X^3 - \frac{3}{2}X^2 + 2X + 1$

   a. X=1; Minimum; Y=11/6
   b. X=1; Maximum; Y=11/6
   c. X=2; Minimum; Y=11/6
   d. X=2; Maximum; Y=5/3
   e. none of the above

5. $Y = \frac{20}{3}X^3 - 60X^2 + 160X + 10$

a. X=2; Minimum; Y=143.333
b. X=2; Maximum; Y=143.333
c. X=4; Minimum; Y=116.666
d. X=4; Maximum; Y=116.666
e. only (b) and (c) of the above

B. Find the first and the second partial derivatives of the following multivariate functions with respect to X.

6. $Y = 2X^3Z^5$

a. $\frac{\partial Y}{\partial X} = 6X^2Z^5$ $\quad \frac{\partial^2 Y}{\partial X^2} = 12XZ^5$

b. $\frac{\partial Y}{\partial X} = 10X^3Z^4$ $\quad \frac{\partial^2 Y}{\partial X^2} = 40X^3Z^3$

c. $\frac{\partial Y}{\partial X} = 3X^2Z^5$ $\quad \frac{\partial^2 Y}{\partial X^2} = 6XZ^5$

d. $\frac{\partial Y}{\partial X} = 6X^3Z^5$ $\quad \frac{\partial^2 Y}{\partial X^2} = 18X^3Z^5$

e. none of the above

7. $Y = 10X^aZ^b - 2X + 3Z$

a. $\frac{\partial Y}{\partial X} = 10aX^{a-1}Z^{b-1} - 2$ $\quad \frac{\partial^2 Y}{\partial X^2} = 10a(a-1)X^{a-2}Z^{b-2}$

b. $\frac{\partial Y}{\partial X} = 10bX^aZ^{b-1} + 3$ $\quad \frac{\partial^2 Y}{\partial X^2} = 10b(b-1)X^aZ^{b-2}$

c. $\frac{\partial Y}{\partial X} = 10aX^{a-1}Z^b - 2$ $\quad \frac{\partial^2 Y}{\partial X^2} = 10a(a-1)X^{a-2}Z^b$

d. $\frac{\partial Y}{\partial X} = 10aX^aZ^b - 2$ $\quad \frac{\partial^2 Y}{\partial x^2} = 10a(a-1)X^{a-1}Z^b$

e. none of the above

8. $Y = 2X^{-3}Z^2 + 5X^2Z - 4X + 20Z$

a. $\frac{\partial Y}{\partial X} = 6X^{-4}Z^2 + 10XZ - 4$ $\frac{\partial^2 Y}{\partial X^2} = -24X^{-5}Z^2 + 10Z$

b. $\frac{\partial Y}{\partial X} = 4X^{-3}Z + 5X^2 + 20$ $\frac{\partial^2 Y}{\partial X^2} = 4X^{-3}$

c. $\frac{\partial Y}{\partial X} = -6X^{-4}Z^2 + 10XZ - 4$ $\frac{\partial^2 Y}{\partial X^2} = 24X^{-5}Z^2 + 10Z$

d. $\frac{\partial Y}{\partial X} = -6X^{-2}Z^2 + 10XZ - 4$ $\frac{\partial^2 Y}{\partial X^2} = 12X^{-1}Z^2 + 10Z$

e. none of the above

9. $Y = \frac{X^2}{Z^2} + 3X^5 - 2Z^3 + XZ$

a. $\frac{\partial Y}{\partial X} = \frac{2X}{Z^2} + 15X^6 + Z$ $\frac{\partial^2 Y}{\partial X^2} = \frac{2}{Z^2} + 90X^7$

b. $\frac{\partial Y}{\partial X} = \frac{2}{Z^2}X + 15X^4 + Z$ $\frac{\partial^2 Y}{\partial X^2} = \frac{2}{Z^2} + 60X^3$

c. $\frac{\partial Y}{\partial X} = \frac{2}{Z^2}X + 15X^4$ $\frac{\partial^2 Y}{\partial X^2} = \frac{2}{Z^2} + 60X^3$

d. $\frac{\partial Y}{\partial X} = \frac{X}{2Z^2} + 15X^4 + Z$ $\frac{\partial^2 Y}{\partial X^2} = \frac{1}{2Z^2} + 60X^3$

e. none of the above

10. $Y = 5X^{10} + 10Z^3 + 2X^{-2}Z^{-3} + 20X - 10Z + 100$

a. $\frac{\partial Y}{\partial X} = 30Z^2 - 6X^{-2}Z^{-4} - 10$ $\frac{\partial^2 Y}{\partial X^2} = 60Z + 24X^{-2}Z^{-5}$

b. $\frac{\partial Y}{\partial X} = 50X^9 - 4X^{-3}Z^{-3} + 20$ $\frac{\partial^2 Y}{\partial X^2} = 450X^8 + 12X^{-4}Z^{-3}$

c. $\frac{\partial Y}{\partial X} = 50X^9 + 10Z^2 + 20 - 10Z$ $\frac{\partial^2 Y}{\partial X^2} = 450X^8 + 10Z - 10$

d. $\frac{\partial Y}{\partial X} = 50X^9 + 30Z^2 - 4X^{-3}Z^{-3} + 20 - 10$

$\frac{\partial^2 Y}{\partial X^2} = 450X^8 + 60Z + 12X^{-4}Z^{-3}$

e. only (a) and (b) of the above

C. Find the first and the second partial derivatives of the following multivariate functions with respect to Z.

11. $Y = 2X^3Z^5$

a. $\frac{\partial Y}{\partial Z} = 10X^3Z^4$ $\frac{\partial^2 Y}{\partial Z^2} = 40X^3Z^3$

b. $\frac{\partial Y}{\partial Z} = 2X^3Z^4$ $\frac{\partial^2 Y}{\partial Z^2} = 8X^3Z^3$

c. $\frac{\partial Y}{\partial Z} = 4 \cdot 3 \cdot 2X^3Z^4$ $\frac{\partial^2 Y}{\partial Z^2} = 3 \cdot 24X^3Z^3$

d. $\frac{\partial Y}{\partial Z} = 30X^2Z^4$ $\frac{\partial^2 Y}{\partial Z^2} = 240X^1Z^3$

e. none of the above

12. $Y = 10X^aZ^b - 2X + 3Z$

a. $\frac{\partial Y}{\partial Z} = 10abX^{a-1}Z^{b-1} + 3$ $\frac{\partial^2 Y}{\partial Z^2} = 10ab(a-1)(b-1)X^{a-2}Z^{b-2}$

b. $\frac{\partial Y}{\partial Z} = 10bX^aZ^{b-1} - 2X + 3$ $\frac{\partial^2 Y}{\partial Z^2} = 10b(b-1)X^aZ^{b-2} - 2X$

c. $\frac{\partial Y}{\partial Z} = 3$ $\frac{\partial^2 Y}{\partial Z^2} = 0$

d. $\frac{\partial Y}{\partial Z} = 10bX^{a}Z^{b-1} + 3$ $\frac{\partial^2 Y}{\partial Z^2} = 10b(b-1)X^{a}Z^{b-2}$

e. none of the above

13. $Y = 2X^{-3}Z^2 + 5X^2Z - 4X + 20Z$

a. $\frac{\partial Y}{\partial Z} = 4X^{-3}Z + 20$ $\frac{\partial^2 Y}{\partial Z^2} = 4X^{-3}$

b. $\frac{\partial Y}{\partial Z} = 4X^{-3}Z + 5X^2 + 20$ $\frac{\partial^2 Y}{\partial Z^2} = 4X^{-3}$

c. $\frac{\partial Y}{\partial Z} = -6X^{-2}Z + 5X^2 + 20$ $\frac{\partial^2 Y}{\partial Z^2} = 12X^{-3}$

d. $\frac{\partial Y}{\partial Z} = 4X^{-3}Z + 5X^2 + 20$ $\frac{\partial^2 Y}{\partial Z^2} = 4X^{-3} + 10X$

e. none of the above

14. $Y = \frac{X^2}{Z^2} + 3X^5 - 2Z^3 + XZ$

a. $\frac{\partial Y}{\partial Z} = \frac{-2X^2Z}{Z^4} - 6Z^2$ $\frac{\partial^2 Y}{\partial Z^2} = \frac{6X^2Z^4}{Z^8} - 12Z$

b. $\frac{\partial Y}{\partial Z} = \frac{-2X^2}{Z^3} - 6Z^2 + X$ $\frac{\partial^2 Y}{\partial Z^2} = \frac{6X^2}{Z^4} - 12Z$

c. $\frac{\partial Y}{\partial Z} = \frac{\partial^2 Y}{\partial Z^2}$

d. $\frac{\partial Y}{\partial Z} = \frac{2X^2}{Z^1} - 6Z^2 + X$ $\frac{\partial^2 Y}{\partial Z^2} = \frac{2X^2}{Z^0} - 12Z$

e. none of the above

15. $Y = 5X^{10} + 10Z^3 + 2X^{-2}Z^{-3} + 20X - 10Z + 100$

a. $\frac{\partial Y}{\partial Z} = 50X^9 - 4X^{-3}Z^{-3} + 20$ $\qquad \frac{\partial^2 Y}{\partial Z^2} = 450X^8 + 12X^{-4}Z^{-3}$

b. $\frac{\partial Y}{\partial Z} = 30Z^4 - 6X^{-2}Z^{-2} - 10$ $\qquad \frac{\partial^2 Y}{\partial Z^2} = 120Z^5 + 12X^{-2}Z^{-1}$

c. $\frac{\partial Y}{\partial Z} = 30Z^2 - 6X^{-2}Z^{-4} - 10$ $\qquad \frac{\partial^2 Y}{\partial Z^2} = 60Z + 24X^{-2}Z^{-5}$

d. $\frac{\partial Y}{\partial Z} = 50X^9 + 30Z^2 - 4X^{-3}Z^{-3} - 10$ $\qquad \frac{\partial^2 Y}{\partial Z^2} = 60Z + 12X^{-4}Z^{-2}$

e. none of the above

D. Solve the following constrained optimization problems:

16. The following is the Sales (S) equation based on two types of advertising expenditures – newspaper advertising expenditure (X) and magazine advertising expenditure (Y):

$$S = 200X + 100Y - 10X^2 - 20Y^2 + 20XY$$

Assuming the total advertising budget is restricted to 20 (i.e., X+Y=20), find X and Y that maximize the sales (S), and identify the value of maximum sales (S) and the corresponding λ value. **(2 points)**

a. X=7; Y=13; S=2450; λ=80

b. X=7; Y=13; S=650; λ=80

c. X=13; Y=7; S=650; λ=80

d. X=13; Y=7; S=2450; λ=80

e. none of the above

17. Find the values of X and Y that maximize the following function and a constraint: (Do not solve for the second order conditions.) **(3 points)**

Maximize $S = -60 + 140X + 100Y - 10X^2 - 8Y^2 - 6XY$

Subject to: $20X + 40Y = 200$

Along with X and Y that maximize the sales (S), identify the value of maximum sales (S) and the corresponding λ value.

a. X=5.56; Y=2.22; S=837.97; λ=0.77

b. X=5.56; Y=2.22; S=517.78; λ=0.77

c. X=5.56; Y=2.22; S=517.78; λ=31.08

d. X=2.22; Y=5.56; S=436.15; λ=0.77

e. X=2.22; Y=5.56; S=436.15; λ=31.08

Chapter 4. Applications to Economics

This chapter shows how algebra and calculus are used to solve economic optimization problems. As can be seen, mathematics is used in economics to answer conceptual problems, not necessarily day-to-day real-world operational problems. For example, the following shows what level of output produced and sold will maximize a firm's profits, provided that the firm's cost and demand functions are known[24].

A. Applications to Solving Economic Problems

1. Defining Q to be the level of output produced and sold, let's assume that the firm's total cost (TC) function is given by the following relationship:

$$TC = 20 + 5Q + Q^2$$

Furthermore, assume that the demand for the output of the firm is a function of price P given by the following relationship:

$$Q = 25 - P$$

a. Defining total profit as the difference between total revenue and total cost, express in terms of Q the total profit function for the firm. (Note: **Total revenue (TR) equals price per unit (P) times the number of units sold (Q). That is, TR = P*Q.**)

Answer:

Step 1. Identify the primary function → Because the primary function is not given in this case, we must identify it by defining the profit function (π) as:

$$\pi = TR - TC$$

In order to identify this profit function, we further need to identify TR and TC by following the steps shown below:

a) Identify Total Revenue (TR) in terms of Q only as follows:

Given TR = Price x Quantity Sold = P x Q

[24] Even though one might argue that a firm's cost and demand relationships can not be specified as mathematical functions that require a high degree of precision, it is nonetheless true that any firm must know its production cost and the customers' demand for its products. Therefore, the issue of what quantity will maximize a firm's profit, based on a **presumed** cost and a demand functions, is very relevant for an analysis. This is the main reason why economics seems to provide conceptual and theoretical answers, rather than practical and operational answers, to a business problem.

→ $P = 25 - Q$ from the demand equation[25]

Therefore, $TR = PQ = (25 - Q)Q = 25Q - Q^2$

b) Identify Total Cost[26] → $TC = 20 + 5Q + Q^2$

c) Profit function[27] →

$$\pi = TR - TC = 25Q - Q^2 - (20 + 5Q + Q^2) = -2Q^2 + 20Q - 20$$

b. Determine the output level, Q, where total profits are maximized.

Answer:

Step 2. Find the first derivative of the primary function with respect to Q; set it equal to zero; and solve for Q → That is, find the FOC and solve for Q:

$$\frac{d\pi}{dQ} = \pi' = -4Q + 20 = 0 \quad \rightarrow \quad Q=5$$

Step 3. Check the SOC for a maximum or a minimum.

$$\frac{d^2\pi}{dQ^2} = \pi'' = -4 < 0 \quad \rightarrow \quad \text{Maximum}$$

c. Calculate total profits and selling price at the profit-maximizing output level.

Answer:

$P = 25 - Q = 25 - 5 = 20$ ← from the demand equation

$$\pi = TR - TC = -2Q^2 + 20Q - 20 = -2(5)^2 + 20(5) - 20 = 30$$

d. If fixed costs increase from $20 to $25 in the total cost relationship, determine the effects of such an increase on the profit-maximizing output level and total profits.

---

[25] This is a critical step to expressing the TR function in terms of Q only. This is the best way to identify the price (P) for the sales revenue calculation unless P is given otherwise. Insisting on using TR = PQ and solving it with partial differentiation of P and Q will NOT work unless you use the chain rule.

[26] Note that Total Cost (TC) is comprised of Total Fixed Cost (TFC) and Total Variable Cost (TVC). Because TFC does not vary with the level of Q produced, TFC is 20 in this case. TVC, however, varies with the level of Q produced and thus, is equal to $5Q + Q^2$ in this case.

[27] Note that this profit function is expressed only in terms of one variable, Q, because the total revenue (TR) and total cost (TC) are both written in terms of Q. If you skip this step of expressing TR = f(Q), you will NOT be able to solve the problem of maximum profits or revenues.

Answer:

Step 1: Identify the primary function → the profit function (π) → $\pi = TR - TC$

a) Total Revenue (TR) = Price x Quantity Sold = PQ

→ P = 25 – Q from the demand equation[28]

→ $TR = PQ = (25 - Q)Q = 25Q - Q^2$

b) Total Cost (TC) = $25 + 5Q + Q^2$

c) Profit function →

$$\pi = TR - TC = 25Q - Q^2 - (25 + 5Q + Q^2) = -2Q^2 + 20Q - 25$$

Step 2: Find the first derivative of the primary function with respect to Q; set it equal to zero; and solve for Q → That is, find the FOC and solve for Q:

$$\frac{d\pi}{dQ} = \pi' = -4Q + 20 = 0 \quad \rightarrow \quad Q=5$$

Step 3: Check the SOC for a maximum or a minimum.

$$\frac{d^2\pi}{dQ^2} = \pi'' = -4 < 0 \quad \rightarrow \quad \text{A maximum exists (at Q=5)}$$

Step 4: Calculate any value that is asked to be identified. For example,

Given $Q = 5$ as the profit-maximizing quantity,

$P = 25 - Q = 25 - 5 = 20$ ← the profit-maximizing price

$\pi = TR - TC = -2Q^2 + 20Q - 25$
$= -2(5)^2 + 20(5) - 25 = 25$ ← the maximum profit

2. Definitions of Marginal Revenue and Marginal Cost

[28] Once again, this step allows one to express TR as a function of Q only. This is a very important method to know to identify and solve a profit function or a revenue function.

In economics, the concept of marginality is very important where "marginal" means "extra", "incremental", or "additional".

The additional revenue generated by selling one additional unit of Q is called "Marginal Revenue" and is expressed as $MR = \frac{Change\ in\ TR}{Change\ in\ Q} = \frac{\Delta TR}{\Delta Q} = \frac{dTR}{dQ}$.

The additional cost incurred by selling one additional unit of Q is called "Marginal Cost" and is expressed as $MC = \frac{Change\ in\ TC}{Change\ in\ Q} = \frac{\Delta TC}{\Delta Q} = \frac{dTC}{dQ}$.

An economic theory known as **"the Fundamental Principle of Profit Maximization"** states that a profit-maximizing firm will produce where MR = MC.

If MR > MC, the firm should produce more because each additional unit produced and sold will bring in a larger additional (=marginal) revenue than additional (=marginal) cost.

Similarly, if MR < MC, it should produce less because each additional unit produced and sold will bring in a smaller additional (=marginal) revenue than additional (=marginal) cost. Only when MR = MC, the firm is in equilibrium (or balance) and has a stable operation.

a. Using the cost and demand functions in Exercise Problem above, determine the marginal revenue (MR) and marginal cost (MC) functions.

Answer:

MR = the first derivative of total revenue (TR) with respect to Q.

Given $TR = PQ = (25 - Q)Q = 25Q - Q^2$

$$MR = \frac{dTR}{dQ} = \frac{\Delta TR}{\Delta Q} = 25 - 2Q$$

MC = the first derivative of total cost (TC) with respect to Q.

Given $TC = 25 + 5Q + Q^2$

$$MC = \frac{dTC}{dQ} = \frac{\Delta TC}{\Delta Q} = 5 + 2Q$$

b. Show that the profit-maximizing output level is found where marginal revenue equals marginal cost[29].

Answer:

Given MR = MC,

$$25 - 2Q = 5 + 2Q$$

Therefore, Q = 5

$$\text{At } Q=5, MR=MC=15$$

3. Using the cost and demand functions of:

$$TR = PQ = (25 - Q)Q = 25Q - Q^2$$

$$TC = 20 + 5Q + Q^2$$

suppose that the government imposes a 20 percent *profits tax* on the firm.

a. Determine the new profit function for the firm.

Answer:

$$\text{New Profit} = (TR–TC) - 0.2(TR - TC)$$

$$= (1 - 0.2)(TR–TC) = 0.8(TR–TC)$$

$$= 0.8(-2Q^2 + 20Q - 20)$$

$$= -1.6Q^2 + 16Q - 16$$

b. Determine the output level at which total profits are maximized.

Answer:

Find FOC and set it equal to 0 and solve for Q:

$$\pi' = -3.2Q + 16 = 0 \quad \rightarrow \quad Q=5$$

---

[29] This illustrates the Fundamental Principle of Profit Maximization can be used to solve Q that maximizes the profit. However, it is less useful because it can NOT utilize the second order condition to verify its conclusion of maximum vs. minimum.

Check SOC:

$$\pi'' = -3.2 < 0 \quad \rightarrow \quad \text{Maximum}$$

c. Calculate total profits (after taxes) and the selling price at the profit-maximizing output level. Also, identify the tax paid.

Answer:

$$P = 25 - Q = 25 - 5 = 20$$

$$\pi = -1.6Q^2 + 16Q - 16 = -1.6(5)^2 + 16(5) - 16 = 24$$

$$\text{Tax} = 20\% \text{ of profits} = 0.2 \text{ x } (30) = 6$$

d. Compare the results in parts (b) and (c) with the results in Exercise 1 above.

Answer:

	No Tax	Profits Tax
Q	5	5
P	20	20
π	30	24
Tax Paid	0	6

4. Suppose that the government imposes a 20 percent *sales tax* (that is, a tax on revenue) on the firm. Find the profit-maximizing output, Q; the price, P; the profits level; and tax, using the original total revenue and cost functions shown below:

$$TR = PQ = (25 - Q)Q = 25Q - Q^2$$

$$TC = 20 + 5Q + Q^2$$

Answer:

(1) Identify the primary function → the new profit function (π)

$$\rightarrow \pi = (TR - TC) - 0.2TR = 0.8TR - TC$$

$$= 0.8(25Q - Q^2) - (20+5Q+Q^2)$$

$$= 20Q - 0.8Q^2 - 20 - 5Q - Q^2$$

$$= -1.8Q^2 + 15Q - 20$$

(2) Find the first derivative of the primary function with respect to Q; set it equal to zero; and solve for Q → That is, find the FOC and solve for Q:

$$\frac{d\pi}{dQ} = \pi' = -3.6Q + 15 = 0 \quad \rightarrow \quad Q=4.17$$

(3) Check the SOC for a maximum or a minimum.

$$\frac{d^2\pi}{dQ^2} = \pi'' = -3.6 < 0 \quad \rightarrow \quad \text{Maximum}$$

(4) $Q = 4.17$

$P = 25 - Q = 25 - 4.17 = 20.83$

$\pi = TR - TC = -1.8Q^2 + 15Q - 20 = -1.8(4.17)^2 + 15(4.17) - 20$
$= 11.25$

Tax = 20% of TR = 0.2 x (20.83) x (4.17) = 17.37

An interesting and important observation:

	No Tax	20% Profits Tax	20% Sales Tax
Q	5	5	4.17
P	20	20	20.83
π	30	24	11.25
Tax Paid	0	6	17.37

This summary table shows that an imposition of a sales tax is most inefficient in resource allocation!!! It calls for a smaller Q and a higher P in the market, and a smallest profit to the firm. That is, a sales tax will hurt the consumers via a higher price and a lower consumption, and the producer via a smaller profit. However, the government loves it because of a larger tax collected.

Without this mathematical exercise, it is very difficult to prove if and how much a sales tax is damaging to the consumers and the producers. This is one of the reasons why mathematics is studied and used in economics and other business disciplines.

5. The McChoi Corporation's average variable cost (AVC[30]) function is given by the following relationship (where Q is the number of units produced and sold):

$$AVC = 25{,}000 - 180Q + 0.50Q^2$$

a. Determine the output level (Q) that minimizes the average variable cost (AVC) function.

Answer:

Because the given AVC function is the primary function, we find the FOC by taking its derivative with respect to Q and solve it for Q as follows:

$$\frac{dAVC}{dQ} = AVC' = -180 + Q = 0 \rightarrow \text{Therefore, } Q = 180$$

b. How does one know that the value of Q determined in part (a) *minimizes* rather than *maximizes* AVC?

Answer:

Find the SOC:

$$\frac{d^2AVC}{dQ^2} = AVC'' = +1 > 0 \rightarrow \text{a minimum at } Q=180.$$

6. The Luv-A-Bull, Inc. had learned that its sales revenue (S) is a function of the amount of advertising (measured in units) in two different media - Cable TV(X) and Internet(Y). This is given by the following relationship: (All units are in million dollars.)

$$S(X,Y) = Y^2 - X^2 - 3XY + Y + 18X + 1$$

a. Find the levels of Cable TV (X*) and Internet (Y*) advertising that maximize the firm's sales revenue.

Answer:

---

[30] An Average Variable Cost (AVC) is also known as a per-unit cost and defined as the Total Variable Cost (TVC) divided by output, Q. That is, AVC=TVC/Q. On the other hand, an Average Fixed Cost (AFC) is then defined as: AFC=TFC/Q.

Find the FOCs as follows[31]:

$$\frac{\partial S}{\partial X} = -2X - 3Y + 18 = 0$$

$$\frac{\partial S}{\partial Y} = 2Y - 3X + 1 = 0$$

Solve them simultaneously by multiplying the top equation by 3 and the bottom equation by 2 to eliminate Y:

$$2 \cdot \frac{\partial S}{\partial X} = 2 \cdot (-2X - 3Y + 18) = -4X - 6Y + 36 = 0$$

$$3 \cdot \frac{\partial S}{\partial Y} = 3 \cdot (2Y - 3X + 1) = 6Y - 9X + 3 = 0$$

Add the two equations and find:

$$-13X + 39 = 0 \quad \rightarrow \quad X = 3$$

Substituting X=3 into either of the two FOCs above, we find:

$$-2X - 3Y + 18 = 0$$

$$-2(3) - 3Y + 18 = 0 \quad \rightarrow \quad Y = 4$$

Or

$$2Y - 3X + 1 = 0$$

$$2Y - 3(3) + 1 = 0 \quad \rightarrow \quad Y = 4$$

b. Calculate the firm's sales revenue (S*) at the optimal values of Cable TV and Internet advertising determined in part (a) above.

Answer:

Given $S(X,Y) = Y^2 - X^2 - 3XY + Y + 18X + 1$

$$S^* = 4^2 - 3^2 - 3(3)(4) + 4 + 18(3) + 1 = 30$$

[31] Note that there are two FOCs because of two independent variables of X and Y.

7. Suppose that we can divide our 24-hour day into either sleeping hours (S) or non-sleeping, thus, working hours (W). If the following equation represents our utility (=satisfaction) function, can you identify the optimal amount of sleeping hours (S) and working hours (W) that will maximize our utility (=satisfaction=U)?

$$\text{Max } U = 2S^2 + 4.25SW + 2.0625W^2$$

Answer:

Step 1: Identify the problem as:

$$\text{Max} \qquad U = 2S^2 + 4.25SW + 2.0625W^2$$

$$\text{Subject to} \qquad S + W = 24$$

Now, set up the problem as a Lagrangean Function:

Max L = Primary Objective Function + λ(Converted Constraint)

= U + λ(24 – S – W)

That is,

$$\text{Max } L = 2S^2 + 4.25SW + 2.0625W^2 + \lambda(24 - S - W)$$

Step 2: Find the FOCs as follows[32]:

$$\frac{\partial L}{\partial S} = 4S + 4.25W - \lambda = 0$$

$$\frac{\partial L}{\partial W} = 4.25S + 4.125W - \lambda = 0$$

$$\frac{\partial L}{\partial \lambda} = 24 - S - W = 0$$

When these 3 FOCs are simultaneously solved, you get:

W*=16, S*=8, λ=100, and U*=1200

Step 3: Check the SOC (Second Order Condition) ← Skip this for the case of more than 1 independent variables.

---

[32] Note that there are 3 FOCs because of 3 unknown independent variables of S, W, and λ.

B. Application to Demand Elasticity Calculation

Demand elasticity measures how much the quantity demanded, Q, changes in percentage points, given a percentage change in a single variable, X, that is believed to affect Q.

1. Two Types of Demand Elasticity Formulas

   a. The Point Elasticity Formula:

$$\varepsilon_X = \frac{\%\Delta Q}{\%\Delta X} = \frac{\dfrac{Q_{new} - Q_{old}}{Q_{old}}}{\dfrac{X_{new} - X_{old}}{X_{old}}} = \frac{\dfrac{\Delta Q}{Q_{old}}}{\dfrac{\Delta X}{X_{old}}} = \frac{\Delta Q}{\Delta X} \cdot \frac{X_{old}}{Q_{old}} = \frac{\partial Q}{\partial X} \cdot \frac{X_{old}}{Q_{old}}$$

   This point elasticity formula is used when a change in X is small.

   b. The Arc Elasticity Formula

$$\varepsilon_{\overline{X}} = \frac{\%\Delta \overline{Q}}{\%\Delta \overline{X}} = \frac{\dfrac{Q_{new} - Q_{old}}{\overline{Q}}}{\dfrac{X_{new} - X_{old}}{\overline{X}}} = \frac{\dfrac{\Delta Q}{\overline{Q}}}{\dfrac{\Delta X}{\overline{X}}} = \frac{\Delta Q}{\Delta X} \cdot \frac{\overline{X}}{\overline{Q}} = \frac{\partial Q}{\partial X} \cdot \frac{\overline{X}}{\overline{Q}}$$

   where $\overline{Q} = \dfrac{Q_{new} + Q_{old}}{2}$ and $\overline{X} = \dfrac{X_{new} + X_{old}}{2}$

   This arc elasticity formula is used when a change in X is large.

2. General Interpretation

   Even though the following interpretation is based on the point elasticity formula, the same interpretation is rendered for the arc elasticity formula.

   If $\varepsilon_X = \dfrac{\%\Delta Q}{\%\Delta X} = \dfrac{dQ}{dX} \cdot \dfrac{X}{Q} < 1$, inelastic demand → $\%\Delta Q < \%\Delta X$

   For example, if $\varepsilon_X = 0.5$, this means that if %ΔX = +1, then %ΔQ = +0.5. Likewise, it can also mean that if %ΔX = – 1, then %ΔQ = – 0.5.

   If $\varepsilon_X = \dfrac{\%\Delta Q}{\%\Delta X} = \dfrac{dQ}{dX} \cdot \dfrac{X}{Q} > 1$, elastic demand → $\%\Delta Q > \%\Delta X$

   For example, if $\varepsilon_X = 2$, this means that if %ΔX = +1, then %ΔQ = +2.

Likewise, it can also mean that if %ΔX = – 1, then %ΔQ = – 2.

$$\text{If } \varepsilon_X = \frac{\%\Delta Q}{\%\Delta X} = \frac{dQ}{dX} \cdot \frac{X}{Q} = 1, \text{ unit-elastic demand} \rightarrow \%\Delta Q = \%\Delta X$$

For example, if $\varepsilon_X = 1$, this means that if %ΔX = +1, then %ΔQ = +1.
Likewise, it can also mean that if %ΔX = – 1, then %ΔQ = – 1.

Note1: There is an exception to the above interpretation. As for the own price elasticity of demand, because its value is always negative, reflecting the negative relationship between the price (P) charged and the quantity (Q) bought, we must **find its absolute value** and render the same interpretation as shown above.

Note 2: The point elasticity formula between and the quantity (Q) demanded and the own price (P) charged can, therefore, be written as follows[33]:

$$|\varepsilon_d| = \frac{\%\Delta Q}{\%\Delta P} = \frac{dQ}{dP} \cdot \frac{P}{Q}$$

This convention of quoting the own price elasticity of demand as a positive number by taking its absolute value, however, requires a great care in interpreting its actual meaning. For example, if an own price elasticity is quoted as 2, it truly means –2. Thus, a 1% **increase (decrease)** in its own price (%ΔP) will cause a 2% **decrease (increase)** in its quantity (%ΔQ) sold because

$$|\varepsilon_d| = \frac{\%\Delta Q}{\%\Delta P} = |-2| = 2 \quad \rightarrow \quad |\varepsilon_d| = \frac{\%\Delta Q}{\%\Delta P} = -2 = \frac{-2}{1} = \frac{2}{-1}$$

Therefore, when interpreting a positive own price elasticity, one must re-introduce the negative sign[34].

This convention of using an absolute value of an elasticity number is applicable only to the own price elasticity. All other elasticities are interpreted as they are without taking their absolute values.

---

33 The reason for the absolute value is because there is an inverse (or negative) relationship between ΔP and ΔQ due to the law of demand, which makes the own price elasticity of demand always a negative number. Because people should know that the own price elasticity is always negative, own price elasticities are treated as if they are positive numbers by taking their absolute values. Note that except the own price elasticity of demand, no other elasticity measures are expressed in absolute values.

34 This convention of using an absolute value of an elasticity number is applicable only to the own price elasticity. All other elasticities are interpreted as they are without taking their absolute values.

3. Examples

Example 1: Own Price Elasticity of Demand

Suppose the follow demand relationship between the number of hamburgers (Q) and its per-unit price (P) is identified by your analyst:

Q = 200 – 10P

a. How many hamburgers would be demanded at a price of \$2? \$4? And \$6?

Answer:

When P = 2, Q = 200 – 10*(2) = 180

When P = 4, Q = 200 – 10*(4) = 160

When P = 6, Q = 200 – 10*(6) = 140

b. Using the point elasticity formula of demand:

$$\varepsilon_d = \frac{\%\Delta Q}{\%\Delta P} = \frac{\dfrac{Q_{new} - Q_{old}}{Q_{old}}}{\dfrac{P_{new} - P_{old}}{P_{old}}} = \frac{\dfrac{\Delta Q}{Q_{old}}}{\dfrac{\Delta P}{P_{old}}} = \frac{\Delta Q}{\Delta P} \cdot \frac{P_{old}}{Q_{old}} = \frac{\partial Q}{\partial P} \cdot \frac{P_{old}}{Q_{old}}$$

i) Calculate the point price elasticity of demand at P=\$2.

Answer:

Because when P = \$2, Q = 180, and $\dfrac{dQ}{dP} = -10$

$$|\varepsilon_d| = \frac{dQ}{dP} \cdot \frac{P}{Q} = (-10) \cdot \frac{2}{180} = -0.1111 \rightarrow 0.1111 \rightarrow \text{inelastic}$$

Note that the negative own price elasticity number is expressed as a positive number. Also, since its value is less than 1, it is called "inelastic," meaning that the quantity change (%ΔQ) is less than the price change (%ΔP) in an absolute magnitude.

That is, the generic interpretation for the point own price elasticity of demand is as follows:

If $|\varepsilon_d| = \dfrac{\%\Delta Q}{\%\Delta P} = \dfrac{dQ}{dP}\cdot\dfrac{P}{Q} < 1$, inelastic demand → $|\%\Delta Q| < |\%\Delta P|$

If $|\varepsilon_d| = \dfrac{\%\Delta Q}{\%\Delta P} = \dfrac{dQ}{dP}\cdot\dfrac{P}{Q} > 1$, elastic demand → $|\%\Delta Q| > |\%\Delta P|$

If $|\varepsilon_d| = \dfrac{\%\Delta Q}{\%\Delta P} = \dfrac{dQ}{dP}\cdot\dfrac{P}{Q} = 1$, unit-elastic demand → $|\%\Delta Q| = |\%\Delta P|$

ii) **Calculate the point price elasticity of demand at P=$4.**

Answer:

Because when P = $4, Q = 160

$$|\varepsilon_d| = \frac{dQ}{dP}\cdot\frac{P}{Q} = (-10)\cdot\frac{4}{160} = -0.25 \rightarrow 0.25 \rightarrow \text{inelastic}$$

c. **Given the arc price elasticity of demand as:**

$$|\varepsilon_a| = \frac{dQ}{dP}\cdot\frac{\overline{P}}{\overline{Q}} \quad \text{Note: } \frac{dQ}{dP} = -10$$

where $\overline{P}$= average of two prices, and $\overline{Q}$= average of two quantities

i) Calculate the arc price elasticity of demand between P=$2 and P=$4.

Answer:

$$\overline{P} = \frac{2+4}{2} = 3 \qquad \text{and} \qquad \overline{Q} = \frac{180+160}{2} = 170$$

$$|\varepsilon_a| = \frac{dQ}{dP}\cdot\frac{\overline{P}}{\overline{Q}} = (-10)\cdot\frac{3}{170} = -0.1764 \rightarrow 0.1764 \rightarrow \text{inelastic}$$

ii) Calculate the arc price elasticity of demand between P=$4 and P=$6.

Answer:

$$\overline{P} = \frac{4+6}{2} = 5 \qquad \text{and} \qquad \overline{Q} = \frac{160+140}{2} = 150$$

$$|\varepsilon_a| = \frac{dQ}{dP} \cdot \frac{\overline{P}}{\overline{Q}} = (-10) \cdot \frac{5}{150} = -0.3333 \rightarrow 0.3333 \rightarrow \text{inelastic}$$

d. If a total of 190 hamburgers were sold last year, what would you expect the prevailing price to have been?

Given Q = 200 – 10P → 190 = 200 – 10P

Therefore, P = $1.

Example 2. A Multivariable Demand Equation

Given the following demand equation for Good Old Hotdogs (Q),

$$Q = 100 - 3P - 4I + 5P_c$$

where P = the price of a Hotdog; I = average disposable income of the consumers ($) for Hotdogs; and P_c = the price of a competitor's Hotdogs

Given a generic point elasticity of demand[35] as:

$$\varepsilon_X = \frac{\partial Q}{\partial X} \cdot \frac{X}{Q}$$

a. Calculate the **point** price elasticity of demand at P=$1, given I = $10 and P_c =$2.

Answer: $\varepsilon_d = \frac{\partial Q}{\partial P} \cdot \frac{P}{Q} = (-3) \cdot \frac{1}{67} = -0.0448 \rightarrow |\varepsilon_d| = 0.0448$

where $Q = 100 - 3P - 4I + 5P_c = 100 - 3(1) - 4(10) + 5(2) = 67$

b. Calculate the **point** income elasticity of demand at I = $10, given P=$1 and P_c = 4.

[35] Note that because this is a case of a multivariate equation, the partial differentiation sign of a lower Greek letter, ∂, is used in this formula. X denotes any variable of one's interest such as own price, income, competitor's price, etc.

Answer: $\varepsilon_I = \dfrac{\partial Q}{\partial I} \cdot \dfrac{I}{Q} = (-4) \cdot \dfrac{10}{77} = -0.5195$

where $Q = 100 - 3P - 4I + 5P_c = 100 - 3(1) - 4(10) + 5(4) = 77$

Note: **We are not changing the sign of the income elasticity from negative to positive. In fact, we are keeping the sign (either + or –) as is.** If an income elasticity of a product is negative, we call it an inferior or Giffen good. If it is positive, we call it a superior or normal good[36].

c. Calculate the **point** cross-price elasticity[37] of demand at $P_c = 2$, given I = \$20 and P=\$3.

Answer: $\varepsilon_{P_c} = \dfrac{\partial Q}{\partial P_c} \cdot \dfrac{P_c}{Q} = (5) \cdot \dfrac{2}{21} = 0.4762$

where $Q = 100 - 3P - 4I + 5P_c = 100 - 3(3) - 4(20) + 5(2) = 21$

Note: **We are not to change the sign of the cross-price elasticity, either. In fact, we must keep the sign (either + or –) intact.** If a cross-price elasticity between two products X and Y is negative, we conclude that X and Y are used or consumed together and thus, X is said to be a complement good to Y. If it is positive, we conclude that X is consumed in place of Y and thus, X is said to be a substitute good with Y.

Example 3. An Exponential Demand Function

Given the following demand equation for Good Old Hotdogs (Q),

$$Q = 25P^{-3}I^{2}A^{-1}P_c^{-2}$$

where P = the price of a Hotdogs; I = average disposable income of the consumers (\$) for Hotdogs; A= advertising expense for Hotdogs; and P_c = the price of a competitor's hotdogs,

Given a generic formula for the point elasticity of demand as:

$$\varepsilon_X = \frac{\%\Delta Q}{\%\Delta X} = \frac{\partial Q}{\partial X} \cdot \frac{X}{Q}$$

[36] This is one of the main topics in a managerial economics course such as Econ 555.
[37] Cross-price elasticity measures the change in my Q given the change in the price of others, P_c, such as my competitors.

a. Calculate the (point) price elasticity of demand.

Answer:

$$\varepsilon_d = \frac{\partial Q}{\partial P} \cdot \frac{P}{Q} = (-3) \cdot (25P^{-4} I^2 A^{-1} P_c^{-2}) \cdot \frac{P}{Q} = -3 \rightarrow |\varepsilon_d| = 3$$

because the numerator of $(25P^{-4} I^2 A^{-1} P_c^{-2}) \cdot P = 25P^{-3} I^2 A^{-1} P_c^{-2} = Q$

b. Calculate the (point) income elasticity of demand.

Answer:

$$\varepsilon_I = \frac{\partial Q}{\partial I} \cdot \frac{I}{Q} = (2) \cdot (25P^{-3} I A^{-1} P_c^{-2}) \cdot \frac{I}{Q} = 2$$

because the numerator of $(25P^{-3} I A^{-1} P_c^{-2}) \cdot I = 25P^{-3} I^2 A^{-1} P_c^{-2} = Q$

c. Calculate the (point) advertising elasticity of demand.

Answer:

$$\varepsilon_A = \frac{\partial Q}{\partial A} \cdot \frac{A}{Q} = (-1) \cdot (25P^{-3} I^2 A^{-2} P_c^{-2}) \cdot \frac{A}{Q} = -1$$

because the numerator of $(25P^{-3} I^2 A^{-2} P_c^{-2}) \cdot A = 25P^{-3} I^2 A^{-1} P_c^{-2} = Q$

d. Calculate the (point) cross-price elasticity of demand.

Answer:

$$\varepsilon_{P_c} = \frac{\partial Q}{\partial P_c} \cdot \frac{P_c}{Q} = (-2) \cdot (25P^{-3} I^2 A^{-1} P_c^{-3}) \cdot \frac{P_c}{Q} = -2$$

because the numerator of $(25P^{-3} I^2 A^{-2} P_c^{-3}) \cdot P_c = 25P^{-3} I^2 A^{-1} P_c^{-2} = Q$

Conclusion:

When a demand equation is expressed as an exponential function, the exponent values are the (point) elasticities of the respective variables and thus, each respective elasticity remains constant over all values of P, I, A, etc.

Exercise Problems on
Chapter 4. Applications to Economics

This set of exercise problems has 20 problems, worth a total of 25 points.

A. Calculation of Demand Elasticity

Suppose the follow demand relationship between the number of automobiles (Q) and its per-unit price (P) is identified by your analyst:

$$Q = 300{,}000 - 20P$$

1. How many automobiles would be demanded at a price of $2,000?

   a. 40,000
   b. 80,000
   c. 120,000
   d. 260,000
   e. 300,000

2. The number of automobiles demanded (Q_1) at a price of $4,000 is ________, and the same (Q_2) at a price of $6,000 is ________.

   a. $Q_1 = 220{,}000$; and $Q_2 = 180{,}000$
   b. $Q_1 = 180{,}000$; and $Q_2 = 220{,}000$
   c. $Q_1 = 220{,}000$; and $Q_2 = 220{,}000$
   d. $Q_1 = 180{,}000$; and $Q_2 = 180{,}000$
   e. none of the above

3. Given the equation for the point own price elasticity of demand as:

$$|\varepsilon_d| = \frac{dQ}{dP} \cdot \frac{P}{Q}$$

   Calculate the point own price elasticity of demand at P=$2,000.

   a. -20 → 20
   b. -0.1538 → 0.1538
   c. -130 → 130
   d. -6.5 → 6.5
   e. not calculable

4. Given the equation for the point own price elasticity of demand as in Problem 3 above, calculate the point own price elasticity of demand at P=$4,000.

a. -20 → 20
b. -0.3636 → 0.3636
c. -55 → 55
d. -2.75→ 2.75
e. not calculable

5. Given the arc own price elasticity of demand as:

$$|\varepsilon_a| = \frac{dQ}{dP} \cdot \frac{\overline{P}}{\overline{Q}}$$

where $\overline{P}$= average of two prices

and $\overline{Q}$= average of two quantities

Calculate the arc own price elasticity of demand between P=$2,000 and P=$4,000.

a. -0.3636 → 0.3636
b. -0.1538 → 0.1538
c. -0.25 → 0.25
d. -4 → 4
e. not calculable

6. Given the equation for the arc own price elasticity of demand as in Problem 5 above, calculate the arc own price elasticity of demand between P=$4,000 and P=$6,000.

a. -0.3636 → 0.3636
b. -0.6667 → 0.6667
c. -0.5 → 0.5
d. -2 → 2
e. not calculable

7. If 200,000 automobiles were demanded and sold last year, what was the (per-unit) price of the automobiles?

a. $20,000
b. $15,000
c. $10,000
d. $5,000
e. not calculable

B. Advanced Problems in Demand Elasticity - 1

Given the following demand equation for Love Chocolate Bars (Q),

$$Q = 10 - 5P + 2I - 3P_c$$

where P = the price of a Love Chocolate Bar; I = average disposable income of the consumers ($) for chocolate bars; and P_c = the price of a competitor's chocolate bar

Given a generic elasticity of demand as:

$$\varepsilon_X = \frac{\partial Q}{\partial X} \cdot \frac{X}{Q}$$

8. The point own price elasticity of demand at P=$2, given I = $20 and P_c = 1, is equal to ______.

   a. -0.27 → 0.27
   b. -0.6667 → 0.6667
   c. -3.7 → 3.7
   d. -2.7 → 2.7
   e. none of the above

9. The point income elasticity of demand at I = $10, given P=$2 and P_c = 1, is equal to ______.

   a. -1.176
   b. 1.176
   c. -0.85
   d. 0.85
   e. none of the above

10. The point cross-price elasticity of demand at P_c = 2, given I = $10 and P=$3, is equal to _____.

   a. -3
   b. -1.5
   c. -0.667
   d. 0.667
   e. none of the above

C. Advanced Problems in Demand Elasticity - 2

Given the following demand equation for Love chocolate bars (Q),

$$Q = 10P^{-2}I^{3}A^{-4}P_c$$

where P = the price of a Love chocolate bar; I = average disposable income of the consumers ($) for chocolate bars; A= advertising expense for Love chocolate bars; and P_c = the price of a competitor's chocolate bar

Given a generic elasticity of demand as:

$$\varepsilon_X = \frac{\partial Q}{\partial X} \cdot \frac{X}{Q}$$

11. The point own price elasticity of demand is ______ and the point income elasticity of demand is ______.

    a. -0.2 → 0.2; -3
    b. -0.5 → 0.5; 3
    c. -3.7 → 3.7; -3
    d. -2 → 2; 3
    e. none of the above

12. The point advertising elasticity of demand is ______ and the point cross-price elasticity of demand is ______.

    a. 4; -1
    b. -4; -1
    c. 4; 1
    d. -4; 1
    e. none of the above

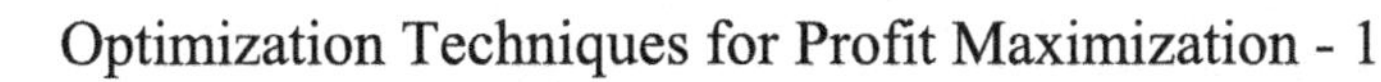

D. Optimization Techniques for Profit Maximization - 1

If a firm sells its product at a fixed price of $121 per unit and has the following total cost function,

$$TC = 0.02Q^3 - 3Q^2 + 175Q + 500$$

13. The profit-maximizing output, Q, is ______ and the corresponding maximum profit level is ______. **(2 points)**

    a. 10; -760
    b. 100; 7600
    c. 90; 4360
    d. 90; -4360
    e. none of the above

14. The marginal revenue (MR) function is ______ and the marginal cost (MC) function is ______.

a. MR = 121; MC = $0.06Q^2 - 6Q + 175$
b. MR = 121Q; MC = $0.06Q^2 - 6Q$
c. MR = 121; MC = $0.02Q^3 - 3Q^2 + 175Q + 500$
d. MR = 121Q; MC = $0.02Q^3 - 3Q^2 + 175Q$
e. none of the above

15. By using the fundamental principle of profit maximizing, the profit maximizing output level, Q, is found to be ______ and the corresponding maximum profit is ______.

a. 10; -760
b. 100; 7600
c. 90; 4360
d. 90; -4360
e. none of the above

E. Optimization Techniques for Profit Maximization – 2

Given the following demand function,

$$Q = 300 - 3P$$

and total cost function,

$$TC = \frac{1}{600}Q^3 - \frac{1}{3}Q^2 + 50Q + \frac{1000}{3}$$

16. The profit-maximizing output, Q, is ______ and the maximum profit level is ______. **(2 points)**

a. 10; -165
b. 10; 165
c. 30; 1121.67
d. 100; 3000
e. none of the above

17. The marginal revenue function is ______ and the marginal cost function is ______. **(2 points)**

a. MR = 3; MC = $\frac{1}{200}Q^2 - \frac{2}{3}Q + 50$

b. MR = $100 - \frac{2}{3}Q$; MC = $\frac{1}{200}Q^2 - \frac{2}{3}Q + 50$

c. $\text{MR} = -\frac{2}{3}$; $\text{MC} = \frac{1}{200}Q^2 - \frac{2}{3}Q + 50$

d. $\text{MR} = 100 - \frac{2}{3}Q$; $\text{MC} = 0.005Q^2 - 6Q + 50$

e. none of the above

18. By using the fundamental principle of profit maximizing, the profit maximizing output level, Q, is found to be ______ and the corresponding maximum profit is ______. **(2 points)**

a. 10; -165
b. 10; 165
c. 30; 1121.67
d. 100; 3000
e. none of the above

F. Optimization Techniques for Profit Maximization – 3

19. A firm has a fixed cost of \$5000 and per unit cost of production of \$2. Also, the firm has the following demand function,

Demand: Q = 10,000 – 1,000P

This firm's total revenue (TR) function is ______ and total cost (TC) function is _______.

a. $\text{TR} = 10Q - \frac{Q^2}{1000}$; $\text{TC} = 2Q$

b. $\text{TR} = 10Q - \frac{Q^2}{1000}$; $\text{TC} = 5000$

c. $\text{TR} = 10Q - \frac{Q^2}{1000}$; $\text{TC} = 2Q + 5000$

d. $\text{TR} = 10{,}000Q - \frac{Q^2}{1000}$; $\text{TC} = 2Q + 5000$

e. none of the above

20. The profit-maximizing output, Q, is ______ and the maximum profit level is ______. **(2 points)**

a. 400; -\$1,960
b. 4000; +\$11,000
c. 400; +\$24,000
d. 4000; +\$13,000
e. none of the above

Part 2. Business Statistics

There are 15 chapters in this part on business statistics. Chapters 1 through 4 discuss the basic concepts, terms, and formulas associated with data collection and presentation. Chapters 5 through 8 discuss the concept of probability and various probability distributions. Chapter 9 describes sampling distributions. Chapter 10 shows the construction of confidence intervals while Chapter 11 elaborates the methods of hypothesis testing. Chapter 12 extends the hypothesis testing to the equality of two independent sample means and the equality of two independent sample variances. Chapters 13 and 14 integrate all these topics into one of the most powerful tools of business data analysis, known as regression analysis. Chapter 15 introduces the concept of nonparametric testing via the Chi-square test.

Chapter 1. Introduction

An Introduction to Statistics:

Statistical analysis is needed because

1. There are lots of data but not enough analyses.

2. Data are facts; analyses are information; and implications are knowledge. Facts deal with the past (history); analyses, the present; and the implications, the future.

3. Businesses demand that you have sufficient knowledge to make good decisions that influence or determine the future.

4. Because the future impact of any decision can not be known with a 100% certainty, modern business managers wish to know the risk involved with their decisions. Because they often believe that a risk that is not measured can not be controlled, they need to quantify and analyze the operational, financial, and managerial data in their possession. For this reason and purpose, statistics is an essential tool for modern business managers.

The roles of statistics:

1. Collect, summarize, and present the data → Descriptive Statistics → Record keeping

2. Analyze and make an inference about the population that produced the data → Inferential Statistics → Prediction with a probability

A. Definitions

1. Population vs. sample

a. Population = the entire members of one's interest

→ These entire members of one's interest can be either finite (thus, known) or infinite (thus, unknown) to an analyst at the start of an analysis.

For example, the entire number of stocks in the S&P 500 Stock Index or the Dow Jones Industrial Average are finite and known up front. On the other hand, on-going processes such as manufacturing light bulbs and automobiles, or treating patients with new drugs, etc. do not have a finite and known size of a population.

b. Sample = a portion or a subset of a population that yields data for an analysis

→ The 20 stocks chosen from the S&P 500 Stock Index, the 10 stocks chosen out of 30 stocks in the Dow Jones Industrial Average, 400 light bulbs chosen out of a day's production line, and the entire U.S. population, etc.

2. Parameter vs. statistic

a. Parameter = a numerical measure of a population characteristic → the population mean (=average), the population variance, the population standard deviation, etc. → **by convention, denoted by Greek letters such as μ, σ, β, etc.**

b. Statistic = a numerical measure of a sample characteristic → the sample mean (=average), the sample variance, the sample standard deviation, etc. **→ by convention, denoted by English letters such as $\overline{X}$, S, b, etc. → A (sample) statistic is an estimate of a (population) parameter.**

3. Parametric vs. nonparametric statistics[38]:

The study of statistics can be divided into parametric vs. nonparametric.

[38] The main focus of this book is parametric statistics. Only one chapter (Chapter 15) will be devoted to nonparametric statistics, covering the topic of a Chi-square distribution.

a. Parametric statistics studies issues related to a sample that is assumed to have come from a population with a defined mean and a defined variance. → It often utilizes a normal distribution, t-distribution, F-distribution, etc.

b. Nonparametric statistics does not require a sample to have a defined mean and/or a defined variance → It most often utilizes a Chi-square distribution.

4. The Fundamental Axiom of Parametric Statistics

The fundamental axioms of parametric statistics are that:

a. if a population mean and a population variance are known, then we can assess the probability of an individual observation or a sample mean occurring → **deductive inference**

b. if they are unknown, then we can collect a sample and based on that, assess the probability of an individual observation, a sample mean, or a population mean occurring → **inductive inference**

Therefore, in parametric statistics, it is critically important to recognize the presence of any information on the population mean and the population variance (or standard deviation). **This is because which statistical tool to use depends critically on the availability and knowledge of population parameters**[39].

5. Deductive vs. Inductive Statistics

Deduction is a method to infer specific information from general information. If 70% of your employees are male, for example, then **repeated random selections** of 10 employees for a working group within your company would most likely yield an average of 7 males and 3 females. This inference about the composition of a specific group from that of the entire group (= population) is called deductive analysis. **The deductive analysis is used to understand how samples are generated from a population, given that population parameters are known.**

On the other hand, induction is a method to infer general information from specific information. This is often the case when population parameters are

[39] Often, because the population mean is unknown, an analyzer collects data to estimate it by calculating a sample mean. In this process, a sample variance (or a sample standard deviation) is obtained. However, if a population variance (or a population standard deviation) is known, it is considered superior to a sample variance (or a sample standard deviation). Therefore, the population variance (or the population standard deviation) must be used in place of the sample variance (or sample standard deviation) unless their validity is in doubt.

unknown. Under this situation, samples are collected to give information about the population. Even though the samples will yield incomplete information about the population, we must use the information as best as we can. For example, if a presidential election survey shows that 55% of people favor Candidate ABC, we can infer from this sample information that 55% of entire voters would favor Candidate ABC. Therefore, **inductive analysis shows how samples can be used to infer the nature and characteristics of a population.**

Of course, this type of inference – be it deductive or inductive – will not be correct 100% of the time. Therefore, we must introduce a level of confidence, expressed in terms of probability of being correct, in making this type of statements. What level of confidence one wishes to instill in such a statement is one of the key topics in statistics.

6. The pitfalls of Statistical Inferences:

Even though statistics can be a very useful analytical tool for modern businesspeople, over-extension or mis-application of its principles can result in detrimental consequences. Following are some of various pitfalls of statistical inferences.

a. Making generalizations about individuals from observations about groups → It can be proven that "Men are taller than women" → However, this may not be true for all individual cases.

b. Making conclusions about a large population from a small sample → "My friend, Joe, lived to be 100 years old while being a chain smoker all his life" → Therefore, I know that smoking is not that harmful.

c. Blindly believing a statistically significant result as the only truth → "Ten cups of coffee a day is harmful to your health" → Is 9 cups OK, then?

d. Misinterpretation of a statistical conclusion → "There is a 70% chance of rain today" → Does this mean that it will rain today or not? → Do you need to take an umbrella with you?

B. Types of Data

1. Primary vs. secondary sources of data

a. Primary data = the data that were collected by the researcher for his/her own use → marketing survey data, focus group data, etc.

b. Secondary data = the data that were not collected by the researcher but provided by a third or a fourth parties → government economic and business data, public financial data, etc.

2. Categorical vs. numerical data (or variables)

a. Categorical data = yes/no data = qualitative data = data based on an answer that goes into a category → Are you a male or a female? Married or not? Which do you prefer to eat – apples, bananas, or others?

b. Numerical data = measurement data = numerically measured data = data based on numerical values → How tall are you? How old are you? How many books do you have?

i. discrete data (or variables) = from a counting process → How many cars/books/phones do you have?

ii. continuous data (or variables) = from a measuring process[40] → How old/heavy/tall are you? → What is an average income of the U.S. households? → However, one can put them into groups, classes, categories, cells, or intervals and present them as if they are categorical data!

C. Data Sources

1. Primary Data Sources

a. Surveys and Interviews
b. Focus Groups → controlled experiments
c. Market Experiments

2. Secondary Data Sources

a. Public sector → Government → the largest source of secondary data
b. Private sector → Companies → financial data for investors
c. Internet Web Browsing

[40] Often, any measurement that comes with a decimal point is considered as continuous data. For example, even if one's annual income is measured in dollars such $50,000, it is considered as a continuous data because it can further be expressed in cents.

Exercise Problems on Chapter 1. Introduction

This set of exercise problems has 10 problems, worth a total of 10 points.

1. Statistics is a discipline that involves the _______.

   a. collection and presentation of data.
   b. effectiveness of quantitative decision making under certainty
   c. analysis of data and drawing inferences from them.
   d. all of the above.
   e. only (a) and (c) of the above.

2. A population is the collection of ____ members of a specific group under study whereas a sample is the collection of _____ members of that group.

   a. all; some
   b. some; all
   c. important; useful
   d. only (a) and (c) of the above
   e. only (b) and (c) of the above

3. A parameter is a _____ measure of a _____ characteristic.

   a. numerical; sample
   b. qualitative; sample
   c. numerical; population
   d. qualitative; population
   e. accurate; universal

4. A statistic is a _____ measure of a _____ characteristic.

   a. numerical; sample
   b. qualitative; sample
   c. numerical; population
   d. qualitative; population
   e. accurate; universal

5. A sample size can be _____ the population size.

   a. larger than
   b. smaller than
   c. larger than or equal to
   d. smaller than or equal to
   e. equal to

6. Statistics can be classified into two types of _____ statistics and _____ statistics.

   a. parametric; inferential
   b. descriptive; inferential
   c. qualitative; quantitative
   d. all of the above
   e. only (b) and (c) of the above

7. Estimation of population characteristics based on sample data and testing of hypotheses thereof are considered as a part of ______ statistics.

   a. analytical
   b. important and useful
   c. descriptive and quantitative
   d. inferential
   e. all of the above

8. Data collected via surveys are classified as _____ data whereas data collected via published public sources are classified as ______.

   a. biased; unbiased
   b. secondary; primary
   c. primary; secondary
   d. only (a) and (b) of the above
   e. only (a) and (c) of the above

9. Discrete and continuous data are the two types of ____ data.

   a. numerical
   b. categorical
   c. descriptive
   d. only (a) and (c) of the above
   e. only (b) and (c) of the above

10. Which of the following are typical examples of categorical data?

   a. height and age
   b. occupation and religion
   c. number of students and their enrollment status
   d. all of the above
   e. only (a) and (c) of the above

Chapter 2. Data Collection Methods

Secondary data are often used in analyzing various business environments and making many decisions thereof. However, someone must have spent the time and effort to collect the primary data to make it available to users as a secondary data. Thus, the collector of a primary data must take extreme care to ascertain its accuracy and validity. In order to prevent the GIGO situation[41] from happening, it is essential for us to understand the various types of sampling (=data collection) methods and their strengths and weaknesses.

A. The Nature and Reasons for a Sample

1. The Nature of a Sample

Even if the nature and the characteristics of a population are known, a sample can still be taken to assess the likelihood or probability of the sample being drawn from such a population[42]. In many situations, however, the nature and the characteristics of a population are not known. In order to know and learn about a population, a sample is collected and analyzed[43]. A sample is a collection of a specific number of observations, n, that are selected without any bias or distortion of a population's characteristics[44].

2. Reasons why a sample is used instead of a population

a. less time-consuming to collect data → because of a smaller number
b. less costly → cost effective
c. custom-made → meets your specific need
d. a population size may not be known → an infinite size is possible

B. Types of Sampling Methods

Sampling methods are classified as either biased or unbiased. If at all possible, a sampling method should be unbiased to yield more confident inferences.

1. Probability (=Unbiased) Sampling Techniques

Because a selection of an observation and thus, a sample thereof, involves some known rule of probability in selecting an observation, these methods are called a probability(-based) sample and said to be unbiased.

---

41 GIGO stands for Garbage-In-Garbage-Out, indicating that if inaccurate input is used (=garbage-in), the consequential output is inevitably inaccurate and useless (=garbage-out).
42 As studied in Chapter 1, this is known as deductive statistical analysis.
43 This is known as inductive statistical analysis.
44 When an entire population is sampled, it can also be called a sample. The definition of a sample and a population overlaps in this case. An example of this case is the U.S. census that takes place every 10 years.

a. Random samples → the probability of an observation being picked into a sample is equal for all observations → based on a random number generator or table, a random sample is picked → =RAND( ) or =RAND( )*100 or =RAND( )*(H-L)+L where H=a high number and L= a low number

Strengths: accurate and unbiased

Weaknesses: can be costly in time and money to assure a true randomness.

b. Systematic samples → If a sample size of n is to be chosen out of a population of N, then every k-th observation should be picked where k=N/n → a uniform probability is assumed and used.

e.g., Light-bulb manufacturing or cookie production

Strengths: less costly in time and money

Weaknesses: can be biased → every Monday (or 9:00 a.m.) may be a bad day (or hour), for example.

c. Stratified samples → Artificially divide the population into mutually exclusive strata and then, select samples randomly out of each of them → It assumes that each stratum is a good representation of the population.

e.g., Morning vs. Afternoon classes can be 2 strata and if 60% of students are in the Morning classes and 40%, in the Afternoon classes, one can select 60%(40%) of the sample observations from the Morning (Afternoon) classes

e.g., To calculate the consumer price index in the U.S., the Bureau of Labor Statistics conducts a stratified sample of 90,000 items from approximately 20,000 stores in 85 geographical areas (=strata) in the 50 states.

Strengths: well-balanced samples → number of females vs. males; morning production vs. afternoon production; etc. on the basis of age, income, occupation, etc.

Weaknesses: can be biased if the strata are not well established

d. Cluster samples →a variant form of the stratified sampling method → identify clusters that are believed to be representative of the population characteristics and then pick clusters randomly to form a sample.

e.g., given students (=random samples) in various classrooms (=clusters), select some classrooms randomly → this assumes that each classroom is reflective of the total student population.

e.g., given geographical regions (=clusters), select a random sample of regions → this assumes that each region is reflective of the population.

Strengths: convenient and less costly

Weaknesses: can be biased if each of the clusters does not represent the population. → i.e., each cluster must be a miniature version of the population.

Note: Most often, when the population size is large, a random systematic stratified sample can be taken → i.e., all these methods can be combined to produce an ideally unbiased sample → e.g., a presidential election poll.

Note: **The Law of a Large Number**[45] states that if the sample size increases, then the accuracy of the sample mean being equal to the population mean increases → Note that the sample mean will be equal to the population mean if the sample size equals the population size.

Note: **The Central Limit Theorem** states that as the sample size increases, the sampling distribution of the mean is (approximately) normally distributed → this theorem allows the use of the standardized normal distribution of Z for any probability distribution, provided that the sample size is large enough → a sample size greater than 30 is often considered "large" enough for the Central Limit Theorem to work[46]

2. Non-probability (=Biased) Sampling Techniques

Because a selection of an observation and thus, a sample thereof, does not involve any rule of probability in selecting an observation, these methods are called a non-probability(-based) sample and said to be biased.

a. Convenience samples → responses that are conveniently obtained →e.g., responses from the front-row students

b. Judgment samples → responses from experts/professionals → e.g., the number of illegal immigrants in the U.S.

---

[45] It also means that if the number of trials increases, then the empirical probability will come closer to the theoretical probability.

[46] When a sample size is infinite, the Z-distribution is applicable. When a sample size is small, such as less than 100, the t-distribution is preferred. However, due to simplicity and convenience, many empirical analysts will say that 30 or more observations is large enough to invoke the Central Limit Theorem .

c. Quota samples → responses based on a quota only → when a quota is met, no more responses will be collected → e.g. If 50% male response and 50% female response are to be obtained, the sampling effort stops when the quota is achieved. → e.g., If 45% Republicans, 45% Democrats, and 10% Independents are to be chosen for a political issue survey, the justification for doing so has to be very clear. Otherwise, it may make the results of a quota sample obsolete and inappropriate.

d. Chunk or group samples → a form of a convenience sample → instead of individuals, chunks or groups of individuals are surveyed → e.g., asking students in GSB420 class only, instead of all MBA students at DePaul.

e. Self-selected samples → responses from volunteers → e.g., students who raise their hands to give answers.

It is important to note that biased samples are neither always useless nor necessarily inaccurate. In fact, they can be very useful and accurate. The main point, however, is that they are more prone to yield erroneous inferences than unbiased sampling methods.

3. Errors in Survey Samples

Even if extreme cares are given in conducting a survey to obtain an accurate primary data, it is often inevitable to find some types of errors present in its collection procedure. Some of the prominent sampling errors are as follows:

a. Coverage error → selection bias

→ An error that results when certain members of a population is omitted from being sampled or partially represented in the sample.

→ e.g., Some families who do not have phones (or those who have unlisted telephone numbers) will never have a chance to be surveyed by telephone because their names will not appear in the phone directory.

→ Some surveyors may intentionally avoid a certain group from being sampled → e.g., do not ask the opposition's responses.

b. Non-response error → non-response bias

→ An error that results when no responses are given by the survey respondents. Often when a questionnaire is too long or complicated, the response rate will be low. → Given a choice of a 3 minute survey vs. a 30 minute survey, the shorter survey tends to have a higher response rate. → Prizes or awards are often promised before an interview is given to minimize this error → Phone call surveys are often done during dinner time to minimize the non-response error.

→ Interviewers' appearances such as good looks and neat dresses will minimize the non-response error → Female models are better than grubby males in getting a higher response rate.

c. Measurement error

→ An error that results when respondents give inaccurate answers to questions surveyed → Often known as a Hawthorne effect, a halo effect, or a Bradley effect[47] that brings about wrong , unintended, or exaggerated responses → e.g., ego- and power-related questions will be of concern → a Viagra survey, a wealth survey, or a racially charged survey is prone to this type of error.

d. Sampling error

→ An error that results when a chance difference exists in being chosen → e.g., in drawing a number from a bowl, those at the bottom may have a less chance of being picked than those at the top → mix the numbers well.

→ This error is the hardest to control because all surveys are subject to this error.

4. Validity of Survey Results

The validity of any survey results depends critically on the sampling method used. It is very unusual for surveyors to make mistake in tabulating surveyed data. However, there can be many issues with how the data has been collected and analyzed. Thus, if one is to find a fault with a survey result, one must evaluate the following:

a. The validity of the data collection/survey method.

b. The fairness of the presentation method → any attempt to exaggerate audio or visual effect?

For many real-world examples on this topic, please read "Freakonomics" by Steven D. Levitt and Stephen J. Dubner, 2005. (ISBN: 0-06-073132-X)

---

[47] It is named after Tom Bradley, a black candidate, who unsuccessfully ran for Governor of California in 1982 against a white candidate, a Republican George Deukmejian. Exit polls indicated Bradley had won but in reality, he lost by a narrow margin. A rumor had it that Collin Powel decided not to run for Presidency in 1996 against the incumbent Bill Clinton due to this Bradley Effect. Because there is a tendency by voters not to reveal their true actions when surveyed, this type of measurement error occur frequently. Thus, it is also known as the Wilder effect or the Dinkins effect for various election survey mishaps that had occurred in the past.

Exercise Problems on
Chapter 2. Data Collection Methods

This set of exercise problems has 8 problems, worth a total of 8 points.

1. Which of the following is classified as a biased sampling technique?

   a. random sampling
   b. stratified random sampling
   c. quota sampling
   d. cluster sampling
   e. none of the above is biased

2. The judgment sampling technique is a/an _____ sampling method which involves _______.

   a. biased; the opinions of experts
   b. biased; the random selection of interviewees
   c. unbiased; the opinions of experts
   d. unbiased; the random selection of interviewees
   e. biased; a minimum qualification of high school diploma

3. In surveying for samples, many types of errors can be made. If a number is to be drawn out of a bucket to determine a winner, a _____ error can often be made.

   a. coverage
   b. non-response
   c. sampling
   d. measurement
   e. halo effect

4. In surveying for samples, many types of errors can be made. If a person answers questions on a survey in an exaggerated or untruthful way by being aware of the interviewer, the survey results may suffer from _____.

   a. a measurement error
   b. a halo effect
   c. a Hawthorne effect
   d. all of the above
   e. none of the above

5. Which of the following sampling error is the most difficult to get rid of?

   a. coverage
   b. non-response
   c. sampling

d. measurement
e. halo effect

6. If you use in your report the answers given by the people in the front row of a meeting room, your report may

   a. be flawed because it used a convenience sample.
   b. not be flawed because it is based on your convenience.
   c. be flawed because it used a probability-based sample.
   d. not be flawed because it accurately showed the real world.
   e. only (a) and (c) of the above.

7. A fishbowl is often is used in a friendly raffle to hold many tickets from which a winning ticket is drawn. This method of selecting a winning number may suffer from a _____ error.

   a. coverage
   b. self-selection
   c. sampling
   d. measurement
   e. only (a) and (b) of the above

8. Low-income families who do not have a telephone at home can often be omitted from a telephone survey. Therefore, it can create an ethical issue to the surveyor as to whether these families should be included in the survey or not. Which of the following is/are the most likely source(s) of this ethical issue?

   a. coverage error
   b. non-response error
   c. measurement error
   d. all of the above
   e. none of the above

Chapter 3. Data Presentation Methods

The methods to present categorical data can also be used for numerical data. The essence of data presentation should lie on the convenience of summarizing the data, the easiness of understanding its contents, and the correctness of the information that are conveyed. In this regard, data can be presented in the form of graphs, pictures, charts, tables, etc.

A. Presentation of Categorical Data

Example>

If an interviewer found that 100 people ate cereals and 60 people ate fruits for breakfast while 40 people ate nothing for breakfast, this information can be presented in many different forms.

1. The Summary Table

This categorical data can be presented in a table as follows:

	Observed Frequency	Relative Frequency	Percentage (%) Frequency	Cumulative Frequency
Cereals →	100	100/200=0.5	50%	0.5 → 50%
Fruits →	60	60/200 = 0.3	30%	0.8 → 80%
Nothing →	40	40/200 = 0.2	20%	1.00 → 100%
Total →	200	200/200=1.00	100%	

Note the definitions of an observed frequency, relative frequency, percentage frequency, and cumulative frequency.

a. an observed frequency is the raw number of occurrences observed for a particular category.

b. a relative frequency is the ratio of an observed frequency divided by the total number of observations.

c. a percentage frequency is the relative frequency multiplied by 100.

d. a cumulative frequency is the sum of relative frequencies or percentage frequencies prior to and including the cell of one's interest.

2. The Bar Chart

The categorical data is most often presented in bar charts that are either vertical or horizontal bar charts. The following shows a vertical bar chart.

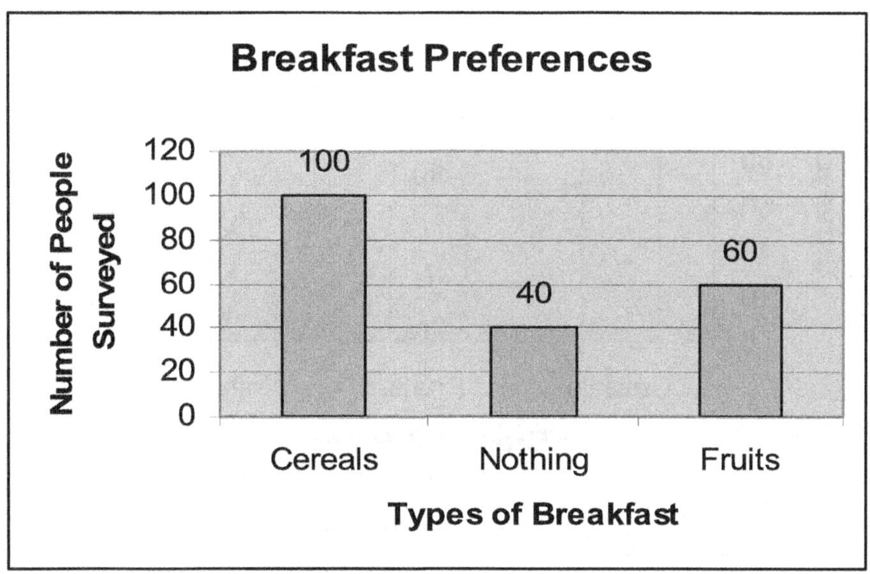

3. The Pie Chart

Categorical data is equally frequently presented in a pie chart as in a bar chart.

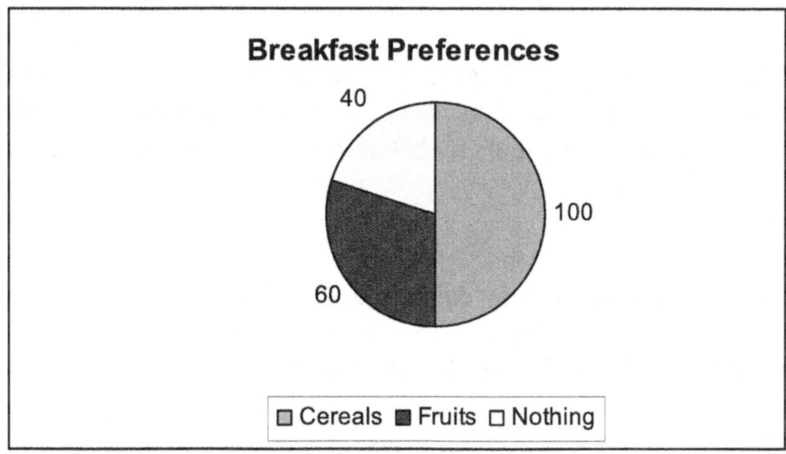

4. The Pareto Diagram

A Pareto diagram is essentially the same as a bar chart, except that the categories are identified in the order of their respective importance. That is, to draw a Pareto diagram, one must order the categories by their importance in terms of frequencies, thus, making it very easy to identify the important vs. not-so-important categories.

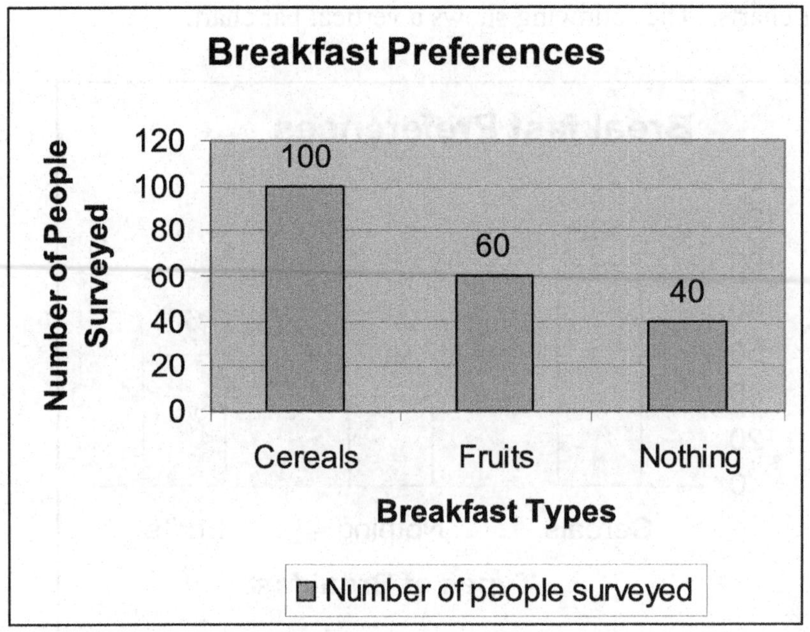

As shown above, "cereals" stands out as the most important whereas "nothing" shows up as least important. Thus, a Pareto diagram tries to portray visually the importance of each category.

B. Presentation of Numerical Data

Numerical data, if they can be properly grouped, can be presented by using the presentation methods for categorical data as shown above. However, they allow many unique methods of their own that can better summarize their characteristics.

Example>

Assume that the following is the survey data on DePaul MBA students' ages:

25, 32, 27, 29, 21, 31, 34, 28, 26, 45

This data can be presented in the following ways:

1. The Ordered Array

→ Rank order the data by magnitude → smallest to largest

21, 25, 26, 27, 28, 29, 31, 32, 34, 45

2. The Stem-and-Leaf Display

A stem-and-leaf display can be constructed either vertically or horizontally. The following shows a vertical stem-and-leaf display, which is more popular than a horizontal stem-and-leaf display.

a. a single stem-and-leaf display

```
2|1, 5, 6, 7, 8, 9
3|1, 2, 4
4|5
```

The vertically displayed values of 2, 3, and 4 represent the group of 20s, 30s, and 40s, and serve as the stem. The single digits associated with each of 20s, 30s, and 40s are written onto the same row and serve as members of a leaf. The visual size of a leaf indicates the frequency of observations.

b. a double stem-and-leaf display

Suppose that we have collected the following ordered array of ages for another group: 22, 23, 23, 27, 27, 30, 30, 32, 34, 50

When this data set is to be presented with the previous data set, the corresponding double stem-and-leaf display can be drawn as follows:

```
7, 7, 3, 3, 2|2|1, 5, 6, 7, 8, 9
   4, 2, 0, 0|3|1, 2, 4
             |4|5
            0|5|
```

3. The Frequency Distribution Table

a. Relative frequency distribution table

Class[48]	First Group	Second Group
20 – 29	6 (0.6)	5 (0.5)
30 – 39	3 (0.3)	4 (0.4)
40 – 49	1 (0.1)	0 (0)
50 – 59	0 (0)	1 (0.1)
Total	10 (1.0)	10 (1.0)

[48] Some statistics books and computer software may use the term, "bin," instead of "class." The approximate width of each class or bin is often determined by rounding up a value obtained from (Range÷k) where k = the number of desired classes or bins. By convention, however, the bin width is often chosen at an increment of 5 or 10, starting from 0.

b. Cumulative frequency distribution table

Class	First Group	Second Group
20 – 29	6 (0.6) → 0.6	5 (0.5) → 0.5
30 – 39	3 (0.3) → 0.9	4 (0.4) → 0.9
40 – 49	1 (0.1) → 1.0	0 (0) → 0.9
50 – 59	0 (0) → 1.0	1 (0.1) → 1.0
Total	10 (1.0)	10 (1.0)

These tables show observed, relative, and cumulative frequencies as was the case with categorical data.

4. The Histogram ≠ A Bar Chart

A histogram is a plot of a probability distribution for numerical data. It is almost the same as a bar chart, except that there is no space between bars. Unfortunately, the Excel Spreadsheet can not draw a histogram and thus, relative and cumulative frequencies are shown in the following double side-by-side bar chart.

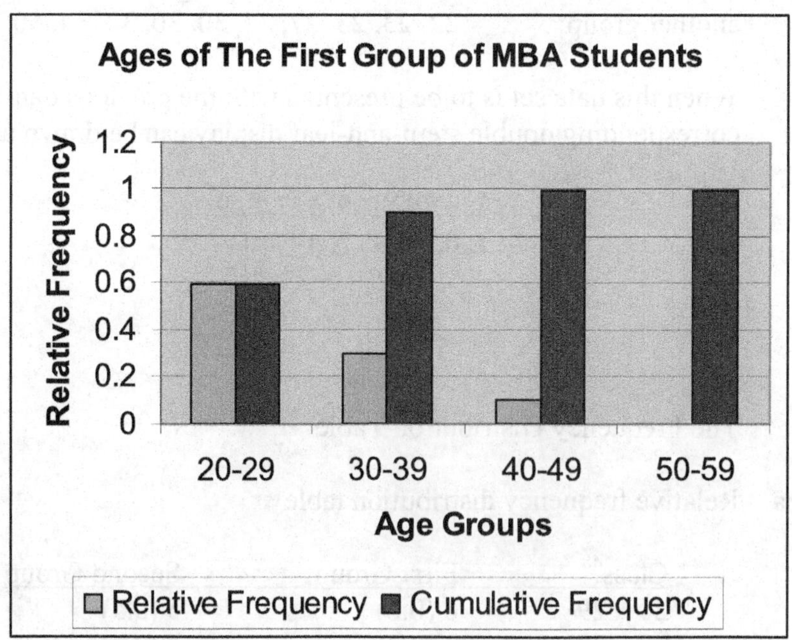

5. The Polygon

Polygons are a line chart of relative or cumulative frequencies.

a. the percentage polygon
→ a line chart of relative frequencies.

b. the cumulative percentage polygon = ogive
→ a line chart of cumulative frequencies.

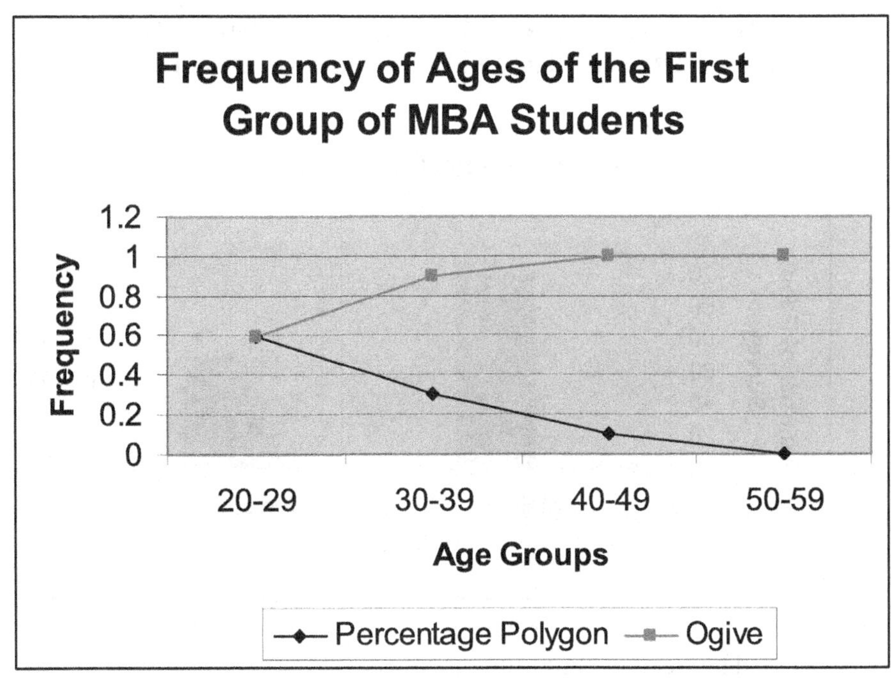

6. Pictograms

Pictograms are bar charts where bars are replaced by pictures that whimsically characterize the nature of the data → e.g., crude oil production can be depicted by the various sizes of oil wells gushing out oil → rail shipment amounts, by the sizes of railcars → obesity, by the sizes of people → whimsical pictograms add humor and fresh perception but may distort the true nature of the data due to its whimsicalness.

C. Presentation of Two or More Variables

Example>

A survey of 200 people found that 20 of 120 males were smokers whereas only 10 of 80 females were.

1. The Contingency Table = Cross-Classification Table

	Smoker	Non-smoker	Total
Male	20	100	120
Female	10	70	80
Total	30	170	200

2. The Side-by-Side Bar Chart

Horizontal vs. Vertical Side-by-Side Bar charts are often used to compare two or more data sets. The following shows a vertical side-by-side bar chart.

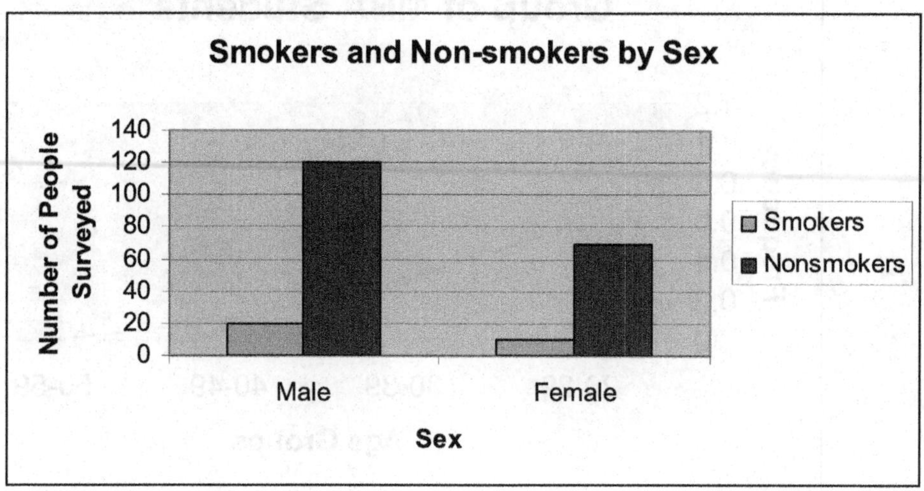

3. The Scatter Diagram

A scatter diagram shows two or more characteristics or variables being plotted against each other on a graph → no lines or bars are used → consequently, only scattered dots are shown on the graph.

Example> Given the following data set,

Individual (i):	1	2	3	4	5
Age (A_i):	20	40	30	60	50
Income (I_i):	\$30K	\$50K	\$40K	\$55K	\$60K

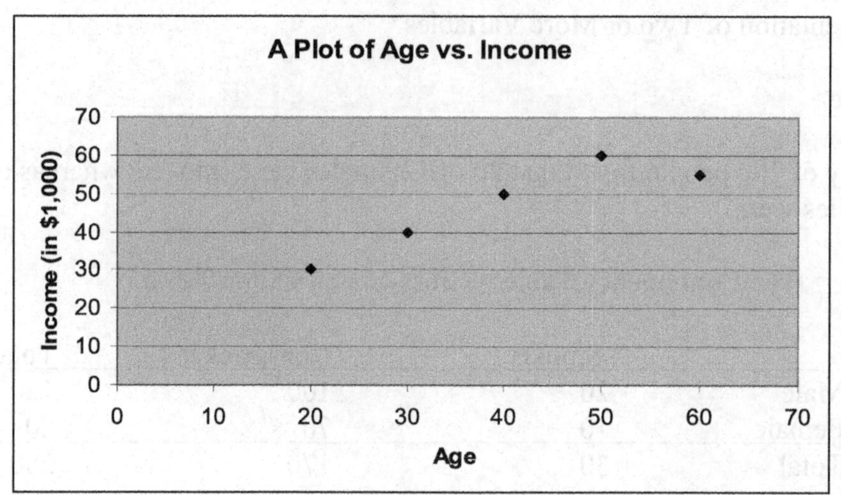

4. The Time-Series Plot → Values of a variable are plotted over a time period scale.

Time Period (t):	1	2	3	4	5
Income (I_t):	\$30K	\$50K	\$40K	\$55K	\$60K

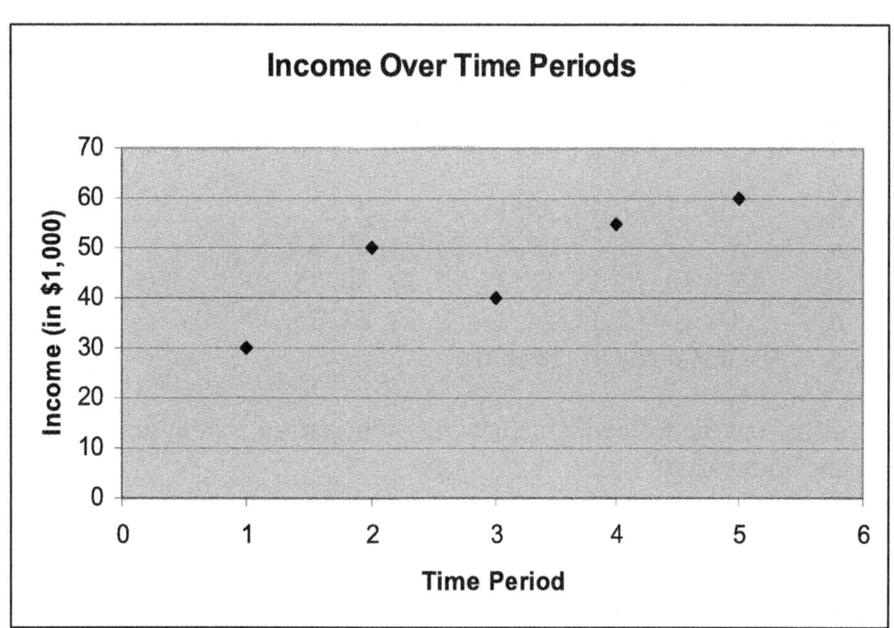

Exercise Problems on
Chapter 3. Data Presentation Methods

This set of exercise problems has 14 problems, worth a total of 14 points.

Given the following data set, X, answer Questions 1 through 5.

X = 25, 31, 10, 29, 15, 24, 27, 35, 39, 17

1. Which of the following represent the ordered array of the above data?

   a. X = 25, 31, 10, 29, 15, 24, 27, 35, 39, 17
   b. X = 17, 15, 10, 29, 27, 25, 24, 39, 35, 31
   c. X = 10, 13, 15, 17, 25, 27, 29, 30, 35, 39
   d. X = 10, 15, 17, 24, 25, 27, 29, 31, 35, 39
   e. all and any of the above

2. Which of the following depicts the stem part of a stem-and-leaf diagram based on the above data?

   a. 1, 2, 3
   b. 0, 1, 2, 3
   c. 0, 1, 2, 3, 4
   d. 0, 1, 4, 5, 7, 9
   e. none of the above

3. What is the range for this data set?

   a. 25 – 17 = 8
   b. 17 – 25 = –8
   c. 39 – 10 = 29
   d. 10 – 39 = –29
   e. none of the above

4. If you had chosen a number of classes to be 6, what would be the most appropriate class interval (=width)?

   a. 3
   b. 4
   c. 5
   d. 6
   e. 7

5. What are the class midpoints (=the averages of the two values defining the class interval) for the 6 classes?

a. 12.5; 17.5; 22.5; 27.5; 32.5; 37.5
b. 12; 17; 22; 27; 32; 37
c. 10; 15; 20; 25; 30; 35
d. 15; 20; 25; 30; 35; 40
e. none of the above

Given the following data about people's preference for soft drinks, answer Questions 6 through 8.

	Number of people who prefer
Coke	30
Pepsi	35
Coffee	25

6. The data shown and presented above is an example of a ____.

a. frequency distribution
b. relative frequency distribution
c. bar table
d. Pareto display
e. all of the above

7. If relative frequency is calculated for the above data in the order of Coke, Pepsi, and Coffee, it would approximately be:

a. 30; 35; 25
b. 0.30; 0.35; 0.25
c. 30%; 35%; 25%
d. 33%; 39%; 28%
e. 0.33; 0.39; 0.28

8. The data shown is a good example of ______ data.

a. numerical
b. categorical
c. relative frequent
d. tabulated and insightful
e. ordered and arrayed

9. Presenting data via the ordered array has following advantages and strengths:

a. the range can be easily shown and calculated.
b. extreme outliers can be easily identified.
c. works equally well with a categorical data set and a numerical data set.
d. all of the above
e. only (a) and (b) of the above

10. A scatter diagram shows a plot of one variable against a/an/the _____ variable whereas a time series diagram shows the plot of one variable against a/an/the _____ variable.

a. categorical; time
b. time; categorical
c. other; time
d. time; other
e. horizontal; vertical

11. The major weakness of a pictogram is that it ______:

a. is too simplistic and thus, does not challenge the intelligence of analysts enough.
b. can distort the truth by using different sizes of bars or pies, for example.
c. is becoming less popular among businessmen due to its complexity of advanced application software.
d. all of the above
e. only (a) and (c) of the above

12. Which of the following can be considered as a good guideline for developing good graphical presentation?

a. the scale on the vertical axis should begin at zero whenever possible.
b. all axes should be properly labeled.
c. the simplest possible graph should be used whenever possible.
d. all of the above
e. only (b) and (c) of the above

13. Which of the following is most often used for categorical data presentation?

a. histogram
b. polygon
c. Pareto diagram
d. all of the above
e. only (a) and (b) of the above

14. If you are to examine possible relationships between two numerical variables, you would use the ________ diagram.

a. scatter
b. time series
c. histogram
d. only (a) and (c) of the above
e. only (b) and (c) of the above

Chapter 4. Statistical Descriptive Measures

This chapter introduces many important statistical measures that describe the central tendency, dispersion, skewness, and peakedness of a numerical data set. Because these measures are the foundation of numerical data analysis - thus, the base for statistical analysis - one must know what they are and how they are calculated.

A. Standard Measures of Central Tendency

The three fundamental measures of central tendency are the mean, the median, and the mode.

1. The Mean = The Expected Value = The Average

a. Population vs. sample mean

* Population mean = the arithmetic average of a population
→ always expressed with a Greek letter such as μ, μ_x, and μ_y.

e.g., If the following 10 observations make up a population,

X_i = 9, 10, 10, 9, 10, 11, 11, 10, 10, and 10

then, the population size, N, is 10 and the population mean is:

$$\mu_X = \frac{\sum_{i=1}^{N} X_i}{N} = \frac{9+10+10+9+10+11+11+10+10+10}{10} = 10$$

* Sample mean = the arithmetic average of a sampled data
→ always expressed with an English letter with a bar on top such as $\overline{X}$ or $\overline{Y}$

e.g., If the following 5 observations were sampled from the above population,
X_i = 10, 9, 10, 11, and 10

then, the sample size, n, is 5 and the sample mean is:

$$\overline{X} = \frac{\sum_{i=1}^{n} X_i}{n} = \frac{10+9+10+11+10}{5} = 10$$

b. Arithmetic vs. geometric mean

* Arithmetic mean = an equal weight is given to all observations as is the case of the sample mean shown above.

* Geometric mean = a geometric weight is given to all observations
→ often used in calculation of a rate of return on investment.

e.g., If you invested $100 in Year 1, saw it grow to $200 in Year 2, and realized a balance of $100 at Year 3, what is the annual rate of return for this 3 year period?

Arithmetic Mean = (the sum of 1st and 2nd Year Returns)/2

$$= \frac{100\% + (-50\%)}{2} = 25\%$$

where the 1st year return = ($200 – $100)/$100 = 1 → 100%
and the 2nd year return = ($100 – $200)/$200 = –0.5 → –50%

Geometric Mean = $\sqrt[2]{(1+1)\cdot(1-0.5)} = 1$ → This implies that the rate of return is zero because 1 represents the principal invested[49].

Given these two measures of an average, it is obvious that the geometric mean is the more accurate measure of a rate of return than an arithmetic mean. However, in other general situations where an average is to be calculated, the arithmetic average is often the preferred one.

e.g., Suppose that a sample data has the following 5 observations:

10, 9, 10, 11, 10

Its arithmetic mean or simply, the sample mean, is calculated as:

$$\overline{X} = \frac{\sum_{i=1}^{n} X_i}{n} = \frac{9+10+10+10+11}{5} = 10$$

Or by an Excel Command of "=average(data)"

Its geometric mean[50] is calculated as:

[49] This method of calculating a geometric mean based on (1 + rate of return) is used in finance to calculate an investment return over an investment period.
[50] Because we are not calculating a rate of return in this example, we do not subtract 1 from this average. If it were a rate of return from an investment, we should subtract 1, the principal investment, from this average.

$$\overline{X_g} = \sqrt[n]{\prod_{i=1}^{n} X_i} = \sqrt[5]{9\times10\times10\times10\times11} = 9.98$$

Or by an Excel Command of "=geomean(data)"

2. The Median = the mid-point value

   The median value is identified by the following procedures:

   a. First, rank-order the data from the smallest to the largest value.

   b. Second, the median value is identified as:

      Median = the [(n+1)/2]-th observation value in the data set.

      e.g., Given n=5 in the above example,

      $$\text{Median value} \rightarrow \frac{5+1}{2} = 3 \rightarrow 3^{rd}\ \text{Observation} \rightarrow 10$$

      Or by an Excel Command of "=median(data)"

   c. If n = odd, the median is the [(n+1)/2]-th observation value;

      If n = even, the median is the **average of the two nearest** observation values from the [(n+1)/2]-th observation.

      Example 1>

      If we have a sample of 9, 10, 10, 10, 11, 11, 11, and 11, then, n=8 and thus, the median would be:

      $$\text{Median value} \rightarrow \frac{8+1}{2} = 4.5 \rightarrow \text{an average of } 4^{th} \text{ and } 5^{th}$$

      $$\text{observations} \rightarrow \frac{10+11}{2} = 10.5 \leftarrow \text{the median}$$

      Example 2>

      If we have a sample of 9, 10, 11, 12, 14, 14, then, n=6 and thus, the median would be:

$$\text{Median value} \rightarrow \frac{6+1}{2} = 3.5 \rightarrow \text{an average of } 3^{rd} \text{ and } 4^{th}$$

$$\text{observations} \rightarrow \frac{11+12}{2} = 11.5 \leftarrow \text{the median}$$

3. The Mode = the most frequently observed value

   In Example 1> above, there are four 11s. Thus, 11 is the mode.

   In Example 2> above, there are two 14s. Thus, 14 is the mode.

   Or by an Excel Command of "=mode(data)"

   If there are two modes in a data set, it is said to have a bimodal distribution. If there are three modes, it is a trimodal distribution. If there are many modes, it is a multimodal distribution.

4. The Sensitivity to an Outlier[51]

   a. Suppose that we have the following data: 9, 10, 10, and 11.

   We find:

   Arithmetic mean = (9+10+10+11)/4 = 10

   Geometric mean[52] = $(9\cdot10\cdot10\cdot11)^{1/4} = \sqrt[4]{9\cdot10\cdot10\cdot11} = 9.975$

   Median = the average of the 2nd and 3rd observations = 10

   b. Suppose that we now have an outlier of 100 to the above situation. That is, we have: 9, 10, 10, 11, and 100.

   We find:

   Arithmetic mean = (9+10+10+11+100)/5 = 28

   Geometric mean = $(9\cdot10\cdot10\cdot11\cdot100)^{1/5} = \sqrt[5]{9\cdot10\cdot10\cdot11\cdot100}$

   = 15.82

   Median = the 3rd observation = 10

   As seen above, the arithmetic mean changes from 10 to 28; the geometric mean, from 9.975 to 15.82; and the median, unchanged. Therefore, it is noted that the arithmetic mean is most sensitive to an outlier and the median is least sensitive to it. The geometric mean falls in between the two measures of central tendency.

---

[51] An outlier is an extreme or unusual value observed in a data set.

[52] We need not subtract 1 from this average because these are not rates of return from investment.

B. Atypical Measures of Central Tendency

1. Quartiles

a. First Quartile, Q1 = a value that represents the 25th percentile = 25% of the values are smaller than Q1 and 75% of the observations are larger than Q1.

Q1= the [(n+1)/4]-th observation found in the ordered array if it is a whole number; or an average to two adjacent observations if it is a fraction of 0.5; or any fractional value round up or down to the next largest integer[53].

Given 9, 10, 11, 12, 13, 14, 15, and 16,

because n=8, Q1 value is found at → $\frac{8+1}{4} = 2.25$ -th observation → the closest integer value = 2 → the 2nd observation → Therefore, Q1=10.

b. Third Quartile, Q3 = a value that represents the 75th percentile = 75% of the values are smaller than Q3 and 25% of the observations are larger than Q3.

Q3 = the [3(n+1)/4]-th observation found in the ordered array if it is a whole number; or an average to two adjacent observations if it is a fraction of 0.5; or any fractional value round up or down to the next largest integer[54].

Given 9, 10, 11, 12, 13, 14, 15, and 16,

because n=8, Q3 value is found at → $\frac{3(8+1)}{4} = 6.75$ -th observation → the closest integer value = 7 → the 7th observation → Therefore, Q3 = 15.

c. Second Quartile, Q2 = a value that represents the 50th percentile = 50% of the values are smaller than Q2 and 50% of the observations are larger than Q2 = the Median.

---

[53] In practice, however, to gain additional precision, a fractional value would be interpreted as a percentage and that percentage is used for adjusting the difference between the nearest two values. For example, when n=8, the Q1 value is found as the 2.25-th value. If the 2nd value is 10 and the 3rd value is 11, then their difference is 1 (=11-10). Since 25% of that difference is 0.25, the Q1 value is identified as 10.25 (=10+0.25), instead of a rounded value to the nearest observation of 10. However, different computer software companies provide their own versions of defining the first quartile value. It is, therefore, important to know what the software does.

[54] The same argument as the case of Q1 applies. That is, given n=8, the Q3 value is found as the 6.75-th value. If the 3rd value is 14 and the 4th value is 15, then their difference is 1 (=15-14). Since 75% of that difference is 0.75, the Q3 value is identified as 14.75 (=14+0.75), instead of a rounded value to the nearest observation of 15.

Q2 = the [(n+1)/2]-th observation found in the ordered array if a whole number or an average to two adjacent observations if it is a fraction of 0.5.

Given 9, 10, 11, 12, 13, 14, 15, and 16,

because n=8, Q2 value is found at → $\frac{8+1}{2}$ = 4.5 -th observation →

the average of 4th and 5th observations → 12.5 ← the median

2. Quintiles

When the observations are divided into 5 groups, they are called quintiles. The first quintile represents the lower 20% of the observations and the fifth quintile, the highest 20% of the observations.

3. Percentiles

When the observations are divided into 100 groups, they are called percentiles. The first quintile represents the lower 1% of the observations and the one-hundredth quintile, the highest 1% of the observations

For example, in the ACT or SAT or GMAT exams, the ranking can be shown in percentile → If you ranked at the 99th percentile, it means that your score is higher than 98% of the people who took the exam → Conversely, only 1% of the people scored higher than you.

C. Measures of Dispersion or Spread or Variation

Let's assume that the following is the data obtained via a census:

9, 10, 10, 10, and 11

Therefore, we can assume that the population size (N) is the same as the sample size (n). That is, N = n = 5.

Given this information, we can calculate the following diverse measures of dispersion:

1. The Range = the Highest Value – the Lowest Value

   Thus, the Range = 11 – 9 = 2

2. The Interquartile Range = the Middle Fifty = Mid-Spread = Q3 – Q1

Q3 value is found at → $\frac{3(5+1)}{4} = 4.5$ -th observation → the average of the 4th and 5th observations.

$$\text{Therefore, Q3} = \frac{10+11}{2} = 10.5$$

Likewise, Q1 value is found at → $\frac{5+1}{4} = 1.5$ -th observation → the average of the 1st and 2nd observations. Therefore, Q1 = 9.5.

Therefore, the Interquartile Range = Q3 – Q1 = 10.5 – 9.5 = 1

3. The Mean Absolute Deviation = MAD

$$MAD = \frac{\sum |X_i - \overline{X}|}{n} \qquad \text{where Deviation} = X_i - \overline{X}$$

Because $\overline{X} = 10$,

$$MAD = \frac{\sum |X_i - \overline{X}|}{n} = \frac{|9-10| + |10-10| + |10-10| + |10-10| + |11-10|}{5} = \frac{2}{5} = 0.4$$

4. The Sum of Squared Deviations = the Sum of Squares

$$SS = \sum_{i=1}^{n} (X_i - \overline{X})^2 = (9-10)^2 + (10-10)^2 + (10-10)^2 + (10-10)^2 + (11-10)^2 = 2$$

5. The Mean Squared Deviation = the Mean Squares → Population Variance = σ^2

$$\sigma^2 = MS = \frac{SS}{N} = \frac{2}{5} = 0.4$$

6. The population standard deviation = σ

$$\sigma = \sqrt{\sigma^2} = \sqrt{0.4} = 0.6324$$

7. The Sample Variance, S^2

$$S^2 = \frac{\sum(X-\overline{X})^2}{n-1} = \frac{SS}{n-1} = \frac{2}{5-1} = \frac{2}{4} = 0.5$$

8. The Sample Standard Deviation, S

$$S = \sqrt{S^2} = \sqrt{0.5} = 0.7071$$

Note 1:

The divisor, (n – 1), used in the sample variance formula is called the degrees of freedom. The reason why 1 is subtracted from the sample size, n, is because 1 observation is lost for calculating the (arithmetic) mean.

Note 2:

If a population (or sample) variance is known, its corresponding population (or sample) standard deviation is also known because a standard deviation is a square root of the variance. Likewise, if a population (or sample) standard deviation is known, its corresponding population (or sample) variance is also known because a variance is a standard deviation squared.

Note 3:

A population (or a sample) variance is measured in a specific unit squared. Thus, an intuitive interpretation of a squared unit as a measure of variability is not straightforward. Thus, a variance is not often used as a measure of variability or spread. A standard deviation is due to its clear straightforward intuition about the unit[55].

Note 4:

Any measure of spread shown above indicates the degree of variability, volatility, or fluctuation about the arithmetic mean. Therefore, the larger the standard deviation (or the variance, for example), the greater the variability of the data points about the mean. As such, a larger standard deviation can mean a larger variability, risk, or uncertainty associated with a data set.

Example 1>

The mean and the standard deviation of hotel room rates in France is 200 euros and 10 euros, respectively. The mean and the standard deviation of hotel room

[55] Consequently, the mean and the standard deviation are measured in the same unit and thus, it is very intuitive to combine a measure of central tendency with a measure of a spread to understand how a population (or a sample) is distributed. For example, the larger the standard deviation, the greater the spread about the mean.

rates in Germany is 200 euros and 8 euros, respectively. Which hotel room rates – France vs. Germany. – vary (or fluctuate) more?

Answer:

Given the same mean of 200 euros, the lower standard deviation in German hotel rates indicates that German hotel rates vary less than French hotel rates. That is, the hotel rates in Germany fluctuates less than in France. Therefore, you can expect to pay less diverging hotel room rates in Germany than in France.

Example 2>

The mean and the standard deviation of hotel room rates in France is 200 euros and 10 euros respectively. The mean and the standard deviation of hotel room rates in Germany is 150 euros and 10 euros, respectively. Which hotel room rates – France vs. Germany. – vary (or fluctuate) more?

Answer:

Because standard deviations of German and French hotel rates are the same, it is easy to jump to a conclusion that both countries' hotel rates vary the same. Even though this is true in an absolute term, if you look at the ratio or percentage of variability to the mean, it may not be the case.

For example, the variability per the mean for France is 10/200 = 0.05 = 5% whereas the same for Germany is 10/150 = 0.0667 = 6.67%. That is, German rates are higher than French when measured from their respective means. That is, in relation to their respective average hotel room rates, German rates vary more than French.

Example 3>

The mean and the standard deviation of hotel room rates in Europe is 200 euros and 10 euros respectively. The mean and the standard deviation of hotel room rates in the U.S. is $150 and $8 respectively. Which hotel room rates – Europe vs. the U.S. – vary (or fluctuate) more?

Answer:

Unless exact exchange rates are known for the entire study period, the degree of variability – 10 euros vs. $8 – can NOT be compared because of the difference in units measured. It's like comparing an orange to an apple.

In order to resolve issues raised in Examples 2 and 3 above, new measures of relative dispersion are introduced here below:

D. Relative Measures of Dispersion

1. The Coefficient of Variation (CV) → a measure of the spread or variability with respect to the size of the mean → a unit-less measure → allows a comparison of variability measured in different units → not necessarily expressed in % but often expressed in %.

Example 1> An Extension of Example 2> above.

Hotel room rates in France: S = 10 euros and $\overline{X}$ = 200 euros,

$$\text{Then,} \quad CV = \frac{S}{\overline{X}} = \frac{10\ euros}{200\ euros} = 0.05 \rightarrow 5\%$$

Hotel room rates in Germany.: S = 10 euros and $\overline{X}$ = 150 euros,

$$\text{Then,} \quad CV = \frac{S}{\overline{X}} = \frac{10\ euros}{150\ euros} = 0.0667 \rightarrow 6.67\%$$

Therefore, the German hotel room rates are (slightly) more variable (or volatile) than the French hotel room rates in relation to the mean.

Example 2> An Extension of Example 3> above.

Hotel room rates in Europe: S = 10 euros and $\overline{X}$ = 200 euros,

$$\text{Then,} \quad CV = \frac{S}{\overline{X}} = \frac{10\ euros}{200\ euros} = 0.05 \rightarrow 5\%$$

Hotel room rates in the U.S.: S = \$8 and $\overline{X}$ = \$150,

$$\text{Then,} \quad CV = \frac{S}{\overline{X}} = \frac{\$8}{\$150} = 0.0533 \rightarrow 5.33\%$$

Therefore, the U.S. hotel room rates are (slightly) more variable (or volatile) than the European hotel room rates in relation to the mean.

Example 3> Given the following information about the light-bulb life and the car speed, determine which is more variable in relation to their means.

Light-bulb Life: S = 0.7071 hour and $\overline{X}$ = 10 hours,

$$\text{Then,}\quad CV = \frac{S}{\overline{X}} = \frac{0.7071\,hour}{10\,hours} = 0.07071 \rightarrow 7\%$$

Car Speed: S = 3 miles per hour and $\overline{X}$ = 50 miles per hour,

$$\text{Then,}\quad CV = \frac{S}{\overline{X}} = \frac{3\,mph}{50\,mph} = 0.06 \rightarrow 6\%$$

Therefore, the light-bulb life is more variable (or volatile) than the car speed in relation to the mean.

Example 4> Given the following information about the monthly rates of return on stock investment in IBM and MSFT, determine the riskier stock of the two:

	Average Monthly Rate of Return	Standard Deviation of Monthly Returns
IBM	0.2%	1%
MSFT	0.1%	0.8%

Answer:

CV for IBM $= \frac{S}{\overline{X}} = \frac{1}{0.2} = 5$ whereas CV for MSFT $= \frac{S}{\overline{X}} = \frac{0.8}{0.1} = 8$.

Therefore, MSFT is riskier than IBM.

Another way of interpreting this result[56] is:

CV of 5 for IBM means that IBM requires 5 units of risk for a 1 unit of return whereas CV of 8 for MSFT means that MSFT requires 8 units of risk for a 1 unit of return.

2. The Z-score, Z-value, or Z-statistic

$$Z = \frac{X_i - \mu}{\sigma}$$ ← a measure of how far X_i is from μ after being adjusted by the size of the population standard deviation, σ.

[56] In investment literature, a similar concept known as a Sharpe Ratio (SR) is defined as: $SR = \frac{E(R - R_f)}{\sqrt{Var(R - R_f)}}$

where R = return of the stock and R_f = return of the benchmark portfolio. This ratio has an opposite interpretation of the coefficient of variation. However, they are similar in that both try to measure the return in relation to the risk. For example, the larger the Sharpe Ratio, the more desirable the stock is because it yields a higher return per unit of risk.

3. The t-score, t-value, or t-statistic

$$t = \frac{X_i - \mu}{S}$$ ← a measure of how far X_i is from μ after being adjusted by the size of the sample standard deviation, S.

Note that all these 3 relative measures of dispersion are unit-less and thus, cross-comparison across any variability is possible.

E. Measures of Shapes[57]

1. The Skewness = the Skewedness = the 3rd Moment = k

Whether a distribution of data points is skewed (either to the left or to the right) or not is an important fact to be recognized. The most ideal shape that is easy to analyze should have no skewness and look like a bell-shaped curve, symmetrical about its mean. **Any curve that shows this general characteristic is called "normal" or a "normal curve."** The following formula, known as "(Ronald A.) Fisher's Skewness," is used to calculate a coefficient of skewness[58], k:

$$k = \frac{\sum (X-\mu)^3}{n\sigma^3}$$

a. Symmetrical or normal when k = 0

Example> Assuming that μ = 10 and $\sigma = 0.6324$ from the previous data set, we can calculate its skewness as:

$$k = \frac{\sum (X-\mu)^3}{n\sigma^3} = \frac{(9-10)^3 + (10-10)^3 + (10-10)^3 + (10-10)^3 + (11-10)^3}{5 \cdot (0.6324)^3} = 0$$

Therefore, the data is judged to have no skewness; or it is symmetrical about its mean.

b. Negative or left-skewed when k < 0

→ the mode, median, and mean are all to the right-side of the distribution and the tail is to the left-side of the distribution.

---

[57] There is a complicated issue of how to calculate these measures of shapes – skewness and peakedness - using sample data. Therefore, different statistical software such as Excel, SAS, Minitab, etc. often yield different results.

[58] There are many variants of this formula. For example, the simplest formula for skewness is often represented by: $k' = \sum (X-\mu)^3 / n$

Example> Negative skewness is often desired in an achievement test, a qualifying exam, or a certificate exam whose main purpose is to examine the general mastery of topics by test takers. If passing a test assures the very minimum qualification or standard, then an exam should be designed easy enough for many to pass it. That is, when a majority of test takers do well, the mode/median/mean of their test scores should be located closer to the higher scores (or to the right-hand side). This observation is often found in medical doctors' board exam, CPA exam, or Driver's license exam → often referred as "a low ceiling."

c. Positive or right-skewed when k > 0
→ the mode, median, and mean are all to the left-side of the distribution and the tail is to the right-side of the distribution.

Example> Positive skewness in an achievement test indicates "a high floor," meaning a hard exam, which may be desirable for topics that require their detailed mastery by test takers. If passing a test is to assure a very high qualification or standard, then an exam should be designed hard enough to screen out those who are less qualified. That is, when a majority of test takers don't do well, the mode/median/mean of their test scores should be located closer to the lower scores (or to the left-hand side). This observation is often found in medical doctors' specialty exam or CFA exam.

2. The Peakedness = the Kurtosis = the 4th Moment = δ

There are many formulas can measure the peakedness of a distribution[59]. The following is known as the Kurtosis coefficient[60]:

$$\delta = \frac{\sum (X-\mu)^4}{n\sigma^4} - 3$$

a. Same as a normal curve (=mesokurtic) when $\delta = 0$.

b. Flatter than a normal curve (=platykurtic) when $\delta < 0$

c. More peaked than a normal curve (=leptokurtic) when $\delta > 0$

[59] A simplest one, for example, is: $\delta' = \sum (X-\mu)^4 / n$.

[60] This is formula is known as "Fisher's Kurtosis" even though it was originally formulated by Karl Pearson.

In the above case, the Kurtosis coefficient, δ, is calculated as:

$$\delta = \frac{\sum(X-\mu)^4}{n\sigma^4} - 3 = \frac{(9-10)^4 + (10-10)^4 + (10-10)^4 + (10-10)^4 + (11-10)^4}{5 \cdot (0.6324)^4} - 3$$

$$= \frac{2}{0.7997} - 3 = 2.5 - 3 = -0.5$$

This value of -0.5 indicates a slightly platykurtic distribution which means that the data shows a slightly flatter top than a normal distribution curve.

F. Graphical Presentation of Distribution via The Box-and-Whisker Plot

1. The Box-and-Whisker Plot[61]

The whiskers are the line connecting the lowest (L) and the highest (H) values; whereas the boxes are the 1st (Q1) and 3rd quartile (Q3) values. The box is split by the median line.

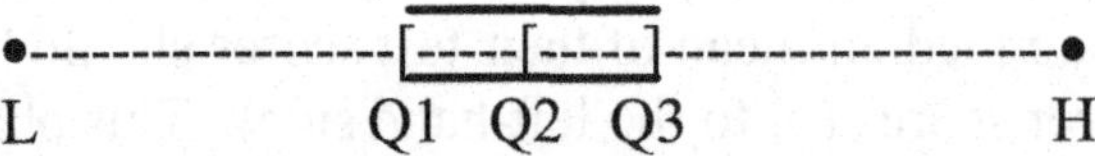

2. The Relationship Between a Box-and-Whisker Plot and a Continuous Probability Distribution Curve.

a. No skewness

If the box is situated in the middle of the whisker, the distribution is bell-shaped, symmetrical, and thus, normal.

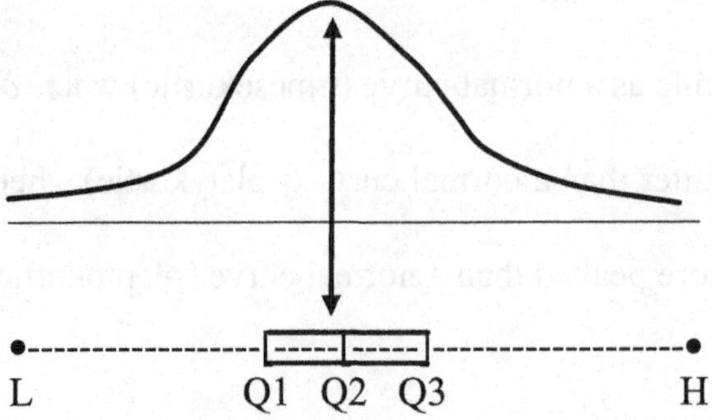

[61] Often, this is called a box plot for simplicity.

In this case, the 3 measures of central tendency – the mean, the mode, and the median – tend to be almost the same[62].

b. The Negative Skewness

If the box is situated to the right-hand side the entire whisker and thus, the left-hand side whisker is longer than the right-hand side whisker, the distribution is left-skewed or negative-skewed.

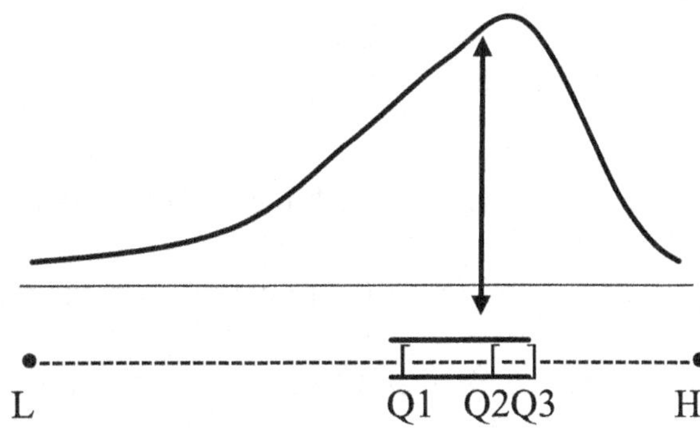

In this case, the mode will be located at the farthest to the right-hand side; then, the median will be to the left-hand side of the mode; and finally, the mean will be located toward the middle of the entire whisker.
→ That is, **Mean < Median < Mode.**

c. The Positive Skewness

If the boxes are situated to the left-hand side the whisker and thus, the left-hand side whisker is shorter than the right-hand side whisker, the distribution is right-skewed or positive-skewed.

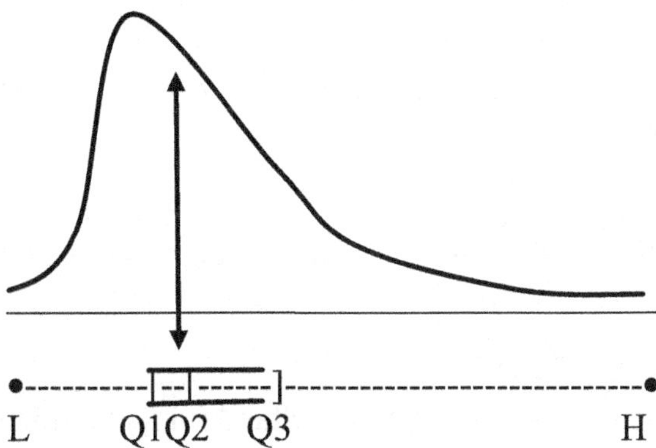

---

62 A note of caution in interpreting the skewness is necessary. When examining actual data for skewness, the case of the mean being exactly the same as the median or the mode is rare. Some differences will often exist. Despite these differences, many distributions can be accepted as a normal symmetrical shape.

In this case, the mode will be located at the farthest to the left-hand side; then, the median will be to the right-hand side of the mode; and finally, the mean will be located toward the middle of the entire whisker.
→ That is, **Mean > Median > Mode.**

G. Measures of Co-movements

1. The Covariance

$$COV(X,Y) = S_{XY} = \frac{\sum (X_i - \overline{X})(Y_i - \overline{Y})}{n-1}$$

a. If S_{XY} is zero → No relationship between X and Y

b. If S_{XY} is positive → X and Y move in the same direction

c. If S_{XY} is negative → X and Y move in the opposite direction

Note: A covariance quantifies the general co-movement between X and Y. However, it can not tell the tightness of the co-movement unless it is combined with variances of X and Y as used in the correlation coefficient described below.

2. The Correlation Coefficient

$$r_{XY} = \frac{S_{XY}}{S_X \cdot S_Y} = \frac{\sum (X_i - \overline{X})(Y_i - \overline{Y})}{\sqrt{\sum (X_i - \overline{X})^2} \cdot \sqrt{\sum (Y_i - \overline{Y})^2}}$$

Where $\overline{X} = \frac{\sum_{i=1}^{n} X_i}{n}$ = Average of X's; $\overline{Y} = \frac{\sum_{i=1}^{n} Y_i}{n}$ = Average of Y's:

$$S_{XY} = \frac{\sum (X_i - \overline{X})(Y_i - \overline{Y})}{n-1}$$ = Sample covariance of X and Y;

$$S_X = \sqrt{\frac{\sum (X_i - \overline{X})^2}{n-1}}$$ = Sample Standard Deviation of X; and

$$S_Y = \sqrt{\frac{\sum (Y_i - \overline{Y})^2}{n-1}}$$ = Sample Standard Deviation of Y

a. If $r_{XY} = 0$ → no linear relationship between X and Y

b. If $r_{XY} = +1$ → perfect positive (upward-sloping) linear relationship → same directional movements between X and Y. That is, if X goes up (down), then Y goes up (down).

c. If $r_{XY} = -1$ → perfect negative (downward-sloping) linear relationship → opposite or inverse directional movements between X and Y. That is, if X goes up (down), Y goes down (up).

d. $-1 \le r_{XY} \le 1$ → the correlation coefficient can not be less than −1 and can not be greater than +1.

Note: A correlation coefficient measures the degree of a LINEAR co-movement between X and Y. It is not capable of measuring a nonlinear (or curvilinear) relationship. That is, when a correlation coefficient between X and Y is equal to zero, we can say for sure that there is no linear relationship between X and Y. However, we can NOT say that there is no nonlinear relationship between X and Y because there can be a circular, parabolic, or higher-degree polynomial relationship. Therefore, it is always a good idea to plot the raw data and visual inspect the relationship between X and Y before calculating their correlation coefficient.

3. Example

Suppose the following data set is given:

i	1	2	3	4	5
X_i	10	20	30	40	50
Y_i	15	10	20	25	30

a. Calculate the Mean (=Average) of X_i's and Y_i's.

$\overline{X} = 30$ and $\overline{Y} = 20$

b. Calculate the sample variance, S_X^2, and standard deviation, S_X, of X_i's.

$$S_X^2 = \frac{(10-30)^2 + (20-30)^2 + (30-30)^2 + (40-30)^2 + (50-30)^2}{5-1} = \frac{1000}{4} = 250$$

Therefore, $S_X = \sqrt{250} = 15.81$

c. Calculate the sample variance and standard deviation of Y_i's.

$$S_Y^2 = \frac{(15-20)^2+(10-20)^2+(20-20)^2+(25-20)^2+(30-20)^2}{5-1} = \frac{250}{4} = 62.5$$

$$\text{Therefore, } S_Y = \sqrt{62.5} = 7.91$$

d. Calculate the sample covariance of X_i's and Y_i's.

$$S_{XY} = \frac{(10-30)(15-20)+(20-30)(10-20)+(30-30)(20-20)+(40-30)(25-20)+(50-30)(30-20)}{5-1}$$

$$\text{Therefore, } S_{XY} = \frac{450}{4} = 112.5$$

e. Calculate the correlation coefficient between X_i's and Y_i's.

$$r_{XY} = \frac{S_{XY}}{S_X \cdot S_Y} = \frac{112.5}{15.81 \cdot 7.91} = 0.8996$$

Because this correlation coefficient is positive and close to 1, we can state that there is a very strong positive (LINEAR) relationship[63] between X and Y → There is a strong positive (LINEAR) co-movement between X and Y → When X increases (decreases), Y tends to increase (decrease).

[63] Because it is a basic knowledge that a correlation coefficient measures a linear relationship, people will often not use the word, "linear," in describing the importance of a correlation coefficient.

Exercise Problems on Chapter 4. Statistical Descriptive Measures

This set of exercise problems has 30 problems, worth a total of 30 points.

Given the following data set, X, answer Questions 1 through 12.

X = 25, 31, 10, 27, 15, 24, 27, 35, 39, 17

1. What is the value of the arithmetic mean of this data set X?

   a. 25
   b. 24.5
   c. 24
   d. 27.78
   e. none of the above

2. What is the value of the geometric mean of this data set X?

   a. 25
   b. 24
   c. 23.30
   d. 27.78
   e. 30

3. The median of this data set X is the ____ observation of the ordered array and its value is _____.

   a. 5th; 15
   b. 5th; 25
   c. 5.5^{th}; 19.5
   d. 5.5^{th}; 26
   e. 6th; 24

4. The mode of this data set X is ____ and the range is ____.

   a. 24; 8
   b. 25; 29
   c. 26; 29
   d. 27; 29
   e. 27; 8

5. The first quartile of this data set X is the ____ observation of the ordered array and its value is _____.

   a. 3rd; 17

b. 3rd; 10
c. 8th; 31
d. 8th; 35
e. none of the above

6. The third quartile of this data set X is the ____ observation of the ordered array and its value is _____.

a. 3rd; 17
b. 3rd; 10
c. 8th; 31
d. 8th; 35
e. none of the above

7. The inter-quartile range of this data set X is often known as the ______ and its value lies between _____.

a. mid-value; 17 and 31
b. middle-fifty; 17 and 31
c. mid-value; 10 and 35
d. middle-fifty; 10 and 35
e. none of the above

8. The mean absolute deviation (MAD) value is _____.

a. 68
b. 70
c. 6.8
d. 7.0
e. none of the above

9. The sum of the squared deviations (SSD) is also known as the _____ and its value is _____.

a. sum of squares; 730
b. sum of deviations; 730
c. sum of squares; 73
d. sum of deviations; 73
e. either (a) or (b) of the above

10. The mean squared deviation (MSD) is also known as the _____ and its value is _____.

a. mean squares; 730
b. mean deviations; 730
c. mean squares; 73

d. mean deviations; 73
e. either (c) or (d) of the above

11. The sample variance is _____ and the sample standard deviation is _____.

a. 73; 8.544
b. 730; 27.018
c. 75.5; 8.689
d. 81.11; 9.006
e. none of the above

12. The skewness of this data set X is measured by the ____ moment about the mean and its value is _____.

a. second; +0.1248
b. third; +0.1248
c. third; –0.1248
d. fourth; +0.1248
e. fourth; –0.1248

13. The skewness of a frequency distribution is usually judged against the value of _____ and the distribution is said to be left-skewed if it is ______.

a. zero; negative
b. zero; positive
c. three; negative
d. three; positive
e. none of the above

14. If a frequency distribution shows a negative skewness, its mode is probably on the ____ hand side and its tail is on the ____ hand side of the distribution.

a. left; right
b. right; left
c. middle; middle
d. upper; lower
e. lower; upper

15. If the results of an achievement test show a negative skewness, it probably means that the test questions were _____ and many people scored _____.

a. difficult; poorly
b. easy; poorly
c. difficult; well
d. easy; well
e. reasonable; accordingly

16. The kurtosis is a measure of _____ of the data frequency distribution and it is often called the _____ moment about the mean.

a. peakedness; 3rd
b. skewness; 3rd
c. peakedness; 4th
d. skewness; 4th
e. peakedness; 5th

17. A box-and-whisker display or plot shows _____ that define(s) the first (1st) and third (3rd) quartile values and _____ that are formed by the lowest and the highest values in the data set.

a. a main body; tails
b. tails; main bodies
c. whiskers; boxes
d. boxes; whiskers
e. none of the above

18. In a left-skewed distribution, the left-hand side whisker is _____ the right-hand side whisker.

a. shorter than
b. longer than
c. equal to
d. prettier than
e. uglier than

Given the following data set for X and Y, answer Questions 19 through 21.

X = 27, 32, 49, 22, 20
Y = 45, 18, 39, 40, 58

19. The mean of X and Y are _____ and _____, respectively.

a. 27; 40
b. 30; 40
c. 40; 27
d. 40; 30
e. 29; 40

20. The correlation coefficient between X and Y is _____.

a. 0.38516
b. -0.38516

c. 1
d. -1
e. unknown

21. Covariance between X and Y is _____.

a. -10,801.0695
b. -103.9281
c. -64.4991
d. 64.4991
e. 103.9281

22. A correlation coefficient of X and Y being _____ means that X and Y move in the _____ direction.

a. positive; opposite
b. positive; same
c. negative; opposite
d. negative; same
e. only (b) and (c) of the above

23. A correlation coefficient can be _____.

a. $-1 \leq$ correlation coefficient $\leq +1$
b. $+1 \geq$ correlation coefficient ≥ -1
c. $-1 <$ correlation coefficient $< +1$
d. $+1 \leq$ correlation coefficient ≤ -1
e. only (a) and (b) of the above

24. A correlation coefficient between X and Y measures a _____ relationship between X and Y.

a. linear
b. nonlinear
c. logarithmic or exponential
d. all of the above
e. only (b) and (c) of the above.

25. Even if a correlation coefficient between X and Y is zero, X and Y can have _____.

a. a linear relationship.
b. a nonlinear relationship.
c. a curvilinear relationship.
d. either (b) or (c) of the above.
e. none of the above.

26. When the data set has an outlier problem, the _____ is often a better measure of central tendency.

a. arithmetic mean
b. geometric mean
c. mode
d. median
e. standard deviation

27. The most common measure of central tendency is the _____ and the most common measure of dispersion is the _____.

a. arithmetic mean; standard deviation
b. geometric mean; standard deviation
c. median; standard deviation
d. median; variance
e. mode; range

28. A coefficient of variation allows a comparison of relative variability of two or more data sets measured in _____ units.

a. absolute
b. relative
c. same
d. different
e. infinite

29. If a correlation coefficient between X and Y is equal to -1, we can conclude that there is a __________ linear relationship.

a. perfect negative
b. imperfect negative
c. negative
d. imperfect
e. perfect

30. If the mean, the mode, and the median are the same, it can be said that the distribution is ______.

a. left skewed
b. right skewed
c. not skewed
d. perfect
e. flat and important

Chapter 5. Probability Theory

The reason why we study the probability theory is that it enables us to concisely summarize theoretical or historical outcomes and consequently, better forecast future outcomes, based on them.

Think about the way that you may answer such questions as:

> Is it possible for someone in this classroom to win a lotto tonight?
> Is it possible to be stricken by lightening? By an airplane?
> Is it possible for your boss to be fired in a week?
> Is it possible for the economy to fall into a recession?
> Is it possible for a person to fall off a third floor and survive?

One can answer "yes" to all of the above questions because they can happen. On the other hand, one can also answer "no" to these questions because their chance of occurring is very small. Instead of simply answering yes or no to any of these questions, truly useful and valuable answers would involve a proper assessment of the probability (or likelihood) of them occurring. For example, the chance of someone winning a lotto tonight is one in a million or the likelihood of our economy falling into a recession this year is 25%. Therefore, it is very important and useful to know how to assess a probability of any event occurring.

A probability of an event occurring can be assessed by a ratio of the number of a desired event occurring divided by all possible outcomes that can occur. Therefore, in order to assess a probability of an event occurring, it is most important to know the frequency of the desired event occurring and the number of all possible outcomes. This chapter will first present some definitions associated with probability theory and the basic rules that are used to count the number of possible events. The concept of marginal and conditional probability will be discussed next. Finally, the Bayesian probability will be introduced[64].

A. Definitions

Before learning about the counting rules, we must be familiar with some definitions that are used in the probability theory.

1. An outcome = anything that can happen → enumerates what is possible →any single characteristic that can be theoretically inferred or empirically obtained.

   Examples:

---

[64] The best way to study the probability theory described herein is not to dwell on the probability formulas but to concentrate on the common sense and intuition behind the formulas. That is, try to use first your common sense and intuition, not the formula, to solve probability questions.

a. In a single toss of a coin, what are the various (possible) outcomes?
→ the (possible) outcomes are a head (H) and a tail (T).

b. In a single toss of 2 coins, what are the various (possible) outcomes?
→ the (possible) outcomes are HH, HT, TH, TT.

Exercises:

a. A single die is rolled. What are the various outcomes?
→ 1, 2, 3, 4, 5, and 6.

b. One coin is flipped and one die is rolled at the same time. What are the various outcomes?
→ (H,1), (H,2), (H,3), (H,4), (H,5), (H,6), and
(T,1), (T,2), (T,3), (T,4), (T,5), (T,6),

2. An event = a group (or collection) of outcomes of one's interest → inquires about what we are interested and thus, wish to know.

   There are two types of events:

   a. A simple (or elementary) event = an outcome = a single characteristic such as two heads → HH → This event is only 1 out of 4 possible outcomes (of HH, HT, TH, and TT) when 2 coins are tossed at the same time or 1 coin is tossed twice.

   b. A joint event = two or more characteristics → "at least one head" when 2 coins are tossed at the same time or 1 coin is tossed twice → HH, HT, TH → There are 3 events in this description of "at least one head."

   When a data is collected on either of these events – a simple or a joint, it is called an "observation."

Examples: Given that a (single) die is rolled,

i. What are the (possible) outcomes?
→ 6 outcomes of getting 1, 2, 3, 4, 5, and 6.

ii. What outcomes make up the event of getting an odd number?
→ Here, the event is made up of all outcomes of "an odd number" → Thus, the possible outcomes are 1, 3, and 5 → Note that an outcome is a member (=element) of an event → An event of getting an odd number has 3 outcomes (or elements) of 1, 3, and 5.

iii. What outcomes make up the event of getting a number less than 3?

→ Here, the event is made up of all outcomes of "a number less than 3" → Thus, the possible outcomes are 1 and 2.

iv. What outcomes make up the event of getting a number less than 1?
→ This event contains no outcomes → Therefore, it is known as a null set.

3. The complement of Event A = A'= all events that are not part of A.
If Event A is having "no head" in a two-coin toss, then the complement of A is "at least one Head' → That is, if A={TT}, then A'={HT, TH, HH}.

4. A sample space = a collection of all possible events (or outcomes)

Examples:

a. In the case of a single coin toss, the sample space is comprised of H and T → These 2 outcomes (or single events or observations) exhaust all possible outcomes → Thus, these comprise the sample space.

b. In the case of a single toss of two coins, the sample space is comprised of HH, TH, HT, and TT → These 4 outcomes (or events or observations) exhaust all possible outcomes → Thus, these comprise the sample space.

Exercises:

a. A single die is rolled. What is the sample space? → 1, 2, 3, 4, 5, and 6 → **a sample space is a collection of all possible outcomes (or events or observations)**.

b. One coin is flipped and one die is rolled at the same time. What is the sample space? → (H,1), (H,2), (H,3), (H,4), (H,5), (H,6), (T,1), (T,2), (T,3), (T,4), (T,5), (T,6),

5. Creation of Events

There are 3 ways to create an event on the basis of other events or outcomes by using the words, "and," "or," and "not."

Assume that we are tossing 2 coins:

a. The Sample Space = All Possible Outcomes:

HH, HT, TH, and TT

b. Events:

i. The event of getting at least one head → two heads **and** one head → HH, HT, TH

ii. The event of getting two heads **or** two tails → HH, TT

iii. The event of getting **no** head → TT

6. A probability = the numerical value representing the chance, likelihood, or possibility that a particular event will occur.

   Assume that we are tossing 2 coins. What is the probability of

   a. getting at least one head?
      → two heads **and** one head out of 4 possible outcomes → HH, HT, TH → Probability of ¾ (= ¼ + ¼ + ¼).

   b. Two head **or** two tails?
      → HH, TT out of 4 possible outcomes → Probability of 2/4 = ½ (= ¼ + ¼).

   c. **No** head?
      → TT out of 4 possible outcomes → Probability of ¼.

7. Summary Exercise Problems

   Given a 6-faceted die with numbers from 1 to 6 on each face,

   a. identify the sample space. → 1, 2, 3, 4, 5, and 6

   b. identify the joint event of all even numbers. → 2, 4, and 6

   c. identify the complement of all even numbers. → 1, 3, and 5

   d. identify the probability of getting an odd number → 3/6 = 1/2.

   e. identify the probability of getting an odd number greater than 4 → 1/6.

   f. identify the probability of getting an odd number greater than 5 → 0/6 = 0.

   g. identify the probability of getting an odd number greater than 4 or an even number less than 4 → the sample space is made up of 5 and 2 → Therefore, the probability = 1/6 + 1/6 = 2/6 = 1/3.

B. The Counting Rules

The following rules provide a general guideline on how to count the number of possible outcomes. Thus, knowing these rules will expedite the process of assessing a probability of an event occurring.

Rule 1:

If any one of k different mutually exclusive and collectively exhaustive events can occur on each of n trials, the number of possible outcomes is: k^n.

Examples:

a. Coin tosses → each toss had 2 outcomes of a Head (H) or a Tail (T)

i. What is the total number of outcomes if a coin is tossed once?
→ Because there is only 1 toss (i.e., n=1), then $2^1 = 2$ → (H, T)

ii. What is the total number of outcomes if a coin is tossed twice?
→ Because there are 2 tosses (i.e., n=2), then $2^2 = 4$
→ (HH, HT, TH, TT)

iii. What is the total number of outcomes if a coin is tossed n times?
→ Because there are n tosses, then 2^n different outcomes.

b. You are to issue an identification number of 4 digits to your employees. How many different numbers can you come up with?

$$10 \times 10 \times 10 \times 10 = 10^4 = 10{,}000$$

Rule 2:

If there are k_1 events on the first trial, k_2 events on the second trial, …, and k_n events on the n-th trial, then the number of possible outcomes is: $(k_1)(k_2)\ldots(k_n)$

Examples:

a. The number of possible license plates → a combination of 3 alphabet letters followed by 3 numbers → 26 x 26 x 26 x 10 x 10 x 10 =17,576,000

b. The number of possible license plates → any combination of 3 alphabet letters and 3 numbers → (26+10) x (26+10) x (26+10) x (26+10) x (26+10) x (26+10) = $(26+10)^6 = 36^6$ = 2,176,782,336

c. You are to issue an identification number of 4 digits to your employees. This time, however, you will not allow any 4 repeated numbers such as 0000, 1111, 2222, or 3333, etc. How many different numbers can you come up with?

10 x 10 x 10 x 10 – 10 = 9,990

d. You are to issue an identification number of 4 digits to your employees. This time, however, you will not allow any repeated numbers such as 1123 or 1213, 1223, etc. How many different numbers can you come up with?

10 x 9 x 8 x 7 = 5,040

Or

$$_nP_X = \frac{n!}{(n-X)!} = \frac{10!}{(10-4)!} = 5{,}040$$ ← Permutation which we will soon study here below.

Rule 3:

The number of ways that all n items can be arranged in order is:

n! = (n)(n-1)(n-2)…(1)

Note 1: The exclamation mark (!) is a mathematical notation, called "factorial."

Note 2: 0! = 1! = 1

Note 3: 5! = 5 x 4 x 3 x 2 x 1= 5 x 4 x 3 x 2 x 1! = 5 x 4 x 3 x 2!
= 5 x 4 x 3! = 5 x 4!

Note 4: 5! = 5 x 4! = 5 x 4 x 3! = 5 x 4 x 3 x 2! = 5 x 4 x 3 x 2 x 1!
= 5 x 4 x 3 x 2 x 1 x 0! = 5 x 4 x 3 x 2 x 1

Note 5: 5!3! = (5 x 4 x 3 x 2 x 1) x (3 x 2 x 1) = 720

Note 6: $\frac{5!}{3!} = \frac{5\cdot4\cdot3\cdot2\cdot1}{3\cdot2\cdot1} = 5\cdot4 = 20$

Example> 4 books of A, B, C, and D are to be put onto a shelf. How many different ways that these books can be arranged?

24 ways (= 4! = 4 x 3 x 2 x 1) as shown below.

ABCD	ABDC	ACBD	ACDB
ADBC	ADCB	BCDA	BCAD
BACD	BADC	BDCA	BDAC
CABD	CADB	CBAD	CBDA
CDAB	CDBA	DABC	DACB
DBAC	DBCA	DCAB	DCBA

Rule 4:

Permutations: The number of ways of arranging X objects selected from n objects **in order** is:

$$_{n}P_{X} = \frac{n!}{(n-X)!}$$

Example> What is the number of **ordered arrangements** of 2 books selected from 4 books?

$$_{4}P_{2} = \frac{4!}{(4-2)!} = \frac{4 \cdot 3 \cdot 2 \cdot 1}{2 \cdot 1} = 12$$

12 as can be verified by the above permutation equation and as shown below.

AB	AC	AD
BA	BC	BD
CA	CB	CD
DA	DB	DC

Rule 5:

Combinations: The number of ways of selecting X objects from n objects, **irrespective order**, is:

$$_{n}C_{X} = \binom{n}{X} = \frac{n!}{X!(n-X)!} = \frac{_{n}P_{X}}{X!}$$

Example 1> What is the number of **unordered arrangements** of 2 books selected from 4 books?

$$_{4}C_{2} = \binom{4}{2} = \frac{4!}{2!(4-2)!} = \frac{_{4}P_{2}}{2!} = \frac{4 \cdot 3 \cdot 2 \cdot 1}{2 \cdot 1 \cdot 2 \cdot 1} = 6$$

6 as can be verified by the above combination equation and as shown below.

AB AC AD
~~BA~~ BC BD
~~CA~~ ~~CB~~ CD
~~DA~~ ~~DB~~ ~~DC~~

Example 2> Given the Illinois Little Lotto where a winning combination of 5 numbers are to be picked from a 39 number pool, calculate the number of different outcomes possible.

$$ {}_{39}C_5 = \binom{39}{5} = \frac{39!}{5!(39-5)!} = \frac{{}_{39}P_5}{5!} = \frac{39\cdot 38\cdot 37\cdot 36\cdot 35}{5\cdot 4\cdot 3\cdot 2\cdot 1} = 575{,}757 $$

C. Rules of Probability

1. Definitions and Types of Probability

a. Classical probability = *a priori* probability = theoretical probability → we know this by theory:

$$ P(i) = P_i = \frac{\textit{number of outcomes contained in Event i}}{\textit{total number of outcomes in the sample space}} $$

i. Given a die, each number has a probability of 1/6 occurring:
The probability of Number 1 occurring = 1/6 = 0.16667
The probability of an odd number occurring = 3/6 = ½ = 0.5

ii. Given a fair coin, ½ for a head and ½ for a tail.

b. Empirical probability = observed probability → a survey shows that 30% of the people are smokers → the probability of a person being a smoker is 0.3.

c. Subjective probability = the probability based on one's experience or hunch → the chance of a having a recession in the next 6 month period is 10%

2. The probability of any event is between 0 and 1, inclusive. That is,

$$ 0 \le P_i \le 1 $$

If $P_i = 0$ → impossible to occur
If $0 < P_i \le 0.25$ → unlikely to occur
If $P_i \approx 0.5$ → a 50-50 chance to occur
If $0.75 \le P_i < 1$ → likely to occur
If $P_i = 1 = 100\%$ → certain to occur

3. The sum of the probabilities of all of the outcomes in the sample space is 1.

$$\sum_{i=1}^{\infty} P_i = 1$$

4. The probability that an event will NOT occur is equal to 1 minus the probability that the event will occur → a complement of an outcome is "not the outcome."

$$P(Complent\ of\ j) = 1 - P_j$$

Also, equivalently

$$P_j = 1 - P(Complent\ of\ j) = 1 - \sum_{i=1}^{\infty} P_i \text{ for } i \neq j$$

and

$$P_j + P(Complent\ of\ j) = P_j + \sum_{i=1}^{\infty} P_i = 1 \text{ for } i \neq j$$

Examples:

i. Given a single die being rolled, what is the probability of NOT rolling 4 or less? That is, what is the probability of rolling 5 and 6?

P(not 1, 2, 3, and 4) = 1 – P(1, 2, 3, and 4) = 1 – 4/6 = 2/6
= P(complement of 1, 2, 3, and 4) = P(5 and 6) = 2/6

ii. If the chance of success is 0.2, what is the probability of no success?

P(no success) = 1 – P(success) = 1 – 0.2 = 0.8 = 80%

iii. In a 3-horse race, if Horse #1 and Horse #2 have a winning probability of 30% each, what is the winning probability of Horse #3?

P(Horse #3 winning) = 1 – P(Horse#3 NOT winning)
= 1 – P(Horses #1 and #2 winning)
= 1 – (0.3 + 0.3) = 0.4

D. The Methods to Present Outcomes and Events:

1. Contingency (=Cross-Classification) Table

2. The Venn Diagram

3. The Decision Tree

E. The Contingency (=Cross-Classification) Table

1. Definitions

A = have an Apple → can change Apple to iPod, for example
A'= do not have an Apple → not A → complement of A

B = have a Banana → can change Banana to DVD, for example
B' = do not have a Banana → not B → complement of B

A or B = $A \cup B$ = A union B

A and B = $A \cap B$ = A intersection B

A + A' = B + B' = Total

2. A Descriptive Situation

a. The following information had been collected:

Do You Have an Apple?	Do You Have a Banana?		
	Yes(B)	No(B')	Total
Yes(A)	100	200	300
No(A')	300	400	700
Total	400	600	1000

b. Construct the probability table as follows by dividing all cells by the grand total number of 1000:

Do You Have an Apple?	Do You Have a Banana?		
	Yes(B)	No(B')	Total
Yes(A)	0.1	0.2	0.3
No(A')	0.3	0.4	0.7
Total	0.4	0.6	1.0

Note: The relative frequencies (or probabilities) are obtained by dividing each cell with the total number of observations.

3. Interpretation of Probability Information Contained in a Contingency Table

a. Marginal Probability

From the table of raw numbers (=frequencies) shown in 2.a. above, we note that (i) "Total" are written on the **margin** of the table; (ii) it shows that 300 people out of 1000 interviewed have Apples → 300/1000 = 0.3 = 30%; and (iii) it shows that 400 people out of 1000 interviewed have Bananas → 400/1000 = 0.4 = 40%. This information is summarized in the table shown in 2.b. above. Thus, these probabilities associated with a category total to the grand total are called a "**marginal**" probability.

b. Joint Probability

Each cell in the above probability table is called a "**joint**" probability because it represents a "joint" event. For example, the 100 people in the upper left-hand corner cell of Table 2.a. above are identified with a joint event of having an apple **and** a banana. Thus, the joint probability of someone having an apple **and** a banana is 100/1000 = 0.1 = $P(A \cap B)$.

Likewise, the 400 people in the lower right-hand corner cell are identified with a joint event of having no apple **and** no banana. Thus, the joint probability of someone having no apple **and** no banana is 400/1000 = 0.4 = $P(A' \cap B')$.

c. Conditional Probability

Now, looking at the body of the table in 2.a. above, we note that:

(i) 100 people out of 300 who have Apples have Bananas → 100/300 = 0.33 = 33% → Similarly, this can be expressed as "the probability of finding a person with a Banana given that she has an Apple" → $P(B|A) = 0.33$ ← **Conditional probability of B given A.**

(ii) 300 people out of 700 who have Apples have Bananas → 300/700 = 0.43 = 43% → Similarly, this can be expressed as "the probability of finding a person with a Banana given that she has NO Apple" → $P(B|A') = 0.43$ ← Conditional probability of B given not A.

(iii) 100 people out of 400 who have Bananas have Apples → 100/400 = 0.25 = 25% → Similarly, this can be expressed as "the probability of finding a person with an Apple given that she has a Banana" → $P(A|B) = 0.25$ ← Conditional probability of A given B.

(iv) 400 people out of 600 who have NO Bananas have NO Apples → 400/600 = 0.67 = 67% → Similarly, this can be expressed as "the probability of finding a person with NO Apple given that she has NO Banana" → P(A'|B') = 0.67 ← Conditional probability of not A given not B.

(v) This type of probability analysis can be applied to other cells in the table.

F. The Venn Diagram

1. The Diagram

The Venn diagram can visually depict the above information as follows:

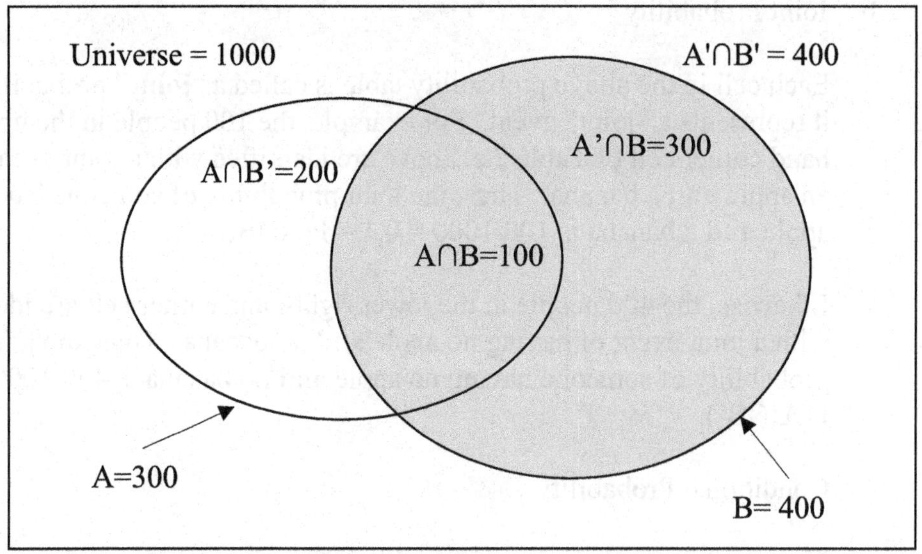

The Universe = 1000 = the grand total number of people surveyed = the entire rectangle
A = 300 = Those who have Apples = Those who belong in the left-hand side circle
B = 400 = Those who have Bananas = Those who belong in the right-hand side circle
A∩B= 100 = Those who have both Apples and Bananas = Those who belong in the middle elongated circle
A∩B'= 200 = Those who have Apples but no Bananas = Those who belong in the left-hand side circle but outside the right-hand side circle
A'∩B= 300 = Those who have Bananas but no Apples = Those who belong in the right-hand side circle but outside the left-hand side circle
A'∩B'= 400 = Those who have neither Apples nor Bananas = Those who are outside the two circles.

2. Matching the Venn diagram with the Contingency Table

→ Finding the meaning for each cell as follows:

Do You Have an Apple?	Do You Have a Banana?		
	Yes(B)	No(B')	Total
Yes(A)	$P(A \cap B)$	$P(A \cap B')$	P(A)
No(A')	$P(A' \cap B)$	$P(A' \cap B')$	P(A')
Total	P(B)	P(B')	1.00

3. Conclusions and Observations:

    a. $P(A) + P(A') = P(B) + P(B') = 1$

    b. $P(A \cap B) + P(A \cap B') = P(A) = 0.3$

    c. $P(A' \cap B) + P(A' \cap B') = P(A') = 0.7$

    d. $P(A \cap B) + P(A' \cap B) = P(B) = 0.4$

    e. $P(A \cap B') + P(A' \cap B') = P(B') = 0.6$

    f. $P(A \cup B) = P(A \cap B) + P(A' \cap B) + P(A \cap B')$
    $= P(B) + P(A) - P(A \cap B)$
    $= 0.1+0.3+0.2 = 0.4+0.3-0.1 = 0.6$

G. The Decision Tree

1. Case 1 → Allows the calculation of Conditional Probability

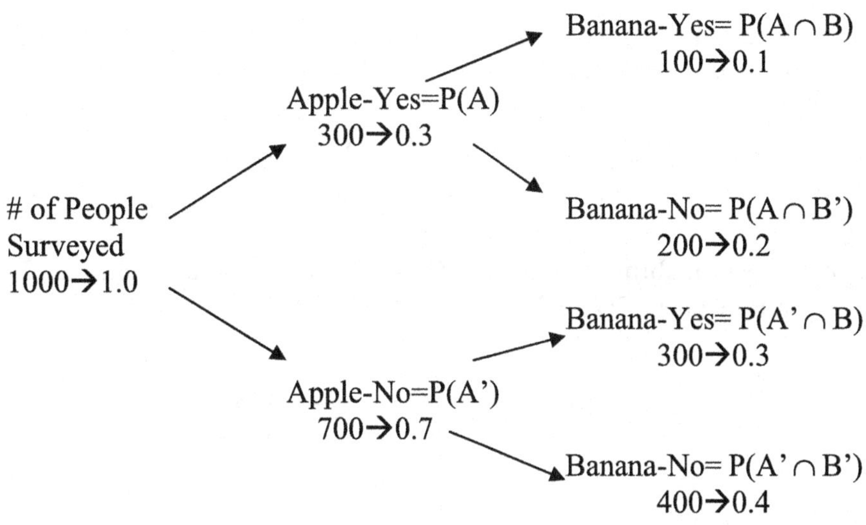

a. The probability of having a Banana given that this individual has an Apple → P(B|A)?

Given A, what is the probability of having B?
That is, P(B|A) = ?

$$P(B|A) = \frac{P(A \cap B)}{P(A)} = \frac{0.1}{0.3} = 0.33$$

b. The probability of having a Banana given that this individual does NOT have an Apple → P(B|A')?

Given A', what is the probability of having B?
That is, P(B|A') = ?

$$P(B|A') = \frac{P(A' \cap B)}{P(A')} = \frac{0.3}{0.7} = 0.43$$

2. Case 2 → Allows the calculation of Conditional Probability

a. The probability of having an Apple given that this individual has a Banana → P(A|B)

Given B, what is the probability of having A?
That is, P(A|B) = ?

$$P(A|B) = \frac{P(A \cap B)}{P(B)} = \frac{0.1}{0.4} = 0.25$$

b. The probability of having an Apple given that this individual does NOT have a Banana → P(A|B')

$$P(A|B') = \frac{P(A \cap B')}{P(B')} = \frac{0.2}{0.6} = 0.33$$

c. The probability of having NO Apple given that this individual does NOT have a Banana → P(A'|B')

$$P(A'|B') = \frac{P(A' \cap B')}{P(B')} = \frac{0.4}{0.6} = 0.67$$

The following decision-tree diagram will help one to see the overall scheme:

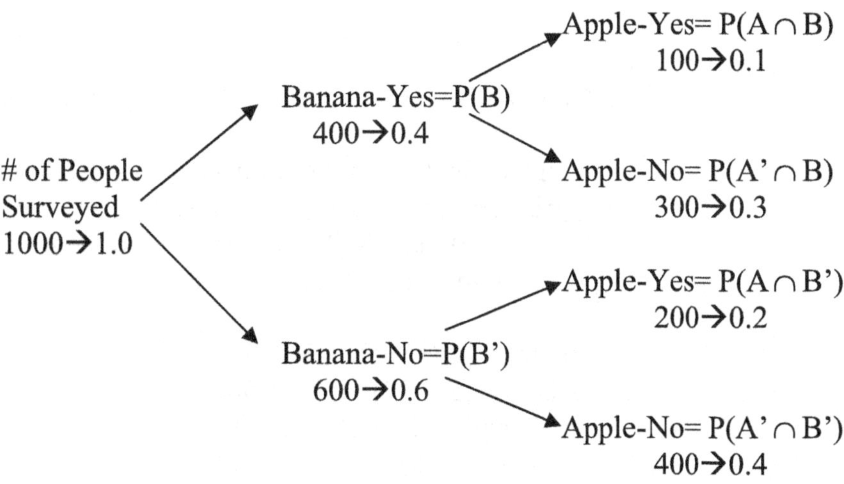

H. Types of Probability

1. Independent probability → mutually exclusive outcomes

a. P(A and B) = P(A)*P(B)

From flipping a coin and throwing a die, what is the probability of getting a head **and** a 1?

$$\text{P(H and 1)} = P(H \cap 1) = \frac{1}{2}\cdot\frac{1}{6} = \frac{1}{12} = 0.0833$$

b. P(A or B) = P(A) + P(B)

From flipping a coin and throwing a die, what is the probability of getting either a head **or** a 1?

$$\text{P(H or 1)} = P(H \cup 1) = \frac{1}{2}+\frac{1}{6} = \frac{8}{12} = 0.667$$

2. Overlapping probability → overlapping outcomes

As shown in the previous case, the overlapped portion has to be deducted:

$$P(A \cup B) = P(A) + P(B) - P(A \cap B)$$

Examples:

a. A card is selected at random from a deck of 52 cards. What is the probability that it is a King or a heart?

P(a King) = 4/52
P(a heart) = 13/52 = ¼
P(a King and a heart) = 1/52

Therefore,

P(a King or a heart) = (4/52) + (13/52) – (1/52) = 16/52 = 4/13

b. A student is selected at random from a class of 50 students which is comprised of 25 female MBA students, 5 female non-MBA students, and 10 male MBA students. Find the probability that the student is a female or a MBA student.

P(a female) = (25+5)/50 = 30/50
P(an MBA) = (25+10)/50 = 35/50
P(a female and an MBA) = 25/50

Therefore,

P(a female or an MBA) = (30/50) + (35/50) – (25/50) = 40/50 = 0.8

3. Dependent probability → conditional outcomes → Conditional Probability

As shown above,

$$P(A|B) = \frac{P(A \cap B)}{P(B)}$$

Alternatively,

$$P(A \cap B) = P(A|B) \cdot P(B) = P(B|A) \cdot P(A)$$
$$= P(B) \cdot P(A|B) = P(A) \cdot P(B|A)$$

Examples:

a. If two cards are selected from a deck and the first card is not replaced. What is the probability of picking two Aces? Two Kings? Two Queens?

P(an Ace and an Ace) = P(an Ace) x P(an Ace given an Ace is picked) = (4/52) x (3/51) = 12/2652 = 1/221 = 0.0045 = 0.45%

b. If two cards are selected from a deck and the first card is not replaced. What is the probability of picking an Ace and a King?

P(an Ace and a King) = P(an Ace) x P(a King given an Ace)
= (4/52) x (4/51) = 16/2652 = 4/663 =0.006 = 0.6%

I. Advanced Dependent/Conditional Probability Theory → Bayes' Theorem

- Developed by Thomas Bayes in the 18th Century.
- Revises previously calculated probabilities when new information is found.

1. The Bayes' Probability Formula

$$P(A|B) = \frac{P(A \cap B)}{P(B)} = \frac{P(B|A) \cdot P(A)}{P(B|A) \cdot P(A) + P(B|A') \cdot P(A')}$$

The above Bayes' formula assumes that P(A) is known prior to an evidence, B, is collected. Based on the new evidence of P(B), the probability of A occurring is revised. Consequently, the Bayes' formula allows one to calculate a conditional probability of A occurring given that B had occurred. That is, P(A|B) is otained.

Note:

P(A|B) is a conditional probability that is determined after empirical or additional evidence, B, is observed → therefore, it is also known as "posterior" or "revised" probability;

P(A) is a marginal probability that is assumed to be known before information on B is collected → Therefore, it is often called a "prior" probability; and

P(B|A)·P(A) = P(A∩B) is a joint probability.

2. Example #1

Suppose that you are interested in knowing the probability of a stock price increase given an interest rate decrease[65]. To assess this probability, you have collected the following historical data:

	Interest Rate		
	Decrease (B)	Increase (B')	Total
Stock Price			
Increase (A)	60	5	65
Decrease (A')	20	55	75
Total	80	60	140

Now, we are to use the Bayes' formula to find out P(A|B) as follows:

[65] Stock price being a firm's future profit stream discounted at an interest rate, this relationship is assumed to be correct.

$$P(A|B) = \frac{P(A \cap B)}{P(B)} = \frac{P(B|A) \cdot P(A)}{P(B|A) \cdot P(A) + P(B|A') \cdot P(A')}$$

where P(B) = 80/140 = 0.5714; P(A) = 65/140 = 0.4643; P(A') = 75/140 = 0.5357 = 1 – P(A) = 1 – 0.4643; P(B|A) = 60/65 = 0.9231; and P(B|A') = 20/75 = 0.2667.

That is,

$$P(A|B) = \frac{0.9231 \cdot 0.4643}{0.9231 \cdot 0.4643 + 0.2667 \cdot 0.5357} = \frac{0.4286}{0.5714} = 0.75$$

Alternatively, you could have identified P(B) = 80/140 = 0.5714 and P(A∩B) = 60/140 = 0.4286, and calculate P(A|B) as:

$$P(A|B) = \frac{P(A \cap B)}{P(B)} = \frac{0.4286}{0.5714} = 0.75$$

This result shows that there is a 75% probability that stock price will increase if an interest rate decreases.

3. Example #2

Suppose the following information is given:

B= Cancer	B'=No cancer
A=Test positive for cancer	A'=Test negative for cancer

And

P(B)=0.05	P(B')=0.95
P(A|B)=0.9	P(A|B') = 0.03

Given the above information, calculate the probability that the cancer is actually present given a positive test result. That is, P(B|A)=?

$$P(B|A) = \frac{(0.9)(0.05)}{(0.9)(0.05) + (0.03)(0.95)} = \frac{0.045}{0.045 + 0.0285} = \frac{0.045}{0.0735} = 0.61$$

Note: P(A|B) + P(A'|B) = 1 → Given P(A|B)=0.9, then P(A'|B) = 0.1.
However, P(A|B) + P(A|B') ≠ 1

J. References

The following can serve as good introductory reference sources on probability theory:

1. Bluman, Allan G. Probability Demystified: A Self-Teaching Guide, McGraw-Hill, 2005.

2. Gonick, Larry, and Smith, Woollcott. The Cartoon Guide to Statistics, HarperCollins, 1993. ISBN: 0-06-273102-5

Exercise Problems on Chapter 5. Probability Theory

This set of exercise problems has 22 problems, worth a total of 25 points.

1. Any single characteristic that is theoretically inferred or empirically proved is called a/an _____ whereas a collection of all possible outcomes is called a/an _______.

   a. outcome; sample space
   b. outcome; complement
   c. sample space; outcome
   d. complement; sample space
   e. event; outcome

2. A probability calculated as a ratio of the number of a particular outcome observed to the total number of outcomes observed is known as _____ probability.

   a. a priori
   b. empirical
   c. subjective
   d. objective
   e. Bayesian

3. A probability value, p, can be

   a. $-1 \leq p \leq 1$
   b. $-1 \leq p \leq 0$
   c. $0 \leq p \leq 1$
   d. $0 < p < 1$
   e. $-1 < p < 1$

4. The total sum of probabilities for all possible outcomes _____.

   a. can be less than 1.
   b. must be less than 1.
   c. must be equal to 1.
   d. can be greater than 1.
   e. must be greater than 1.

5. A complement of Event X means

   a. all events that are part of X.
   b. all events that are not part of X.
   c. partial events that complement X.
   d. partial events that do not complement X.
   e. X is independent of all other events.

The following information is to be used in answering Questions 6 through 10.

Suppose that MBA students at DePaul were surveyed for their preference for GSB 420. The following is their answer:

	Male Students	Female Students
Yes: Love GSB 420	60	40
No: Hate GSB 420	20	30

6. The probability of DePaul's MBA students loving GSB 420 is _____.

   a. 100 b. 70 c. 0.9
   d. 0.7 e. 0.6667

7. Given the total MBA students at DePaul, the probability of DePaul's MBA students who are female and love GSB 420 is _____.

   a. 40 b. 0.4 c. 0.2667
   d. 0.6667 e. 0.5714

8. Given DePaul's female MBA student population, the probability of finding a female MBA student who loves GSB 420 is _____.

   a. 0.8 b. 0.4 c. 0.2667
   d. 0.6667 e. 0.5714

9. Given DePaul's MBA student population, the probability of finding a male who hates GSB 420 or a female MBA student who loves GSB 420 is _____.

   a. 0.8 b. 0.4 c. 0.2667
   d. 0.6667 e. 0.5714

10. Given DePaul's MBA student population, the probability of finding a male student or any student who loves GSB 420 is _____.

    a. 0.8 b. 0.4 c. 0.2667
    d. 0.6667 e. 0.5714

Suppose that survey results show that 42% of small firms introduced at least one new, or significantly improved product, service, design or process in the past year. Among these firms, 54.5% said that they introduced at least one new or improved product; 29.4%, a new or improved service; and 15.5%, either new or improved process or design.

11. On the basis of this information, the probability of any U.S. small firm introducing a new or improved product would be _____.

    a. 0.7711 b. 0.545 c. 0.42
    d. 0.2289 e. unknown

12. It is also reported that three quarters of all small firm owners in the U.S. encouraged their employees to suggest ideas for new products or services and 52% of them gave bonuses or recognition to those who came up with ideas. What is the probability of any small firm owner in the U.S. having paid out a bonus or recognition to the employees?

a. 0.75 b. 0.52 c. 0.39
d. 0.61 e. unknown

13. Assume that 50% of people take Vitamins B and C together. Also assume that 80% of people take Vitamin C every day. What is the probability that a person takes Vitamin B given that the person takes Vitamin C?

a. 0.8 b. 0.5 c. 0.4
d. 0.625 e. unknown

14. If a coin is tossed 3 times in a row, the number of all possible outcomes such as HHH, HHT, or TTT that can result from these 3 tosses is _____.

a. 2 b. 3 c. 4
d. 6 e. 8

15. If you are to create an identification (ID) number for your employees using any 2 numbers from 1 to 9 and any 2 letters in an alphabet, how many different possible ID numbers can you come up with?

a. 234 b. 2,106 c. 6,084
d. 54,756 e. 67,600

16. Permutation calculates the number of ways X objects can be selected from n objects ______ whereas combination calculates the number of ways X objects can be selected from n objects _____.

a. in order; irrespective order b. irrespective order; in order
c. in an orderly fashion; in a confused fashion
d. in a random fashion; in an orderly fashion
e. only (b) and (d) of the above

17. Assume that the Illinois Lottery has hired you as a consultant to design a new pick-4 game where they wish to draw 4 letters out of a bucket that contains 26 letters of an alphabet. Once a letter is picked, it will not be replaced. Identify the probability of picking a winning combination of 4 letters. (Hint: the order of appearance does not matter.)

a. 1 out of 358,800 b. 1 out of 14,950
c. 1 out of 456,976 d. 1 out of 19,040

e. 1 out of 1,000,000

18. Suppose that you have matched first 2 out of 4 letters. Two more letters are to be picked for a winning combination of 4 letters. What is the probability of you being the winner? (3 points)

a. 1 out of 676
b. 1 out of 650
c. 1 out of 600
d. 1 out of 552
e. 1 out of 276

19. Suppose that you own a restaurant that allows a diner to pick any 4 out of 10 foods on the menu, irrespective of order. How many different combinations of foods should your cook be able to prepare?

a. 10,000
b. 5,040
c. 416.67
d. 210
e. none of the above

The following questions are worth 2 points each.

20. In GSB 420, 20% of the students get an A; 30%, a B; 40%, a C; and 10%, a D. On a survey, 90% of students with an A said they studied very hard for the exam; 70% of students with a B said they studied very hard for the exam; 40% of students with a C said they studied very hard for the exam; and 20% of students with a D said they studied very hard for the exam.

The probability of finding a student who said he/she studied very hard for the exam is _____.

a. 0.3
b. 0.27
c. 0.57
d. 2.2
e. 0.55

21. Given the description above, if a student studies very hard for the exam, what is the probability that he/she will receive an A?

a. 0.18
b. 0.57
c. 0.9
d. 0.3157
e. 2.2

22. Assume that 30% of the U.S. citizens are classified as "rich" and the remaining 70% are classified as "poor." Furthermore, assume that 50% of the rich U.S. citizens say that they invest in stocks whereas 10% of the poor U.S. citizens say that they invest in stocks. 20% of the rich and 40% of the poor U.S. citizens do not invest at all. What is the probability that a stock investor is classified as "rich"?

a. 0.3
b. 0.5
c. 0.2
d. 0.68
e. 0.06

Chapter 6. Discrete Probability Distributions

A discrete probability distribution describes the probability of a discrete random variable occurring. A discrete variable is an observation (or a data) that is measured in a discrete unit such as categorical data of yes-or-no and discrete numerical data obtained from a counting process such as a number of customers in line, phone calls received, customer orders taken, etc. A **random variable** is a data set whose values are observed randomly without any managed plans of selecting them.

Among discrete probability distributions, however, the binomial distribution and the Poisson distribution are the most prominent. The binomial distribution allows a probability assessment of an event that has only two outcomes of either yes-or-no, failure-or-success, or one-or-the-other-group. The Poisson distribution, on the other hand, allows the probability assessment of an outcome based on an average number of occurrences observed.

Before discussing these distributions, alternative but very powerful concepts of calculating the mean and the variance of a discrete random variable by the use of the expectations operator are first in order.

A. Definitions

1. The Mean = The Expected Value of a Discrete Random Variable = E(X) = The First Moment about the Origin = μ

$$\mu = E(X) = \sum_{i=1}^{m} X_i \cdot P(X_i)$$

where X_i = the i-th observation of X = the i-th X value
and $P(X_i)$ = the probability of X_i occurring

Example 1>

Given 10, 9, 10, 11, and 10 as a population, the arithmetic average (=the population mean) is calculated as:

$$\mu = \frac{\sum_{i=1}^{N} X_i}{N} = \frac{9+10+10+10+11}{5} = 10$$

Note that there are one 9, three 10s, and one 11, out of 5 observations.

Alternatively, therefore, the mean (=the expected value) of X can be calculated as:

$$\mu = \frac{1}{5}\cdot 9 + \frac{3}{5}\cdot 10 + \frac{1}{5}\cdot 11 = 10 \rightarrow \mu = \sum_{i=1}^{m} P(X_i)\cdot X_i = E(X)$$

where $P(X_i)$ is the probability of X_i occurring. This calculation emphasizes the relative frequency (=probability) of 9, 10 and 11 occurring in the population. → This method is useful if only the probabilities of certain outcomes to occur are known. That is, you need not know all the observation values as is the case of calculating the arithmetic average.

Note: The expected value of X, E(X), is the mean of X, μ.

Example 2>

Howie Mandel's "Deal or No Deal" TV Game Show[66]:

Given the following suit cases of $5; $300; $50,000; and $1 million are to be opened, the Banker offers $165,000. Will you take it? Deal? Or No Deal? If you base your decision purely on the expected value concept, will you make a deal or not?

Answer:

The answer to this question will critically depend on your risk preference. However, if you are to make a decision solely based on the mathematical expectation, you should reject the deal because the offer is far lower than the expected value of the game as shown below:

$$E(X) = \frac{5 + 300 + 50000 + 1000000}{4} = \frac{1050305}{4} = 262576.25$$

In reality, the contestant took the deal and stopped the game. Was he irrational?

Example 3>

A survey showed that Americans exercise an average of 20 minutes daily. However, only 20% of Americans exercise daily. How long do you think these exercising Americans exercise every day?

---

[66] This game starts with 26 brief cases with values, ranging between $0.01 and $1 million, that are written inside each brief case. To start the game, the contestant initially picks 6 brief cases to be opened, except the one he/she keeps to his/her side and the number of brief cases to be opened decreases as the game progresses.

Answer:

Because E(X) = $0.2 \times X + 0.8 \times 0 = 20$, X = 100 minutes.

Example 4>

How would you determine the annual premium of a $500,000 life insurance policy for a 29-year-old male? Suppose that 5 out of 10,000 29-year old male die each year.

Answer:

Identify the probability of a 29-year-old male dying in one year → That is, the mortality rate for a 29-year-old male is 0.0005 (=5/10,000). Then, calculate the break-even point for the premium paid by the insured and the death benefit to be received by the beneficiary as follows:

Noting that a payment is a negative number to the insured individual whereas a receipt is a positive number, we can identify:

Break-even = E(Cash Flow) = 0 = (–X)*(1 – 0.0005) + $500,000 (0.0005)

Solving for X, the insurance premium, we find: X = $250.13 per year.

2. The Variance, σ^2 = The 2nd Moment about the Mean

$$\sigma^2 = E(X^2) = \sum_{i=1}^{m} P(X_i) \cdot [X_i - E(X)]^2 = \sum_{i=1}^{m} P(X_i) \cdot [X_i - \mu]^2$$

e.g., Given a population of 10, 9, 10, 11, and 10 and thus, the population size (N) = 5, the population variance, σ^2, is calculated as:

$$\sigma^2 = \frac{\sum_{i=1}^{N}(X_i - \mu)^2}{N} = \frac{(9-10)^2 + 0^2 + 0^2 + 0^2 + (11-10)^2}{5} = \frac{2}{5} = 0.4$$

Alternatively, the population variance of X can be calculated as:

$$\sigma^2 = \frac{1}{5} \cdot (9-10)^2 + \frac{3}{5} \cdot (10-10)^2 + \frac{1}{5} \cdot (11-10)^2 = \frac{2}{5} = 0.4$$

because $\sigma^2 = E(X^2) = \sum_{i=1}^{m} P(X_i) \cdot [X_i - \mu]^2$

3. The Standard Deviation, σ

$$\sigma = \sqrt{\sigma^2} \qquad \rightarrow \sigma = \sqrt{0.4} = 0.6324$$

Note 1: The standard deviation is a square root of the variance. Thus, without the variance, the standard deviation can not be calculated.

Note 2: Traditionally, a standard deviation is written as a positive number even though $(\pm\sigma)^2 = \sigma^2$ clearly shows the existence of a negative standard deviation.

B. The Binomial Distribution

The binomial (probability) distribution measure the probability of success (=yes) and failure (=no) in a sample of n observations → Only 2 outcomes are possible[67].

1. Assumptions

a. Each trial can have only two mutually exclusive outcomes or can be grouped into two mutually exclusive outcomes. Thus, each outcome can be classified as either a success or a failure.

b. The number of trials, n, is known.

c. The outcomes of each trial are independent of each other.

d. The probability of a success (or a failure) remains the same for each trial.

Examples:

a. Does a 3 coin toss follow a binomial distribution?

Yes. In a 3 coin toss (either tossing 3 coins at the same time or tossing 1 coin three times), (a) outcomes are classified as either a head (=success) or a tail (=failure); (b) there is a fixed number of trials - 1 trial of 3 coins or 3 trials of 1 coin; (c) each coin toss is independent of one another; and (4) the probability of a head remains the same at 0.5. Thus, all conditions (=assumptions) of a binomial distribution is fulfilled in this case.

[67] If three outcomes are possible, then the probability distribution is called the trinomial probability distribution.

b. Can rolling a die be considered as following a binomial probability distribution?

Yes, if 2 (for example) is a success and the rest of the numbers such as 1, 3, 4, 5, and 6 are grouped as a failure. The success probability is, then, 1/6 whereas the failure probability is 5/6 (= 1 – 1/6).

2. The Binomial Probability Function

$$P(X) = {}_nC_X \cdot p^X \cdot (1-p)^{n-X} = \frac{n!}{X!(n-X)!} \cdot p^X (1-p)^{n-X}$$

a. The Mean of the Binomial Distribution

$$\mu = E(X) = np$$

b. The Variance of the Binomial Distribution

$$\sigma^2 = np(1-p)$$

c. The Standard Deviation of the Binomial Distribution

$$\sigma = \sqrt{np(1-p)}$$

3. The Methodology of Applying the Binomial Distribution

Step 1: Assess if the binomial distribution is appropriate by examining the number of outcomes possible. If appropriate, identify the probability of success, p, and the probability of failure, (1 – p). If we let q = (1 – p), then the above equations can be simplified somewhat. For example, $\sigma^2 = npq$ and $\sigma = \sqrt{npq}$.

Step 2: Identify the number of trials, n, and the number of successes, X.

Step 3: Apply these values to the binomial probability function to obtain the binomial probability, P(X).

C. Examples

1. What is the probability of having 2 heads in 3 coin tosses?

Note: p = probability of a head in a single coin toss = 0.5
Number of Heads = successes = X = 2
Number of trials = number of observations = n = 3

a. Numerical Solution:

$$P(X=2) = {}_3C_2 \cdot (0.5)^2 \cdot (1-0.5)^{3-2} = \frac{3!}{2!(3-2)!} \cdot 0.5^2 (1-0.5)^{3-2}$$

$$= 3 \times 0.125 = 3 \times \frac{1}{8}$$

b. Enumerative Solution:

i. What is the number of possible outcomes from 3 coin tosses?

$2 \times 2 \times 2 = 2^3 = 8$ ways

Alternatively, we find the 8 ways to be:

HHH	(HHT)	(HTH)	HTT
(THH)	THT	TTH	TTT

ii. What is the probability of 2 heads in 3 coin tosses?

As shown above, 3 cases of 2 heads out of 8 possible outcomes are found by enumeration.

Alternatively, we can use the rule of combination to count the number of outcomes because the order of a head appearing does not matter → unordered → use the rule of combination as follows:

$$C_{n\,X} = \frac{n!}{X!(n-X)!} = \frac{3!}{2!(3-2)!} = 3$$

Therefore,

P(2 heads)=3 cases x probability of 1/8 $= 3 \times \frac{1}{8} =$ 3 x 0.125 = 0.375

2. On average, how many times will you see a head in these 3 trials of coin tosses? What is the corresponding standard deviation?

$$\mu = np = 3 \cdot 0.5 = 1.5$$
$$\sigma = \sqrt{np(1-p)} = \sqrt{3 \cdot 0.5 \cdot (1-0.5)} = 0.866$$

3. What is the probability of having 1 or 2 heads in 4 coin tosses?

Note: p = probability of a head in a single coin toss = 0.5
Number of Heads= number of successes = X = 1 **or** 2
Number of trials = number of observations = n = 4

Solution:

$$P(X=1)=P(1)=\frac{4!}{1!(4-1)!}\cdot 0.5^1(1-0.5)^{4-1}=0.25$$

$$P(X=2)=P(2)=\frac{4!}{2!(4-2)!}\cdot 0.5^2(1-0.5)^{4-2}=0.375$$

Therefore,

P(1 or 2) = P(0<X<3)= P(1) + P(2) = 0.25 + 0.375 = 0.625

Note 1: You can use the Excel Spreadsheet command of "**=BINOMDIST(X,n,p,true)**" for cumulative probability and "**=BINOMDIST(X,n,p,false)**" for relative probability.

Note 2: You can look up the binomial probability value from the following partial Binomial Probability Table → the probability, p, can be found in the first row, and n and X are found under their respective columns. For example, given p=0.5, n=4, and X=1, you should find P(X=1) = 0.25.

						p				
n	X	0.1	0.2	0.3	0.4	0.5	0.6	0.7	0.8	0.9
2	0	0.8100	0.6400	0.4900	0.3600	0.2500	0.1600	0.0900	0.0400	0.0100
	1	0.1800	0.3200	0.4200	0.4800	0.5000	0.4800	0.4200	0.3200	0.1800
	2	0.0100	0.0400	0.0900	0.1600	0.2500	0.3600	0.4900	0.6400	0.8100
3	0	0.7290	0.5120	0.3430	0.2160	0.1250	0.0640	0.0270	0.0080	0.0010
	1	0.2430	0.3840	0.4410	0.4320	0.3750	0.2880	0.1890	0.0960	0.0270
	2	0.0270	0.0960	0.1890	0.2880	0.3750	0.4320	0.4410	0.3840	0.2430
	3	0.0010	0.0080	0.0270	0.0640	0.1250	0.2160	0.3430	0.5120	0.7290
4	0	0.6561	0.4096	0.2401	0.1296	0.0625	0.0256	0.0081	0.0016	0.0001
	1	0.2916	0.4096	0.4116	0.3456	0.2500	0.1536	0.0756	0.0256	0.0036
	2	0.0486	0.1536	0.2646	0.3456	0.3750	0.3456	0.2646	0.1536	0.0486
	3	0.0036	0.0256	0.0756	0.1536	0.2500	0.3456	0.4116	0.4096	0.2916
	4	0.0001	0.0016	0.0081	0.0256	0.0625	0.1296	0.2401	0.4096	0.6561
5	0	0.5905	0.3277	0.1681	0.0778	0.0313	0.0102	0.0024	0.0003	0.0000
	1	0.3281	0.4096	0.3602	0.2592	0.1563	0.0768	0.0284	0.0064	0.0005
	2	0.0729	0.2048	0.3087	0.3456	0.3125	0.2304	0.1323	0.0512	0.0081
	3	0.0081	0.0512	0.1323	0.2304	0.3125	0.3456	0.3087	0.2048	0.0729
	4	0.0005	0.0064	0.0284	0.0768	0.1563	0.2592	0.3602	0.4096	0.3281
	5	0.0000	0.0003	0.0024	0.0102	0.0313	0.0778	0.1681	0.3277	0.5905
6	0	0.5314	0.2621	0.1176	0.0467	0.0156	0.0041	0.0007	0.0001	0.0000
	1	0.3543	0.3932	0.3025	0.1866	0.0938	0.0369	0.0102	0.0015	0.0001
	2	0.0984	0.2458	0.3241	0.3110	0.2344	0.1382	0.0595	0.0154	0.0012
	3	0.0146	0.0819	0.1852	0.2765	0.3125	0.2765	0.1852	0.0819	0.0146
	4	0.0012	0.0154	0.0595	0.1382	0.2344	0.3110	0.3241	0.2458	0.0984
	5	0.0001	0.0015	0.0102	0.0369	0.0938	0.1866	0.3025	0.3932	0.3543
	6	0.0000	0.0001	0.0007	0.0041	0.0156	0.0467	0.1176	0.2621	0.5314

4. On average, how many times will you see a head in these 4 trials of coin tosses? What is the corresponding standard deviation?

$$\mu = np = 4 \cdot 0.5 = 2$$

$$\sigma = \sqrt{np(1-p)} = \sqrt{4 \cdot 0.5 \cdot (1-0.5)} = 1$$

5. Applications

   a. Assume that Julie can hit the bull's eye 60% of the time. If she shoots 3 bullets, what is the probability that she will hit the bull's eye 2 times?

Answer: Given n = 3, X = 2, and p = 0.6,

$$P(X=2) = P(2) = \frac{3!}{2!(3-2)!} \cdot 0.6^2 (1-0.6)^{3-2} = 0.432$$

   b. When taking a 10-question true-false quiz, a student guesses the answers by flipping a coin. What is the probability that she will get 6 out of 10 answers correctly?

Answer: Given n = 10, X = 6, and p = 0.5,

$$P(X=6) = P(6) = \frac{10!}{6!(10-6)!} \cdot 0.5^6 (1-0.5)^{10-6} = 0.205 \approx 20.5\%$$

→ one chance out of 5 trials

   c. A basketball player can make free throws 80% of time. What is the probability that he will make next 3 free throws in a row?

Answer: Given n = 3, X = 3, and p = 0.8,

$$P(X=3) = \frac{3!}{3!(3-3)!} \cdot 0.8^3 (1-0.8)^{3-3} = 0.512 \approx 51.2\%$$

→ one chance out of 2 situations to make 3 consecutive free throws.

Note: The same result can be found as:

$0.8 \times 0.8 \times 0.8 = 0.8^3 = 0.512$ → 51.2%

The reason for this identical outcome is based on the assumption that all three free throws are independent from one another.

d. Assume that a sharp shooter can hit the bull's eye 90% of the time. If she shoots 100 bullets, what are the mean (=average) and the standard deviation of the number of her hitting the bull's eye?

Answer: Given n = 100 and p = 0.9, $\mu = np = 100 \cdot 0.9 = 90$ and

$$\sigma = \sqrt{np(1-p)} = \sqrt{100 \cdot 0.9 \cdot (1-0.9)} = 3$$

e. If 5% of the people who take aspirin get a headache, what is the probability that one out of 6 people who take the aspirin will get a headache?

Answer: Given n = 6, X = 1, and p = 0.05,

$$P(X=1) = \frac{6!}{1!(6-1)!} \cdot 0.05^1(1-0.05)^{6-1} = 0.2321 \approx 23.21\%$$

D. The Poisson Distribution

The Poisson distribution measures the probability of an event occurring within a time block or a specified space → note that the event is comprised of discrete units such as a number of people, a number of phone calls, a number of books ordered, etc.

The Poisson Distribution is named after its originator, Simeon D. Poisson (1781-1840), and used to calculate a probability associated with situations where an event of one's interest occurs over a specific time period, geographical space, or volume, etc.

1. The Poisson Probability Function

$$P(X) = \frac{e^{-\lambda} \cdot \lambda^X}{X!} \quad where \quad e = 2.718281828 \approx 2.7183$$

a. The Mean of the Poisson Distribution: $\mu = E(X) = \lambda$

b. The Variance of the Poisson Distribution: $\sigma^2 = \lambda$

c. The Standard Deviation of the Poisson Distribution: $\sigma = \sqrt{\lambda}$

2. The Methodology of Applying the Poisson Distribution

Step 1: Assess if the Poisson distribution is appropriate by examining the number of outcomes possible. If appropriate, identify the average number of occurrences, λ. → no probability of successes or failures is needed.

Step 2: Identify the number of successes, X. → the number of trial, n, is not needed.

Step 3: Apply these values to the Poisson probability function to obtain the probability, P(X).

Always you are assumed to know the value of lambda, λ, which is what is expected or observed based on one's experience or data analysis[68]. That is, lambda, λ, is the mean or the average value of the variable of one's interest. **Given that you know the value of lambda, λ, you now want to know the probability of an event, X, occurring.**

Note:

The binomial and the Poisson distributions are like close cousins because both deal with discrete probability. Therefore, the Poisson distribution can yield a probability value that is quite close to the binomial distribution when $n \geq 20$ and the success rate, $p \leq 0.05$. **However, unlike the binomial distribution, the Poisson distribution deals with the number of occurrences, not yes-or-no issues.**

3. Examples

a. Suppose that you know that there are on average 10 people standing in line for a coffee at Starbuck's during the rush hour. What is the probability that there will be no one in line this morning?

Answer:

Hint 1: There is no probability of successes or failures given and the number of trials, n, is not given → the Binomial Distribution is **not** appropriate → However, because the number of people is measured in a discrete unit, the Poisson Distribution is appropriate.

Hint 2: Therefore, we identify λ=10 and X=0.

Therefore, $P(0) = \dfrac{e^{-10} \cdot 10^0}{0!} = 0.0000454$

b. What is the probability of less than 3 people in line?

Answer:

Hint: P(X<3) = P(0) + P(1) + P(2)

Therefore, we first calculate: $P(0) = \dfrac{e^{-10} \cdot 10^0}{0!} = 0.0000454$

[68] In some business situations, one can hypothesize its value based on prior experiences.

$$P(1) = \frac{e^{-10} \cdot 10^1}{1!} = 0.000454 \qquad \text{and} \qquad P(2) = \frac{e^{-10} \cdot 10^2}{2!} = 0.00227$$

Therefore, P(X<3) = 0.0000454+0.000454+0.00227=0.00277

Note 1: You can use the Excel Spreadsheet command of **"=POISSON(X,lambda,true)"** for cumulative probability and **"=POISSON(X,lambda,false)"** for relative probability

Note 2: You can look up the Poisson probability value from the following partial Poisson Probability Table → Some selected Lambda values are listed in the first row and X values under the X column. For example, given λ=10, and X=2, you may find P(2) = 0.0023, which is a rounded number of 0.00227. → Lambda and X values not listed in this table must be calculated by using the above Excel command for a relative Poisson probability.

The following Poisson probability distribution table is produced by the Excel command of **"=POISSON(X,lambda,false)"** for relative probability

	Lambda									
X	1	2	3	4	5	6	7	8	9	10
0	0.3679	0.1353	0.0498	0.0183	0.0067	0.0025	0.0009	0.0003	0.0001	0.0000
1	0.3679	0.2707	0.1494	0.0733	0.0337	0.0149	0.0064	0.0027	0.0011	0.0005
2	0.1839	0.2707	0.2240	0.1465	0.0842	0.0446	0.0223	0.0107	0.0050	0.0023
3	0.0613	0.1804	0.2240	0.1954	0.1404	0.0892	0.0521	0.0286	0.0150	0.0076
4	0.0153	0.0902	0.1680	0.1954	0.1755	0.1339	0.0912	0.0573	0.0337	0.0189
5	0.0031	0.0361	0.1008	0.1563	0.1755	0.1606	0.1277	0.0916	0.0607	0.0378
6	0.0005	0.0120	0.0504	0.1042	0.1462	0.1606	0.1490	0.1221	0.0911	0.0631
7	0.0001	0.0034	0.0216	0.0595	0.1044	0.1377	0.1490	0.1396	0.1171	0.0901
8	0.0000	0.0009	0.0081	0.0298	0.0653	0.1033	0.1304	0.1396	0.1318	0.1126
9	0.0000	0.0002	0.0027	0.0132	0.0363	0.0688	0.1014	0.1241	0.1318	0.1251
10	0.0000	0.0000	0.0008	0.0053	0.0181	0.0413	0.0710	0.0993	0.1186	0.1251
11	0.0000	0.0000	0.0002	0.0019	0.0082	0.0225	0.0452	0.0722	0.0970	0.1137
12	0.0000	0.0000	0.0001	0.0006	0.0034	0.0113	0.0263	0.0481	0.0728	0.0948
13	0.0000	0.0000	0.0000	0.0002	0.0013	0.0052	0.0142	0.0296	0.0504	0.0729
14	0.0000	0.0000	0.0000	0.0001	0.0005	0.0022	0.0071	0.0169	0.0324	0.0521
15	0.0000	0.0000	0.0000	0.0000	0.0002	0.0009	0.0033	0.0090	0.0194	0.0347
16	0.0000	0.0000	0.0000	0.0000	0.0000	0.0003	0.0014	0.0045	0.0109	0.0217
17	0.0000	0.0000	0.0000	0.0000	0.0000	0.0001	0.0006	0.0021	0.0058	0.0128
18	0.0000	0.0000	0.0000	0.0000	0.0000	0.0000	0.0002	0.0009	0.0029	0.0071
19	0.0000	0.0000	0.0000	0.0000	0.0000	0.0000	0.0001	0.0004	0.0014	0.0037
20	0.0000	0.0000	0.0000	0.0000	0.0000	0.0000	0.0000	0.0002	0.0006	0.0019

c. What is the probability of more than 9 people in line?

Answer:

$P(X>9) = P(10) + P(11) + \ldots\ldots. + \infty$

$= 1 - P(X \le 9) = 1 - [P(0)+P(1)+\ldots+P(9)] = 1 - 0.458 = 0.542$

Note 1: $P(0) + P(1) + \ldots + P(9) = 0.458$
is obtained from the Poisson probability table shown in the above Poisson Probability Table.

Note 2: Does this $P(X>9) = 0.542$ make an intuitive sense?
Yes, given that the mean is 10 which may be the median, the probability of greater than 9 should be larger than 0.5 such as 0.542.

d. What is the probability of between 3 and 9 people in line?

Answer: Because "between 3 and 9" technically excludes 3 and 9,
$P(3 < X < 9) = P(4) + P(5) + P(6) + P(7) + P(8) = 0.3301$

Alternatively, $P(3 < X < 9) = 1 - [P(X \le 3) + P(X \ge 9)]$
$= 1 - [0.0104 + 0.6595]$
$= 1 - [0.6699] = 0.3301$

4. Advanced Applications of the Poisson Distribution

Example 1> Assume that a telemarketing company gets on average 5 orders per 1000 calls it makes. If an employee of the company makes 400 calls today, what is the probability that the employee will get 4 orders or more?

Answer:

Given that there is on average 5 orders per 1000 calls, the probability of getting an order is 0.005 (=5/1000). Thus, if 400 calls are made, then one can expect on average of 2 orders (=400 calls x 0.005). Thus, the value of Lambda, λ, is 2 and the probability can be calculated as follows:

$$P(X \ge 4) = P(X > 3) = 1 - P(X \le 3) = 1 - [P(0) + P(1) + P(2) + P(3)]$$

where $P(0) = \frac{e^{-2} \cdot 2^0}{0!} = 0.1353$ $\quad P(1) = \frac{e^{-2} \cdot 2^1}{1!} = 0.27067$

$P(2) = \frac{e^{-2} \cdot 2^2}{2!} = 0.27067$ and $P(3) = \frac{e^{-2} \cdot 2^3}{3!} = 0.1804$

Therefore, $P(X \geq 4) = P(X>3) = 1 - P(X \leq 3)$

$= 1 - [0.1353+0.2706+0.2706+0.1804] = 1 - 0.8569 = 0.1431$

Note: The key to solving this problem lies with the identification of the lambda value (=the average).

Example 2> Assume that the average number of inquiries to a toll-free telephone number for a computer company is 10 per business hour. Each inquiry takes about 6 minutes on average to be resolved. What is the probability of getting exactly 15 calls per business hour today?

Answer:

$$P(15) = \frac{e^{-10} \cdot 10^{15}}{15!} = 0.0347$$

Note: The information of 6 minutes has no role to play in probability calculation.

Example 3> An order taker at an online shopping company normally has 5 errors per day when filling the orders. What is the probability that a newly hired order taker will have 6 errors or less today (assuming that the new hire has the same skill as an average order taker)?

Answer:

$P(X \leq 6) = P(0) + P(1) + P(2) + P(3) + P(4) + P(5) + P(6)$

$= 0.0067 + 0.0337 + 0.0842 + 0.1404 + 0.1755 + 0.1755 + 0.1462$

$= 0.7622 \rightarrow 76.22\%$

Or alternatively, by using the Excel command of "=poisson(6,5,1)" we find:

0.762183

Note that in Excel command, the last number, 1, indicates a cumulative probability. If we were to calculate a relative probability, it should have been "0".

Exercise Problems on
Chapter 6. Discrete Probability Distributions

This set of exercise problems has 16 problems, worth a total of 20 points.

1. The expected value of a discrete random variable X is the same as the _____.

   a. mean of X
   b. arithmetic average of X
   c. geometric average of X
   d. only (a) and (b) of the above
   e. only (a) and (c) of the above

2. As the Chief Financial Officer (CFO) of your firm, you are discussing the future sales prospect of your firm's products in a board of directors meeting. If you tabulated the following sales data for the last 10 years and used them to forecast sales, what would be your expected sales?

   $1 million sales occurred 3 times.
   $2 million sales occurred 5 times.
   $3 million sales occurred 2 times.

   a. $1.5 million
   b. $1.9 million
   c. $2 million
   d. $2.1 million
   e. $2.5 million

3. What is the standard deviation of your sales estimate (or forecast), using the above information?

   a. $0.4 million
   b. $0.49 million
   c. $0.5 million
   d. $0.6 million
   e. $0.7 million

4. Suppose that one of the board members suggests the following:
   She expects that a 30% chance of booming economy with a possible sales of $4 million, a 50% chance of so-so economy with a possible sales of $3 million, and a 20% chance of bad economy with a possible sales of $0.5 million. Under this scenario, what would be the expected sales?

   a. $4 million
   b. $3 million
   c. $2.8 million
   d. $2.5 million
   e. not calculable

5. Which of the following is the underlying assumption(s) of the binomial probability distribution?

   a. each trial can have only two outcomes.
   b. the number of trials is known.
   c. the outcomes of each trial are independent of each other.
   d. all of the above
   e. only (a) and (c) of the above

6. What is the probability of having 5 heads in 8 coin tosses?

a.	0.0179	b.	0.0039	c.	0.2188
d.	0.6250	e.	none of the above		

7. The mean and the standard deviation of a binomial distribution with p=0.4 and n=20 is _____ and _____, respectively.

a.	10; 4.8	b.	8; 4.8	c.	10; 2.1909
d.	8; 2.1909	e.	12; 2.1909		

8. Given that there is a 30% chance of a head appearing in a coin toss (and thus, there is a 70% chance of a tail appearing), what will be the expected number of heads that will appear if the coin is tossed 30 times?

a.	9	b.	15	c.	21
d.	30	e.	unknown		

9. A fair coin is a coin whose head and tail appears in an equal probability of 50% each. What is the probability of obtaining either 5 heads or 6 heads in 10 tosses of a fair coin? Pick a number as close as it comes by carrying 6 decimal points in calculating this probability.

a.	0.5	b.	0.45	c.	0.3
d.	0.55	e.	0.6		

10. If President George Bush has a 30% approval rating from the U.S. citizens, what will be the probability that you will find 5 Bush supporters in a crowd of 10 people? Pick a number as close as it comes by carrying 6 decimal points in calculating this probability.

a.	0.3	b.	0.2	c.	0.1
d.	0.05	e.	0		

11. The Poisson distribution is used to calculate the probability associated with an event occurring _____.

a.	in a specific time block	b.	in a geographical space
c.	in a specific volume	d.	all of the above
e.	none of the above		

12. The Poisson distribution uses the exponential value of *e* which has a numerical value of _____.

a.	0	b.	1	c.	2.5
d.	2.71828	e.	3.14159		

The following questions are worth 2 points each.

13. Assume that as a quality control manager of a firm that produces chocolate-chip cookies, you believe that the number of chocolate chips on a cookie is distributed as a Poisson distribution. Assume further that your machine is designed to put 10 chocolate chips per cookie. The industry standard accepts 8 or more chocolate chips per cookie. What is the probability that you will find a cookie with exactly 8 chocolate chips?

a.	0.8	b.	0.8944	c.	0.1126
d.	0.2203	e.	0.5		

14. Assume the same situation as above. That is, your machine is designed to put 10 chocolate chips per cookie. The industry standard accepts 8 or more chocolate chips per cookie. What is the probability that you will find a cookie with less than 8 chocolate chips which is below the industry standard?

a.	0.4765	b.	0.8944	c.	0.1126
d.	0.2203	e.	0.5324		

15. Assume the same situation as above. That is, assume that your machine is designed to put 10 chocolate chips per cookie. What is the probability that you will find cookies with between 7 and 12 chocolate chips (i.e., more than 7 chocolate chips and less than 12 chocolate chips) per cookie?

a.	0.5324	b.	0.3639	c.	0.4765
d.	0.8001	e.	0.9		

16. Assume the same situation as above. That is, assume that your machine is designed to put 10 chocolate chips per cookie. What is the probability that you will find a cookie with more than 11 chocolate chips per cookie?

a.	0.2203	b.	0.3032	c.	0.4765
d.	0.6968	e.	0.8		

Chapter 7. The Normal Probability Distribution

In this chapter and the following chapter, we will study two most important continuous probability distributions – the normal distribution and the t-distribution. As you may recall, a continuous probability distribution shows the probability of observing a continuous numerical data. Because data on income, sales, age, etc., for example, are considered continuous numerical data, when we collect and analyze them, we must ask what underlying probability distribution may have generated such data points. In other words, we must ask which of the many continuous probability distributions – such as a uniform distribution, an exponential distribution, a normal distribution, a t-distribution, etc. [69] – the data may have come from. This question is not easy to answer because we would not normally know their underlying probability distribution.

However, when analyzing data, we are most interested in its central tendency that is measured and represented by its (sample) mean. If the mean of any data set is of our interest, then the Central Limit Theorem states that the sampling distribution of the mean[70] is normally distributed if the sample size is sufficiently large[71]. What this means is then, a normal distribution can be used in analyzing a distribution of any mean[72], regardless of the underlying probability distribution the data had come from.

More specifically, if the value of the population standard deviation (or variance) is known, we assume that the data is from a normal distribution. If it is not known and thus, needs to be estimated by the sample standard deviation (or variance), then we assume that the data is from a t-distribution[73].

Let's now examine the characteristics and the uses of the normal probability distribution below and those of the t-distribution in the next chapter.

A. The Background on the Normal Probability Distribution

The normal probability distribution is often referred to as the normal distribution, the normal probability function, or the normal cumulative probability.

Abraham de Moivre, a French mathematician, had developed the concept first in 1733, based on the observed properties of the binomial probability function. However, as

---

[69] Other well-known continuous probability functions include an F-distribution and a Chi-square distribution, which we study at later chapters.
[70] "The sampling distribution of the mean" is a fancy way of saying that various sample means (=averages) can be obtained from a given population and these different means have a distribution of their own. Thus, a more correct expression could have been "a distribution of various means obtained from many samples taken from the same population."
[71] Often a sample size being greater than 30 is considered large.
[72] In combination with the Central Limit Theorem, a well-known statistical property, called the unbiasedness of a sample mean, assures this application. The unbiasedness of a sample mean says that a sample mean is a good (=unbiased) estimate of the population mean.
[73] Also, a t-distribution is often used when a sample size is small - such as less than 30.

shown below in the 10 Deutsche Mark[74], Karl Gauss (1777-1855), a German mathematician, is credited with this probability distribution function because Gauss had a stronger and clearer vision for its usefulness and applicability to astronomy and other scientific fields. For this reason, the normal distribution is often known as the Empirical Rule or the Gaussian (probability) distribution[75].

B. The Characteristics or properties of the Normal Distribution

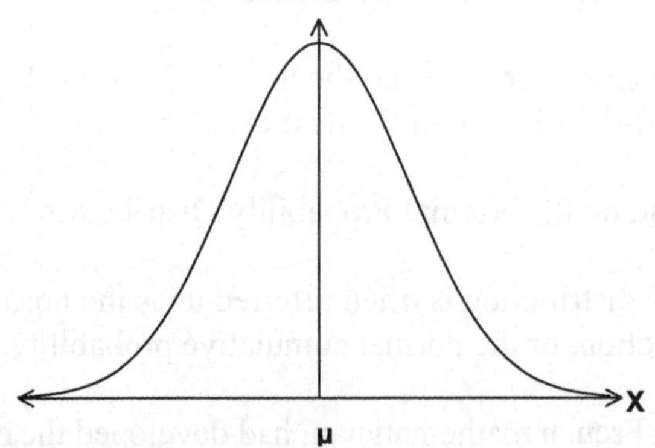

As shown in the above graph of a normal distribution, we can observe its characteristics as follows:

1. It is bell-shaped or symmetrical about the mean, μ.

---

[74] Since the creation of a common currency, euro, in the European Union in January 1999, Deutsche Mark is no longer in circulation.

[75] While the Empirical Rule is based on the empirical observation of a **normally distributed** population, there is a more general rule of thumb for **any** population. It is called the Chebyshev's Theorem, named after a Russian mathematician, Pafnuty Chebyshev (1821-1894).

2. Its measures of central tendency – the mean, median, and mode – are all identical.

3. Its mean and variance (or standard deviation) are known.

4. One standard deviation above (or below) its mean represents a probability of 0.1587 or 15.87% → Therefore, the probability between one standard deviation above and below the mean is 0.6826 (=1 – 2 x 0.1587)

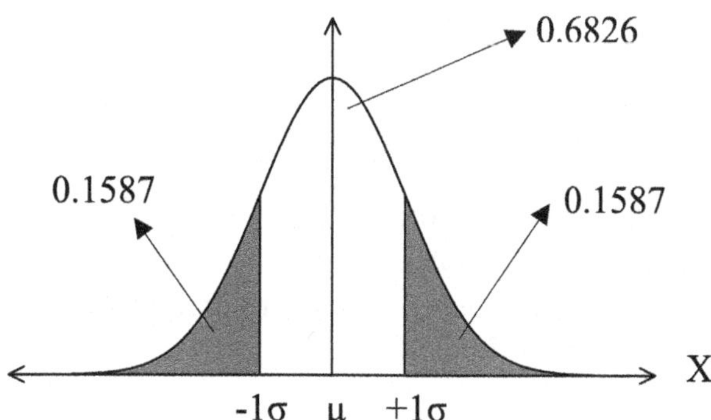

Note:

Any observation that lies outside of ± 1 standard deviation from the mean has a total (summed) probability of 31.74%. → Because many observations will often fall outside this $\pm 1\sigma$ boundary, there will be no unusualness or abnormality when a value is observed outside this boundary → Thus, this value is considered a normal or usual observation.

5. Two standard deviations above (or below) its mean represent a probability of 0.0228 or 2.28% → The probability between two standard deviations above and below the mean is 0.9544 (=1 – 2 x 0.0228)

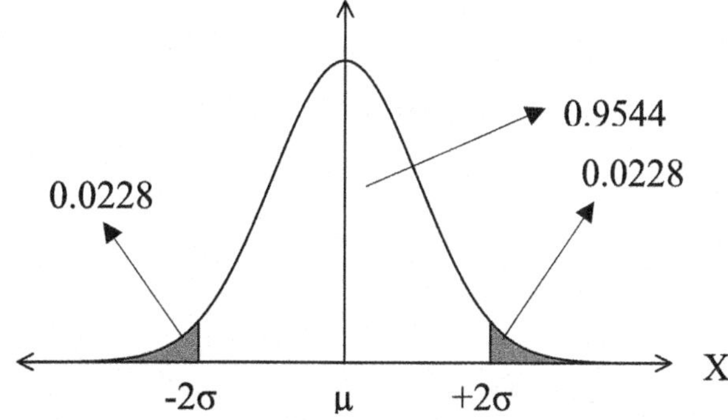

Note:

Any observation that lies outside of ± 2 standard deviations from the mean has a probability of 4.56%. Thus, any value that lies outside this $\pm 2\sigma$ boundary can be considered an abnormal or unusual observation.

6. Three standard deviations above (or below) its mean represents a probability of 0.00135 or 0.135% → The probability between three standard deviations above and below the mean is 0.9973 (=1 – 2 x 0.00135)

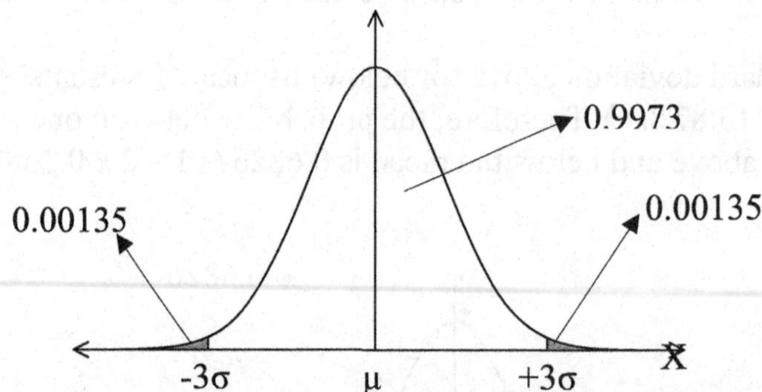

Note: Any observation that lies outside of ±3 standard deviations from the mean has a probability of 0.135%. Thus, any value that lies outside this ±3σ boundary can be considered an extremely abnormal or unusual observation → an outlier.

7. Its inter-quartile range – the "middle fifty" – is equal to ±1.33 standard deviations.

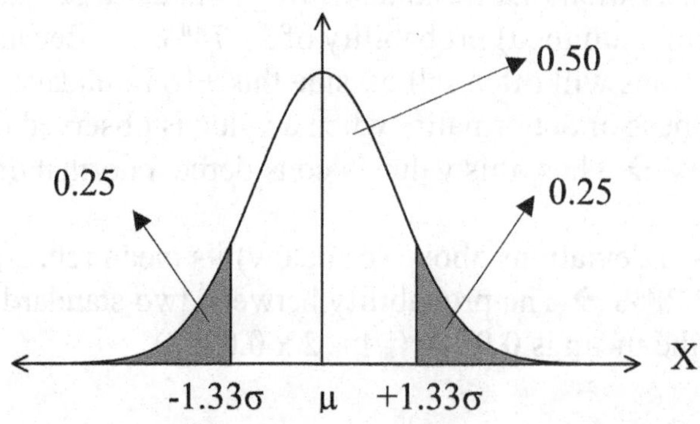

8. It can have a range of X values from -∞ to +∞ → X can take any value.

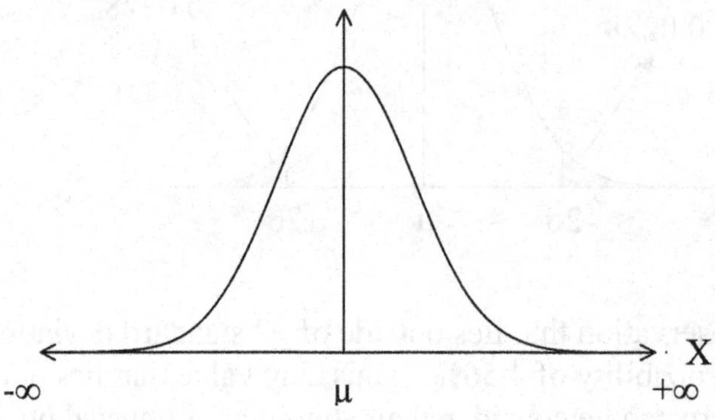

D. The Standard(ized) Normal Probability Density Function

$$f(X) = \frac{1}{\sigma\sqrt{2\pi}} e^{-(1/2)[(X-\mu)/\sigma]^2}$$

This function is the foundation for calculating the cumulative probability of any normal variable, X, given its population mean and population standard deviation are known. The important point is that a normal distribution is defined by only these two parameters of μ and σ.

The above equation can be dissected and simplified as follows:

1. The Standard(ized) normal random variable = the standard(ized) normal variate = the standard normal variate = Z = Z-value = Z-score

$$Z = \frac{X - \mu}{\sigma}$$

2. The Characteristics of the Standard(ized) Normal Variate, Z

   a. When X = μ, Z = 0
   b. The value of the standard deviation of Z = σ_Z = σ =1

3. The Standard(ized) Normal Probability Density Function

$$f(Z) = \frac{1}{\sqrt{2\pi}} e^{-(1/2)Z^2}$$

4. The Cumulative Standard(ized) Normal Probability Density Function

   The sum of standard(ized) normal density function values as shown in the following Z-table are tabulated via the Excel Spreadsheet command of **"=NORMDIST(X,mean,standard deviation,True)"** for a cumulative probability.

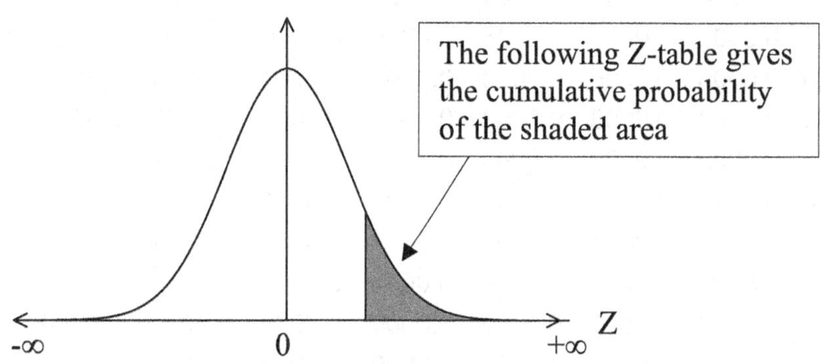

The Cumulative Standard Normal Probability Distribution for Positive Z Values										
Z	0.00	0.01	0.02	0.03	0.04	0.05	0.06	0.07	0.08	0.09
0.0	0.5000	0.4960	0.4920	0.4880	0.4840	0.4801	0.4761	0.4721	0.4681	0.4641
0.1	0.4602	0.4562	0.4522	0.4483	0.4443	0.4404	0.4364	0.4325	0.4286	0.4247
0.2	0.4207	0.4168	0.4129	0.4090	0.4052	0.4013	0.3974	0.3936	0.3897	0.3859
0.3	0.3821	0.3783	0.3745	0.3707	0.3669	0.3632	0.3594	0.3557	0.3520	0.3483
0.4	0.3446	0.3409	0.3372	0.3336	0.3300	0.3264	0.3228	0.3192	0.3156	0.3121
0.5	0.3085	0.3050	0.3015	0.2981	0.2946	0.2912	0.2877	0.2843	0.2810	0.2776
0.6	0.2743	0.2709	0.2676	0.2643	0.2611	0.2578	0.2546	0.2514	0.2483	0.2451
0.7	0.2420	0.2389	0.2358	0.2327	0.2296	0.2266	0.2236	0.2206	0.2177	0.2148
0.8	0.2119	0.2090	0.2061	0.2033	0.2005	0.1977	0.1949	0.1922	0.1894	0.1867
0.9	0.1841	0.1814	0.1788	0.1762	0.1736	0.1711	0.1685	0.1660	0.1635	0.1611
1.0	0.1587	0.1562	0.1539	0.1515	0.1492	0.1469	0.1446	0.1423	0.1401	0.1379
1.1	0.1357	0.1335	0.1314	0.1292	0.1271	0.1251	0.1230	0.1210	0.1190	0.1170
1.2	0.1151	0.1131	0.1112	0.1093	0.1075	0.1056	0.1038	0.1020	0.1003	0.0985
1.3	0.0968	0.0951	0.0934	0.0918	0.0901	0.0885	0.0869	0.0853	0.0838	0.0823
1.4	0.0808	0.0793	0.0778	0.0764	0.0749	0.0735	0.0721	0.0708	0.0694	0.0681
1.5	0.0668	0.0655	0.0643	0.0630	0.0618	0.0606	0.0594	0.0582	0.0571	0.0559
1.6	0.0548	0.0537	0.0526	0.0516	0.0505	0.0495	0.0485	0.0475	0.0465	0.0455
1.7	0.0446	0.0436	0.0427	0.0418	0.0409	0.0401	0.0392	0.0384	0.0375	0.0367
1.8	0.0359	0.0351	0.0344	0.0336	0.0329	0.0322	0.0314	0.0307	0.0301	0.0294
1.9	0.0287	0.0281	0.0274	0.0268	0.0262	0.0256	0.0250	0.0244	0.0239	0.0233
2.0	0.0228	0.0222	0.0217	0.0212	0.0207	0.0202	0.0197	0.0192	0.0188	0.0183
2.1	0.0179	0.0174	0.0170	0.0166	0.0162	0.0158	0.0154	0.0150	0.0146	0.0143
2.2	0.0139	0.0136	0.0132	0.0129	0.0125	0.0122	0.0119	0.0116	0.0113	0.0110
2.3	0.0107	0.0104	0.0102	0.0099	0.0096	0.0094	0.0091	0.0089	0.0087	0.0084
2.4	0.0082	0.0080	0.0078	0.0075	0.0073	0.0071	0.0069	0.0068	0.0066	0.0064
2.5	0.0062	0.0060	0.0059	0.0057	0.0055	0.0054	0.0052	0.0051	0.0049	0.0048
2.6	0.0047	0.0045	0.0044	0.0043	0.0041	0.0040	0.0039	0.0038	0.0037	0.0036
2.7	0.0035	0.0034	0.0033	0.0032	0.0031	0.0030	0.0029	0.0028	0.0027	0.0026
2.8	0.0026	0.0025	0.0024	0.0023	0.0023	0.0022	0.0021	0.0021	0.0020	0.0019
2.9	0.0019	0.0018	0.0018	0.0017	0.0016	0.0016	0.0015	0.0015	0.0014	0.0014
3.0	0.0013	0.0013	0.0013	0.0012	0.0012	0.0011	0.0011	0.0011	0.0010	0.0010
3.1	0.0010	0.0009	0.0009	0.0009	0.0008	0.0008	0.0008	0.0008	0.0007	0.0007
3.2	0.0007	0.0007	0.0006	0.0006	0.0006	0.0006	0.0006	0.0005	0.0005	0.0005
3.3	0.0005	0.0005	0.0005	0.0004	0.0004	0.0004	0.0004	0.0004	0.0004	0.0003
3.4	0.0003	0.0003	0.0003	0.0003	0.0003	0.0003	0.0003	0.0003	0.0003	0.0002
3.5	0.0002	0.0002	0.0002	0.0002	0.0002	0.0002	0.0002	0.0002	0.0002	0.0002
3.6	0.0002	0.0002	0.0001	0.0001	0.0001	0.0001	0.0001	0.0001	0.0001	0.0001
3.7	0.0001	0.0001	0.0001	0.0001	0.0001	0.0001	0.0001	0.0001	0.0001	0.0001
3.8	0.0001	0.0001	0.0001	0.0001	0.0001	0.0001	0.0001	0.0001	0.0001	0.0001
3.9	0.0000	0.0000	0.0000	0.0000	0.0000	0.0000	0.0000	0.0000	0.0000	0.0000
4.0	0.0000	0.0000	0.0000	0.0000	0.0000	0.0000	0.0000	0.0000	0.0000	0.0000

E. How to Read and Use the Z Table

1. The above table shows the cumulative standard normal probability of positive Z values only. The values of the first column and the first row combined define the Z value and the values inside the table show the cumulative probability.

   That is, the probability shown in the above table is for P(Z > k) where **k is found as the sum of Z value in the first column and the Z value in the first row**. For example, if k = 1.51, it is equivalent to k = (1.5) + (0.01). Therefore the cross section of Z = 1.5 and 0.01 will yield P(Z > 1.51) = 0.0655. **Note the first row values represent the second digit below the decimal point of k.**

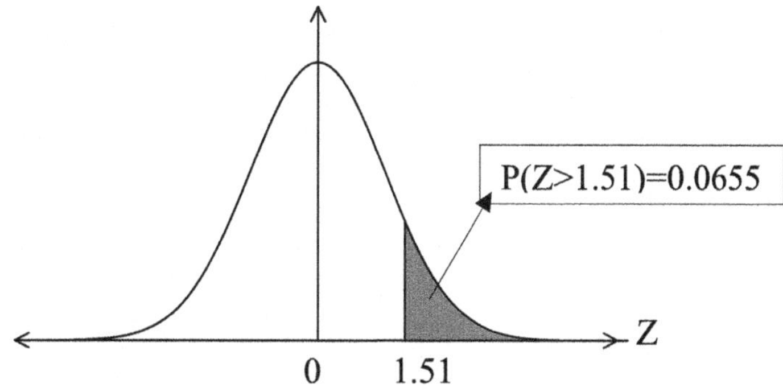

   Because P(Z > 1.51) represents a cumulative probability associated with the **upper (or right-hand-side) tail** of the standard normal distribution, it is often known as an upper-tail probability.

2. Given that P(Z > 1.51) = 0.0655 as shown above

   a. If we wish to know P(Z < 1.51), how can we find its probability?

      Because the Z table does not give the probability of P(Z<1.51), we ingeniously use the concept of the complement to calculate this probability:

   That is,

$$P(Z < 1.51) = 1 - P(Z > 1.51) = 1 - 0.0655 = 0.9345$$

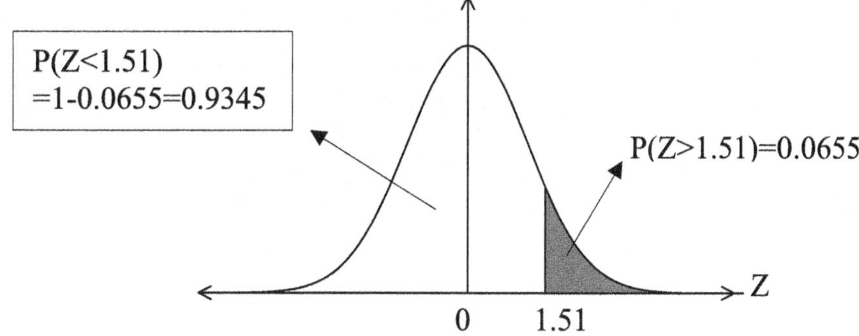

b. If we wish to know $P(Z < -1.51)$, how can we find its probability?

Because the above Z table does not show negative Z values, **we must rely on the knowledge that a normal distribution is symmetrical about the mean**. That is, we should note that $P(Z < -1.51) = P(Z > 1.51) = 0.0655$.

That is,
due to the symmetrical nature of the standard normal distribution, $P(Z < -k) = P(Z > k)$.

e.g., $P(Z > 1.51) = P(Z < -1.51) = 0.0655 = 1 - 0.9345$

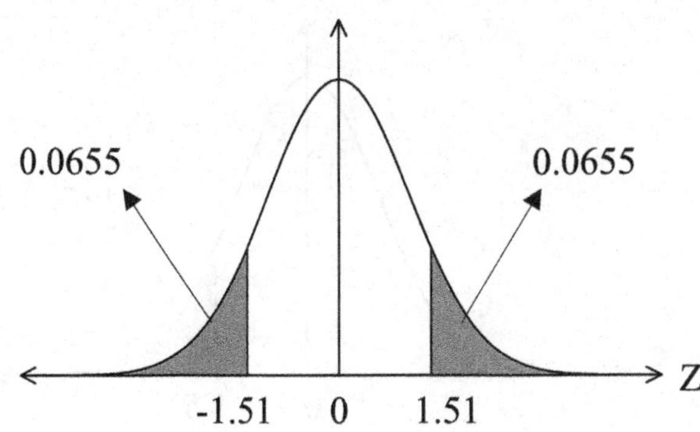

c. If we wish to know $P(-1.51 < Z < 1.51)$, how can we find its probability?

In this case, **we use the knowledge that (1) a normal distribution is symmetrical about the mean and (2) all probabilities sum to 1.**

That is,
because we know that $P(Z < -1.51) = P(Z > 1.51) = 0.0655$,
we find $P(-1.51 < Z < 1.51) = 1 - 2 \times 0.0655 = 1 - 0.1310 = 0.8690$.

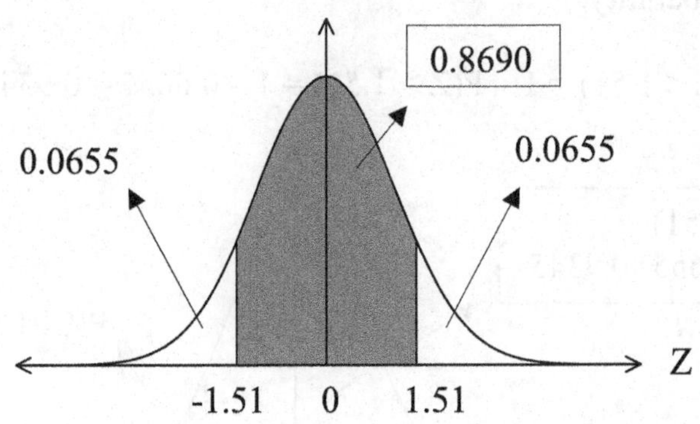

3. What is the probability of Z being greater than 1.96? → That is, P(Z > 1.96)?

   If Z >1.96 , we first find the probability value of P(Z >1.96) from the above Z table by looking up the value at the cross-section of the first column under Z that shows 1.9 and the first row of 0.06.

   That is, P(Z>1.96) = 0.0250 = 2.5%.

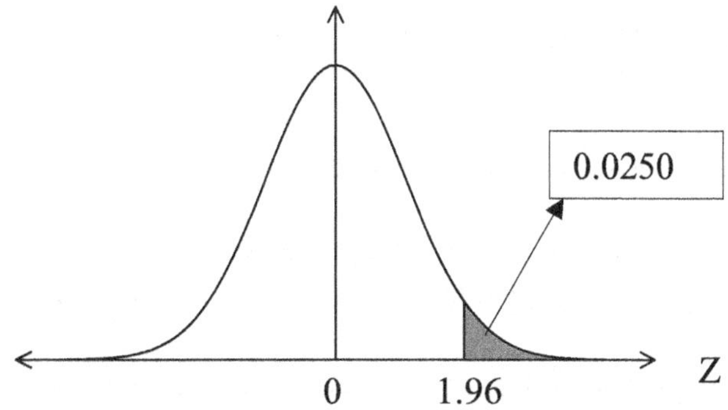

4. What is the probability of Z being between –2 and +2?
   → This can be re-stated as P(–2 < Z < 2) = P(|Z| < 2)?

   In situations like this, it is very useful to draw a normal distribution curve as shown below and **visualize** what probabilities need to be identified. That is, given –2 < Z < 2, it means two standard deviations above and below the mean.

   First, we find P(Z<2) = P(Z<2.00) from the above Z table as 0.0228.
   Next, we utilize the symmetrical nature of the Z distribution to find P(Z<–2.00) → That is, P(Z<–2.00) = 0.0228 → Therefore, the probability between two standard deviations above and below the mean is 0.9544 (=1 – 2 x 0.0228) or 95.44%[76]

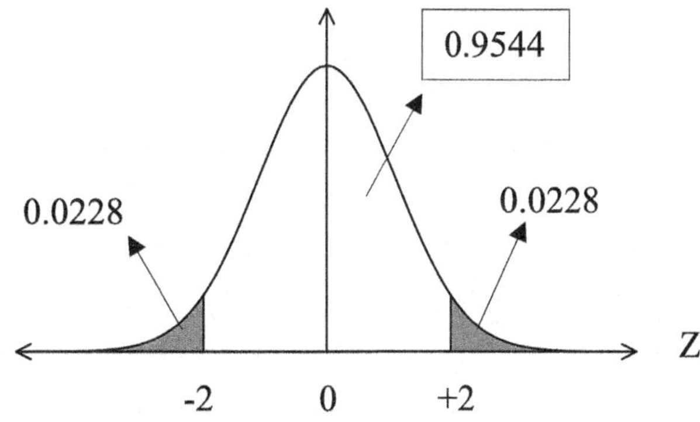

[76] Because 95.44% can be conveniently treated equally with 95%, it is often said that two standard deviations above and below a mean will capture 95% of observations. However, the exact Z- value that will capture 95% of observations is ±1.96, not ±2, as shown in the previous example of P(Z>1.96).

6. The meaning of a 6 sigma management

   Because a Greek letter, sigma (σ), is commonly used to denote the population standard deviation, they are both synonymously used. From the Z table, one can find that the probability of an observation being outside ±3 sigma from the mean is about 2.6 (= 2 x 0.0013) out of 1,000 chance.

   The same for ±6 sigma is about 2 out of 1 billion chance, which means a perfection – especially, in case of manufacturing operation where only 2 defects are found out of 1 billion products produced[77].

F. Comparing a normal variable X with the standard(ized) normal variable Z

1. Suppose that a normal variable X has the population mean of 10 (i.e., μ = 10) and the population standard deviation of 0.6 (i.e., σ = 0.6).

   a. What shape does the assumed probability distribution of the X variable have?

      Because X is a normal variable, it shows a bell-shaped curve as follows:

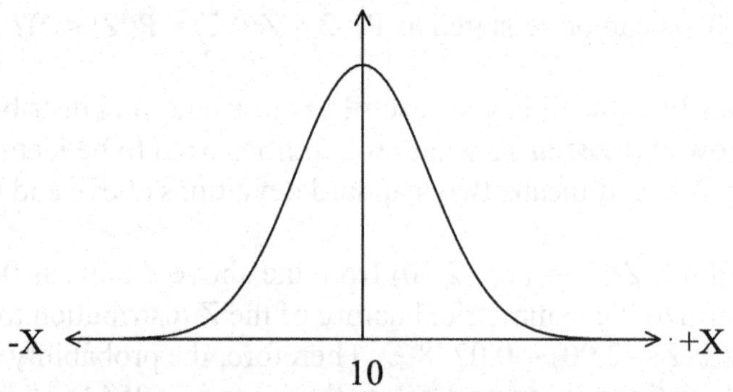

   b. What is the probability of finding a X value that is greater than 11.2?

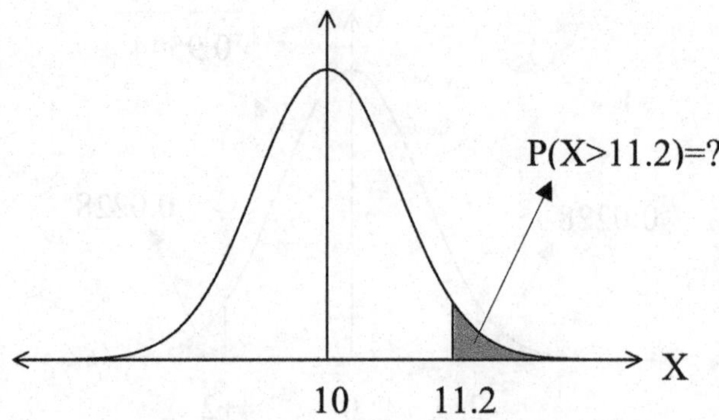

[77] In reality when major businesses such as the Motorola or the General Electric use the term, 6 sigma, for a quality control purpose, they often mean ±4.3 sigma. This lack of accuracy in using the term, 6 sigma, for quality control has been an interesting conversation topic between statisticians and business managers.

There is no probability table to yield P(X>11.2). In order to assess P(X>11.2), we must convert the X value into a Z value by using the following Z-transformation formula, or simply the Z formula:

$$Z_{calculated} = Z_c = \frac{X - \mu}{\sigma}$$

By plugging a desired X value and its mean, μ, into the above Z formula, we can calculate the standardized normal variate, the Z-value → **often called the calculated Z-value**.

That is,

given the Z formula of: $Z_{calculated} = Z_c = \frac{X - \mu}{\sigma}$; X = 11.2; and μ = 10,

the calculated Z-value is: $Z_{calculated} = Z_c = \frac{X - \mu}{\sigma} = \frac{11.2 - 10}{0.6} = 2$

2. What is the probability of finding a Z value that is greater than 2.00?

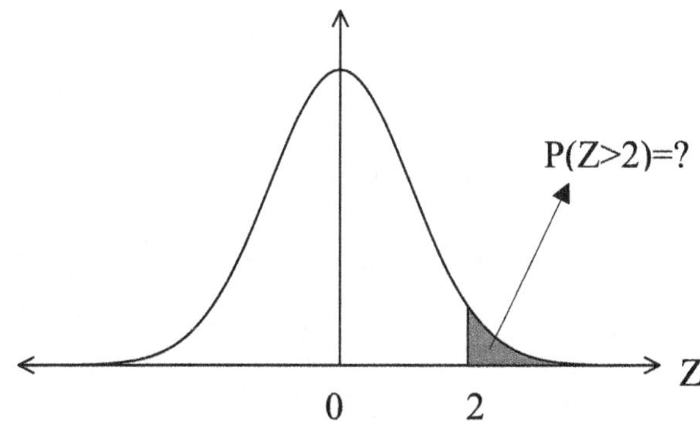

The answer of 0.0228 can be from the Z table by looking at P(Z>2.00), which is the answer to P(X>11.2) because both X and Z have a normal distribution. That is, **finding P(Z>2) is equivalent to finding P(X>11.2)**. This makes sense in that 11.2 is 2 standard deviations above the X mean of 10 whereas 2 is 2 standard deviations above the Z mean of 0.

A Very Important Conclusion:

Note that the distributions of X are Z are essentially the same due to their normality assumption; only their measurement units are different. The Z formula eliminates any unit-of-measurement problem. That is, we can convert the problem expressed in original units (X) into standardized units (Z). Therefore, once we know the probability associated with Z, then we can know the probability associated with X.

G. Applications - 1

Given **a population** of 10, 9, 10, 11, and 10 hours of light bulb life, we can calculate various probabilities associated with the life of a light bulb as follows:

In order to calculate the probability based on the normal distribution, it is necessary to carry out the following steps:

Step 1: Identify or calculate the population mean → $\mu = 10$

Step 2: Identify or calculate the population standard deviation[78], σ.
→ $\sigma = \sqrt{0.4} = 0.63245 \approx 0.6$ for simplicity

Step 3: Identify the value of X that we are interested in calculating probability for. → X = 10.6, for example.

Step 4: Plug these values into the Z formula and calculate the standardized normal variate, the Z-value → **often called the calculated Z-value.**

$$Z_{calculated} = Z_c = \frac{X - \mu}{\sigma}$$

Step 5: Identify the appropriate probability via the Z-probability table or the Excel command.

Let's look at some examples below:

1. Given $\mu = 10$ and $\sigma = 0.6$, calculate the probability that a light bulb will burn more than 10.6 hours.

Solution:

Note that we can identify $\mu = 10, \sigma = 0.6$, and X = 10.6. Therefore, the calculated Z-value can be obtained as follows:

$$Z_{calculated} = Z_c = \frac{X - \mu}{\sigma} = \frac{10.6 - 10}{0.6} = 1$$

Note: Because we are interested in knowing the probability of **X being greater than 10.6,** we must identify the probability of **Z_c being greater than 1** from the Z table or the Excel Spreadsheet. That is, we must recognize that calculating the

[78] When using the Z formula, we always assume that the population standard deviation, σ, is known or calculable. If not, the (Student-) t distribution must be used. The calculation method for the population standard deviation was given in Chapter 4.

probability of X being greater than 10.6 is equivalent to calculating the probability of Z_c being greater than 1. Because this process is used for all normal variables, we call this process the "standardization" process.

Graphically and visually, this process of identifying the probability can be shown below:

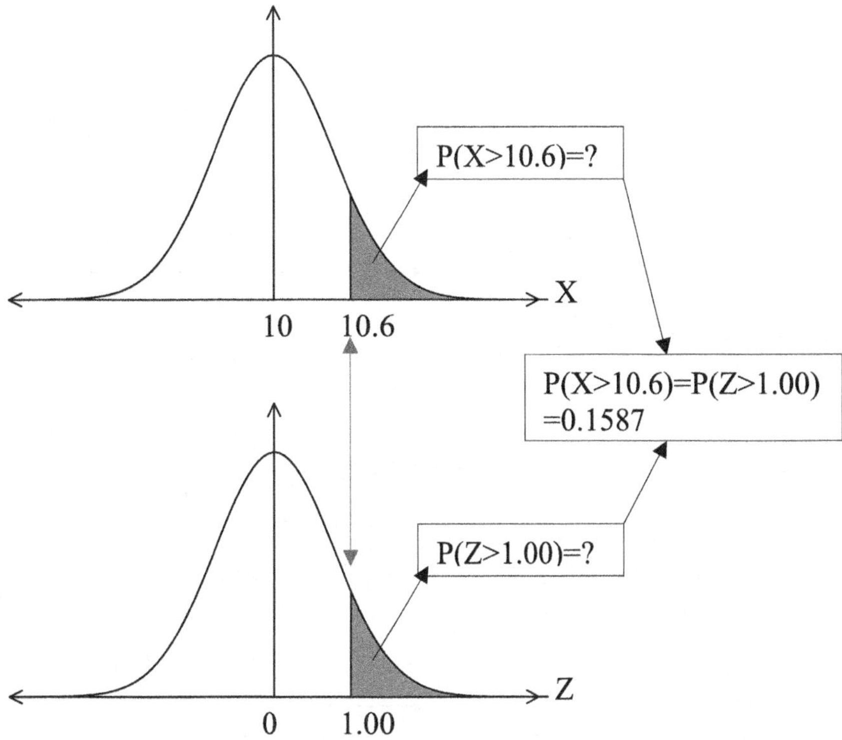

That is, the probability of P(X>10.6) is the same as that of P(Z>1.00) because both ask for the probability being greater than 1 standard deviation above the mean.

Alternatively, this process can be simplified as follows:

$$P(X > 10.6) = P(\frac{X-\mu}{\sigma} > \frac{10.6-\mu}{\sigma}) = P(Z > \frac{10.6-10}{0.6})$$

$$= P(Z > 1) = 0.1587$$

2. Given $\mu = 10$ and $\sigma = 0.6$, calculate the probability that a light bulb will burn less than 9.4 hours.

Solution:

$$P(X<9.4) = P(\frac{X-\mu}{\sigma} < \frac{9.4-\mu}{\sigma}) = P(Z<\frac{9.4-10}{0.6}) = P(Z<-1) = 1 - 0.9413$$

$$= 0.1587$$

We can graphically visualize this problem and solution as follows:

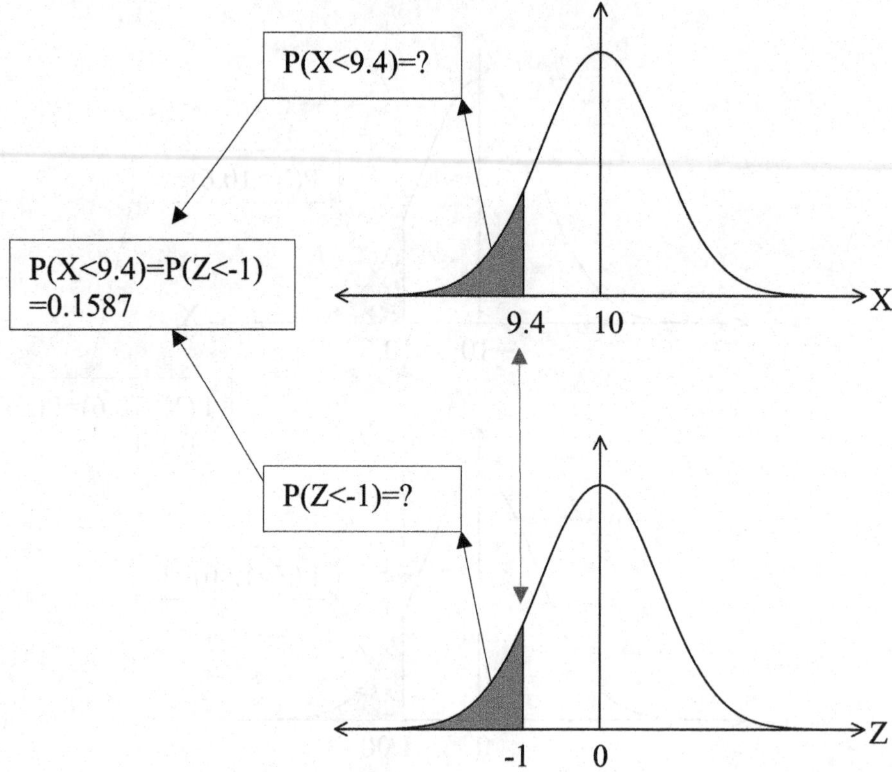

That is, the probability of P(X<9.4) is the same as that of P(Z<-1.00) because both ask for the probability being less than 1 standard deviation below the mean.

3. Given $\mu = 10$ and $\sigma = 0.6$, calculate the probability that a light bulb will burn between 9.4 and 10.6 hours.

Solution:

$$P(9.4 < X < 10.6) = P(\frac{9.4-\mu}{\sigma} < Z < \frac{10.6-\mu}{\sigma})$$

$$= P(\frac{9.4-10}{0.6} < Z < \frac{10.6-10}{0.6})$$

$$= P(-1 < Z < 1) = 1 - (0.1587+0.1587) = 0.6826$$

We can graphically visualize this problem and solution as follows:

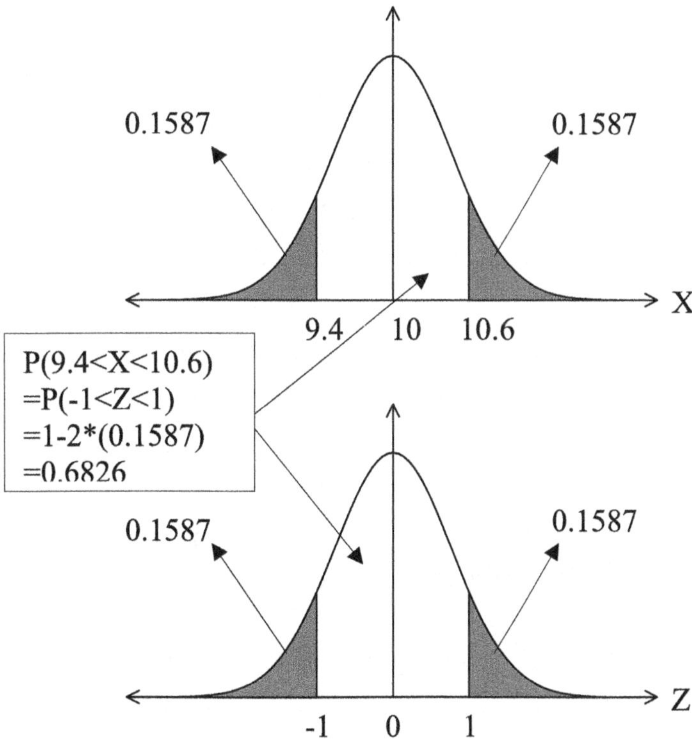

4. Given $\mu = 10$ and $\sigma = 0.6$, calculate the probability that a light bulb will burn less than 8.8 hours.

Solution:

$$P(X<8.8) = P(Z<\frac{8.8-10}{0.6}) = P(Z<-2) = P(Z>2) = 0.0228$$

Note that due to the symmetrical nature of the Z distribution, the answer was obtained by noting $P(Z<-2) = P(Z>2) = 0.0228$

5. Given $\mu = 10$ and $\sigma = 0.6$, calculate the probability that a light bulb will burn more than 9 hours.

Solution:

$$P(X > 9) = P(Z > \frac{9-10}{0.6}) = P(Z > -1.67) = 1 - P(Z<1.67) = 1 - 0.0475 =$$

$$0.9525$$

We can graphically visualize this problem and solution as follows:

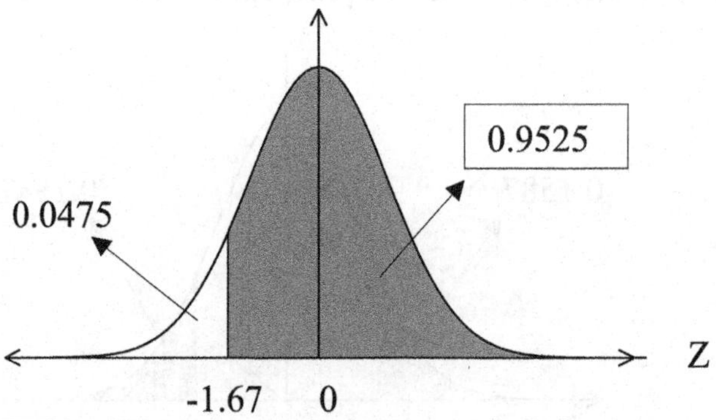

6. Given $\mu = 10$ and $\sigma = 0.6$, calculate the probability that a light bulb will burn between 9.1 and 11.3 hours.

Solution:

$$P(9.1 < X < 11.3) = P(-1.5 < Z < 2.17) = 0.9850 – 0.0668$$
$$= 1 - (0.0668 + 0.0150) = 0.9182$$

We can graphically visualize this problem and solution as follows:

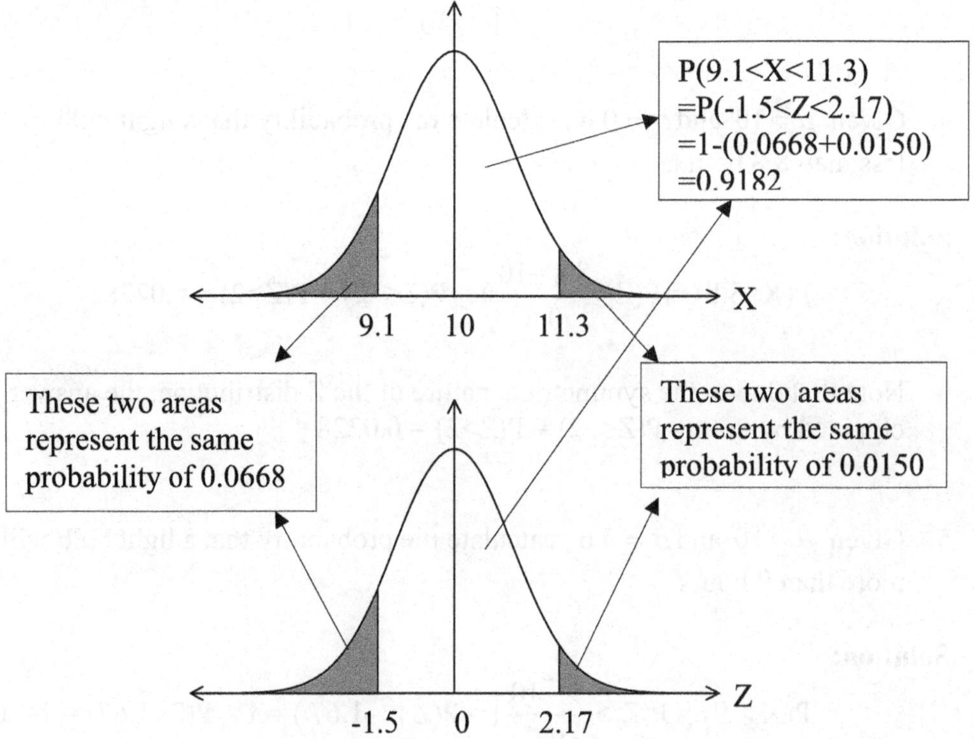

Note:

a. Can you claim that 19 out of 20 light bulbs will burn between 8.8 hours and 11.5 hours?

Solution:

We need to calculate the probability of the light bulbs burning between 8.8 hours and 11.5 hours, based on $\mu = 10$ and $\sigma = 0.6$, as follows:

That is, $P(8.8 < X < 11.5) = P(-2 < Z < 2.5) = 0.9938 - 0.0228 = 0.9710$

This shows that there is a 97.1% probability that the light bulbs will burn between 8.8 and 11.5 hours, which is a higher probability than "19 out of 20." (Note that 19 out of 20 shows a probability of 0.95 or 95%.) Therefore, yes, we can claim that 19 out of 20 light bulbs will burn between 8.8 and 11.5 hours.

b. What is the probability that one of the light bulbs will burn more than 12 hours?

$P(X > 12) = P(Z > 3.33) = 0.00043$

Therefore, 4.3 out of 10,000 bulbs will burn more than 12 hours.

H. Applications – 2

Suppose that you know that the true average (population mean) car speed[79] on Highway I-94 is 60 mph and its population standard deviation is 5 mph. Based on this information, what is the probability that you will find a car on I-94 traveling:

1. faster than 65 mph?

Answer: $P(X>65) = P(Z>\frac{65-60}{5}) = P(Z>1) = 0.1587$

We can graphically visualize this problem and solution as follows:

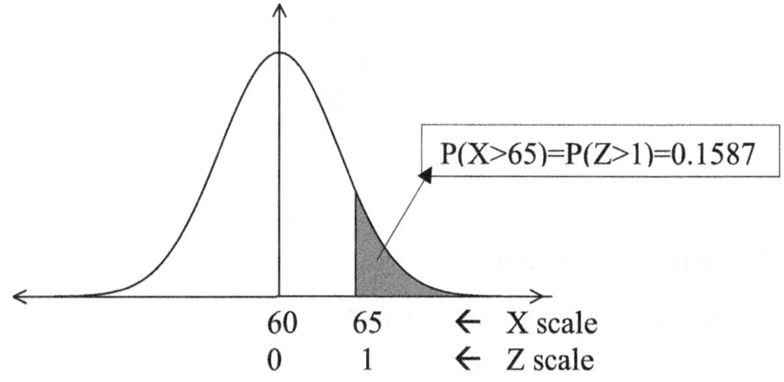

[79] Note that the "true average" means the "population mean." Also, when someone "hypothesizes a true mean," it is called a "hypothesized mean" and treated equally as a "population mean." These are a statistical concept and jargon, which will prove to be very valuable and useful in the later chapters.

2. slower than 55 mph?

Answer: $P(X<55) = P(Z<\frac{55-60}{5}) = P(Z<-1) = P(Z>1) = 0.1587$

We can graphically visualize this problem and solution as follows:

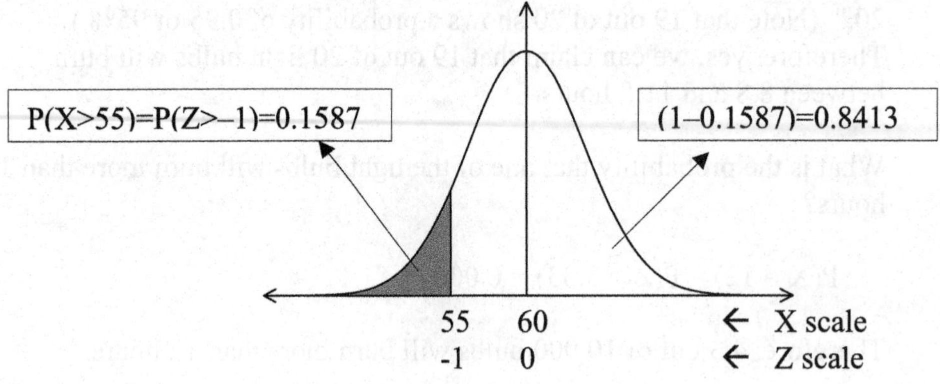

3. between 55 and 65 mph?

Answer: $P(55 < X < 65) = P(-1 < Z < 1) = 1 - 2 \cdot 0.1587 = 0.6826$

We can graphically visualize this problem and solution as follows:

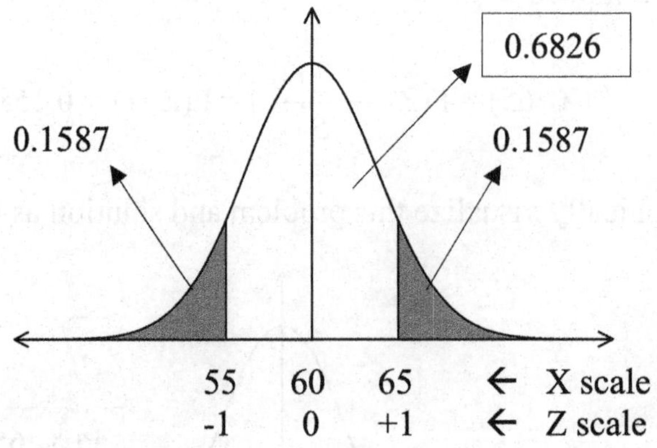

4. faster than 70 mph?

Answer: $P(X > 70) = P(Z>\frac{70-60}{5}) = P(Z>2) = 0.0228$

5. slower than 50 mph?

Solution: $P(X<50) = P(Z<\frac{50-60}{5}) = P(Z<-2) = P(Z>2) = 0.0228$

6. between 57 and 68 hours?

   Solution: $P(57 < X < 68) = P(-0.6 < Z < 1.6) = 1 - 0.2743 - 0.0548 = 0.6709$

I. Advanced Problems:

1. Given the population mean of 60 mph and the population standard deviation of 5 mph, what percentage of cars will violate the maximum speed limit of 55 mph?
   → Can you claim that 80% of cars are violating the speed limit of 55 mph?

   Answer: Yes! Because $P(X>55) = P(Z>-1) = 0.8413$.

2. If you wish to instruct state troopers to catch top 10% of the speeders, what would be the cut-off speed for them to use?

   Answer:
   In order to answer this question, graphically visualizing the problem and solution as follows will be a great help:

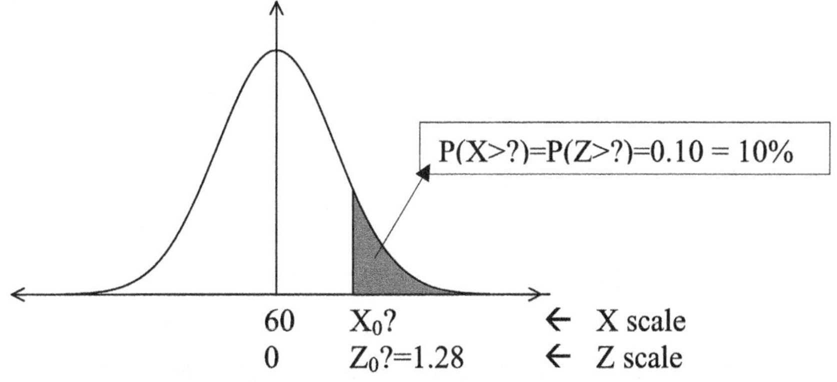

First we must look for Z_0 that satisfies $P(Z > Z_0) = 0.1 = 10\%$

Therefore, from the Z table, we find $Z_0 \approx 1.28$.

Next, we find the value of X, X_0, from the Z formula where $\mu = 60$, $\sigma = 5$, and $Z_0 = 1.28$ as follows:

$$Z_0 = \frac{X_0 - \mu}{\sigma} \rightarrow \quad 1.28 = \frac{X_0 - 60}{5} \quad \rightarrow \quad X_0 = 66.4.$$

Therefore, ticketing the cars with a speed greater than 66.4 mph will constitute the top 10% speeders.

3. If you wish to instruct state troopers to give a warning to the slowest 1% of drivers, what would be the cut-off speed for them to use?

Answer:

Once again we can graphically visualize the problem and solution as follows:

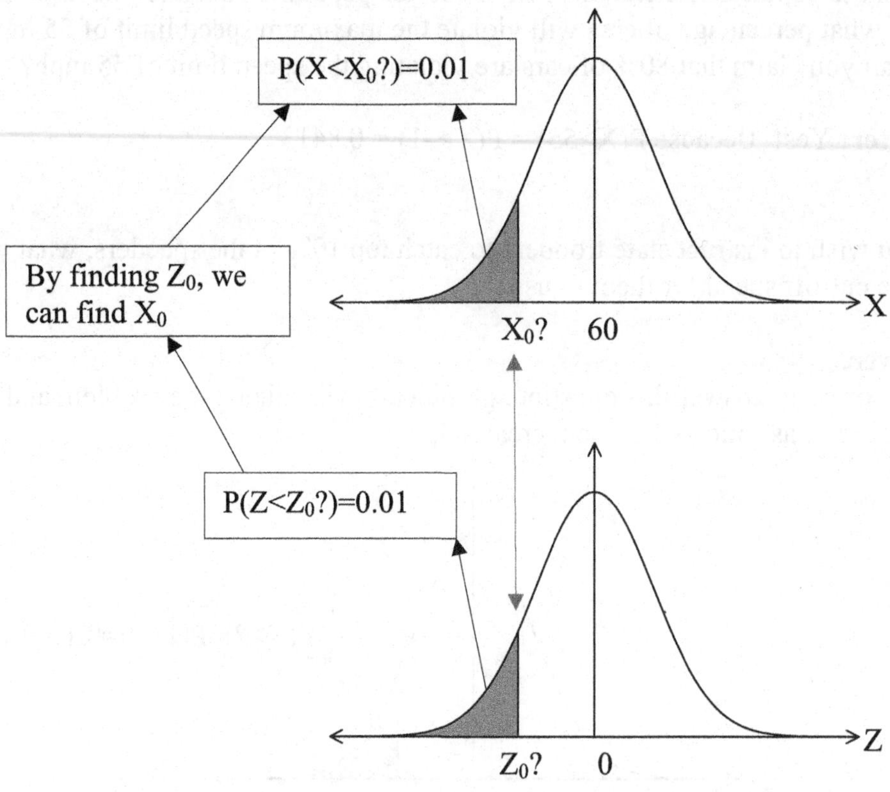

First we must look for Z_0 that satisfies $P(Z < Z_0) = 0.01 = 1\%$

Therefore, from the Z table, we find Z_0 is between –2.32 and –2.33. By interpolation, we assume $Z_0 \approx -2.325$. (Note: we must choose the negative Z value because we know that the slowest 1% falls below the X mean of 60 mph, or equivalently, below the Z mean of 0.)

Next, we find the value of X, X_0, from the Z formula where $\mu = 60$, $\sigma = 5$, and $Z_0 = -2.325$ as follows:

$$Z_0 = \frac{X_0 - \mu}{\sigma} \rightarrow \quad -2.325 = \frac{X_0 - 60}{5} \quad \rightarrow \quad X_0 = 48.375$$

Therefore, giving a warning to the cars with a speed slower than 48.375 mph will constitute the slowest 1% of the drivers.

Exercise Problems on Chapter 7. The Normal Probability Distribution

This set of exercise problems has 13 problems, worth a total of 15 points.

1. The major difference between a discrete and a continuous probability distributions is found if the random variable of one's interest can be measured on a continuum or not. An example of a discrete random variable is _____ and that of a continuous random variable is _____.

   a. height; weight
   b. the number of flight arrivals per day; height
   c. age; weight
   d. all of the above
   e. only (a) and (c) of the above

2. The characteristics of a normal (cumulative probability distribution) curve include _____.

   a. it is bell-shaped
   b. it is symmetrical about the mean
   c. its mean, median, and mode are all equal and same.
   d. all of the above
   e. only (a) and (b) of the above

3. The parameter(s) that define a normal distribution include _____.

   a. the mean
   b. the inter-quartile
   c. the range
   d. only (a) and (b) of the above
   e. only (a) and (c) of the above.

4. The larger the value of a standard deviation is, holding all other factors the same, the _____ the normal probability curve is.

   a. flatter
   b. steeper
   c. narrower
   d. less random
   e. only (b) and (c) of the above

5. The transformation of any normal distribution into a standardized normal distribution, Z, forces the transformed (or standardized) mean value to be _____ and the transformed (or standardized) standard deviation value to be _____.

a. 1; 0
b. 0; 1
c. 0; 0
d. 1; 1
e. none of the above

6. The parameter(s) that define a standardized normal variable, Z, include _____.

a. the sample mean and the sample standard deviation
b. the population mean and the sample standard deviation
c. the sample mean and the population standard deviation
d. the population mean and the population standard deviation
e. none of the above.

7. The values of the mean and one standard deviation of a standardized normal variable, Z, are _____ and _____, respectively.

a. 1; 0
b. 0; 1
c. 0; 0
d. 1; 1
e. none of the above

8. Theoretically speaking, a standardized normal variable, Z, can take a range of values _____.

a. from -1 to +1
b. from -2 to +2
c. from -10 to +10
d. from $-\infty$ to $+\infty$
e. only 0

9. Assume a population that has a mean of 10 and a standard deviation of 3. What is the Z value of a sample observation, X, that has a value of 4?

a. 1
b. -1
c. 2
d. -2
e. 3

10. Based on the cumulative standardized normal probability table, the probability of Z being less than -1.5 is _____ and that of Z being greater than 1.75 is _____.

a. 0.0668; 0.0401
b. 0.0668; 4.01%

c. 6.68%; 0.0401
d. 6.68%; 4.01%
e. all of the above

11. Suppose that you found from the survey of how many bottles of Coke people drink in a day that the population mean is 10 and a population standard deviation is 5. What is the probability of finding a person drinking 5 or less number of bottles of Coke a day?

a. 0.8413 or 84.13%
b. 0.8413 or 8.413%
c. 0.1587 or 15.87%
d. 0.1587 or 1.587%
e. 0.3413

The following problems are worth 2 points each.

12. Given a similar situation as above, you found from another survey of how many bottles of Coke people drink in a day that the population mean is 8 and a population standard deviation is 4. What is the probability of finding people who drink between 5 and 13 bottles of Coke a day?

a. 0.1056
b. 0.2266
c. 0.6678
d. 0.7734
e. 0.8944

13. Given that the population mean is 8 bottles and a population standard deviation is 4 bottles, if you wish to identify the number of bottles of Coke that the top 10% of heavy Coke drinkers will drink, what would it be?

a. 13.12 bottles
b. 14.5 bottles
c. 12 bottles
d. 15.67 bottles
e. 16 bottles

Chapter 8. The t-Probability Distribution

In this chapter, we examine the second most prominent probability distribution known officially as the Student's t-probability distribution[80]. For short, however, it is most often called a t-distribution.

Unlike the standard(ized) normal distribution of Z, the t-distribution normalizes a value based on a sample standard deviation. The reason is because when a population standard deviation is not known, Z-value cannot be calculated. In a situation like this, we must use a sample standard deviation in place of a population deviation, resulting in a new transformation or standardizing formula as follows:

$$t = \frac{X - \mu}{S}$$

where S is the sample standard deviation[81]. **The resulting t-value is often called a t-score, t-statistic, or calculated t-value**. The above t-formula enables one to standardize any random variable whose population standard deviation is unknown[82]. As its formula is very similar to the Z-formula, its main characteristics are very similar as well.

A. The Comparison of the t-distribution in relation to the Z-distribution

The following shows the characteristics of the t-distribution by comparing its similarities and dissimilarities to the Z-distribution:

The similarities are:

(1) The t-distribution is symmetrical about the mean as the Z-distribution is. Therefore, it is bell-shaped.

(2) The mean of a t-distribution is zero (0) as is the Z-distribution.

(3) The value of one standard deviation of a t-distribution is 1 as is the Z-distribution.

(4) The t-distribution is the same as the Z-distribution if the sample size n increases to an infinity.

---

[80] The first most important one is the Z-distribution. The developer of this t-distribution, William S. Gosset (1876-1937), was a British statistician who worked for a brewery, Guinness. Because he was prevented from using his real name in publication of his work due to his company policy, he used a pseudonym, "Student." In describing the t-distribution, always a lower case, t, is used; never the upper case, T.

[81] The formula and the calculation method thereof are shown in Chapter 4.

[82] The t-distribution is a refinement of a Z-distribution, applicable specifically to small sample cases. If a sample size is very large, exceeding 100, the use of either a Z- or a t-distribution may not matter.

Therefore, the general shape of a t-distribution as shown below is very similar to the Z-distribution:

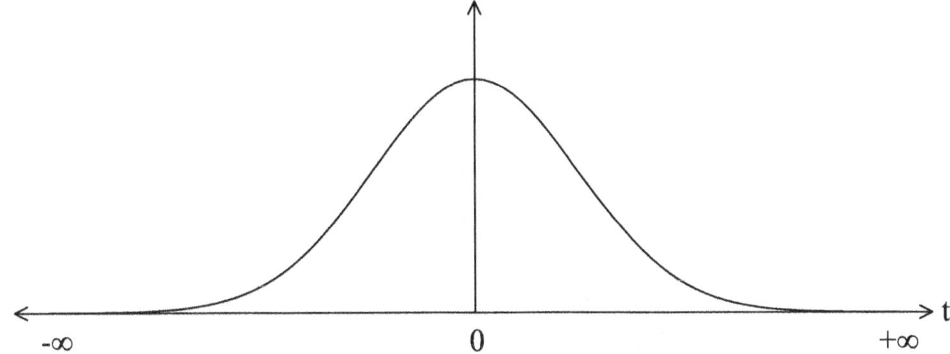

The dissimilarities are:

(1) If the sample size is small, the t-distribution has a fatter tail and flatter top than the Z-distribution.

(2) The t-statistic is sensitive to the degrees of freedom[83] whereas the Z-distribution is not. In the case of a sample mean, t-statistic has (n–1) degrees of freedom.

B. Critical t-Value Table

1. What is a critical t-value?

For an upper-tail probability, a **positive** critical t-value, denoted as $t_{critical}$ or t_{table}, defines a cumulative probability greater than itself as shown below.

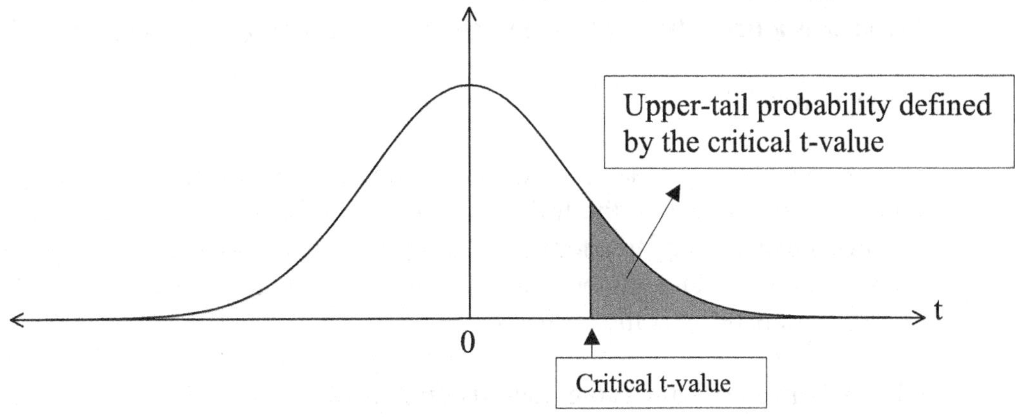

[83] The degree of freedom is the number of observations that are free to vary in determining a parameter. It equals the number of observations minus the number of additional parameters to be calculated. For example, to calculate a sample mean, one needs all n observations. However, when calculating a sample variance or a sample standard deviation, a sample mean is needed and it has to be calculated. Because a sample mean is an estimator of a population parameter, μ, we lose 1 degree of freedom. Thus, the appropriate degrees of freedom for either a sample variance or a sample standard deviation are (n – 1).

For a lower-tail probability, a **negative** critical t-value, denoted as $t_{critical}$ or t_{table}, defines a cumulative probability less than itself as shown below[84].

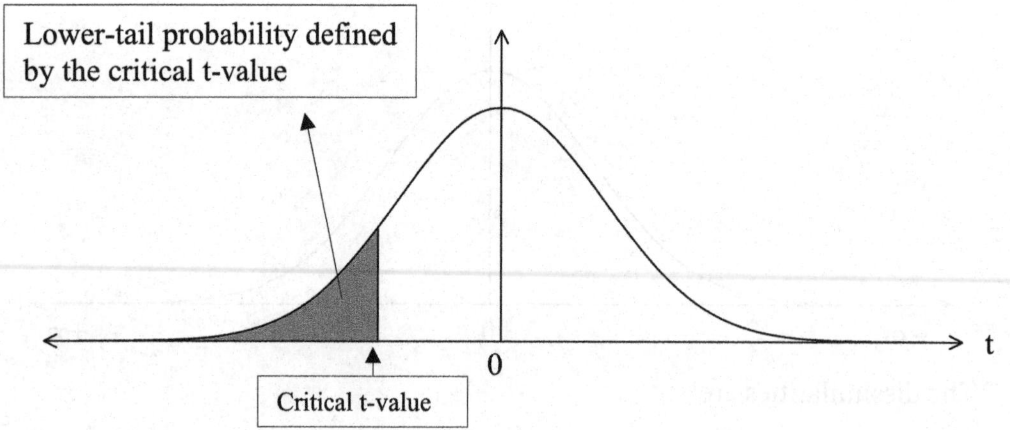

2. How to Calculate Critical t-Values

Due to complexity, t-values are calculated by a computer software such as the Excel Spreadsheet or can be looked up in a table of t-values.

The Excel Spreadsheet automatically calculates the critical values for a two-tail probability via "=TINV(probability chosen, degrees of freedom)." That is, when "=TINV(0.05, 1)" is used in Excel, for example, the critical t-value found implies both an upper probability of 0.05/2 and a lower probability of 0.05/2.

Therefore, its command has to be revised as "=TINV(2*probability chosen, degrees of freedom)" if the critical value for only an upper tail probability is desired. This command is used to generate the following critical (or table) t-values at a probability shown in the first row and (n – 1) degrees of freedom.

3. How to Read the t-Table

The critical t-value associated with specific degrees of freedom and a specific probability is shown in the following table. Unlike the Z-table, the first column shows the degrees of freedom (DF) and the top row shows the probability of the upper tail only. The number found at the intersection of the degrees of freedom and the probability is the critical t-value.

To look up a critical t-value, identify first the degrees of freedom such as 24. Then identify the level of an upper cumulative probability such as 0.05 (=5%). The intersection of these row and column will yield the critical t-value such as 1.7109.

[84] For a two-tail probability, there will be two critical t-values – one negative and one positive. This topic will be studied in detail in Chapters 9, 10, and 11.

Critical t-Values for an Upper-Tail Probability								
Degrees of Freedom	Upper-Tail Probability							
	0.25	0.2	0.15	0.1	0.05	0.025	0.01	0.005
1	1.0000	1.3764	1.9626	3.0777	6.3138	12.7062	31.8205	63.6567
2	0.8165	1.0607	1.3862	1.8856	2.9200	4.3027	6.9646	9.9248
3	0.7649	0.9785	1.2498	1.6377	2.3534	3.1824	4.5407	5.8409
4	0.7407	0.9410	1.1896	1.5332	2.1318	2.7764	3.7469	4.6041
5	0.7267	0.9195	1.1558	1.4759	2.0150	2.5706	3.3649	4.0321
6	0.7176	0.9057	1.1342	1.4398	1.9432	2.4469	3.1427	3.7074
7	0.7111	0.8960	1.1192	1.4149	1.8946	2.3646	2.9980	3.4995
8	0.7064	0.8889	1.1081	1.3968	1.8595	2.3060	2.8965	3.3554
9	0.7027	0.8834	1.0997	1.3830	1.8331	2.2622	2.8214	3.2498
10	0.6998	0.8791	1.0931	1.3722	1.8125	2.2281	2.7638	3.1693
11	0.6974	0.8755	1.0877	1.3634	1.7959	2.2010	2.7181	3.1058
12	0.6955	0.8726	1.0832	1.3562	1.7823	2.1788	2.6810	3.0545
13	0.6938	0.8702	1.0795	1.3502	1.7709	2.1604	2.6503	3.0123
14	0.6924	0.8681	1.0763	1.3450	1.7613	2.1448	2.6245	2.9768
15	0.6912	0.8662	1.0735	1.3406	1.7531	2.1314	2.6025	2.9467
16	0.6901	0.8647	1.0711	1.3368	1.7459	2.1199	2.5835	2.9208
17	0.6892	0.8633	1.0690	1.3334	1.7396	2.1098	2.5669	2.8982
18	0.6884	0.8620	1.0672	1.3304	1.7341	2.1009	2.5524	2.8784
19	0.6876	0.8610	1.0655	1.3277	1.7291	2.0930	2.5395	2.8609
20	0.6870	0.8600	1.0640	1.3253	1.7247	2.0860	2.5280	2.8453
21	0.6864	0.8591	1.0627	1.3232	1.7207	2.0796	2.5176	2.8314
22	0.6858	0.8583	1.0614	1.3212	1.7171	2.0739	2.5083	2.8188
23	0.6853	0.8575	1.0603	1.3195	1.7139	2.0687	2.4999	2.8073
24	0.6848	0.8569	1.0593	1.3178	1.7109	2.0639	2.4922	2.7969
25	0.6844	0.8562	1.0584	1.3163	1.7081	2.0595	2.4851	2.7874
26	0.6840	0.8557	1.0575	1.3150	1.7056	2.0555	2.4786	2.7787
27	0.6837	0.8551	1.0567	1.3137	1.7033	2.0518	2.4727	2.7707
28	0.6834	0.8546	1.0560	1.3125	1.7011	2.0484	2.4671	2.7633
29	0.6830	0.8542	1.0553	1.3114	1.6991	2.0452	2.4620	2.7564
30	0.6828	0.8538	1.0547	1.3104	1.6973	2.0423	2.4573	2.7500
31	0.6825	0.8534	1.0541	1.3095	1.6955	2.0395	2.4528	2.7440
32	0.6822	0.8530	1.0535	1.3086	1.6939	2.0369	2.4487	2.7385
33	0.6820	0.8526	1.0530	1.3077	1.6924	2.0345	2.4448	2.7333
34	0.6818	0.8523	1.0525	1.3070	1.6909	2.0322	2.4411	2.7284
35	0.6816	0.8520	1.0520	1.3062	1.6896	2.0301	2.4377	2.7238
36	0.6814	0.8517	1.0516	1.3055	1.6883	2.0281	2.4345	2.7195
37	0.6812	0.8514	1.0512	1.3049	1.6871	2.0262	2.4314	2.7154
38	0.6810	0.8512	1.0508	1.3042	1.6860	2.0244	2.4286	2.7116
39	0.6808	0.8509	1.0504	1.3036	1.6849	2.0227	2.4258	2.7079
40	0.6807	0.8507	1.0500	1.3031	1.6839	2.0211	2.4233	2.7045
42	0.6804	0.8503	1.0494	1.3020	1.6820	2.0181	2.4185	2.6981
44	0.6801	0.8499	1.0488	1.3011	1.6802	2.0154	2.4141	2.6923
46	0.6799	0.8495	1.0483	1.3002	1.6787	2.0129	2.4102	2.6870
48	0.6796	0.8492	1.0478	1.2994	1.6772	2.0106	2.4066	2.6822
50	0.6794	0.8489	1.0473	1.2987	1.6759	2.0086	2.4033	2.6778
60	0.6786	0.8477	1.0455	1.2958	1.6706	2.0003	2.3901	2.6603
80	0.6776	0.8461	1.0432	1.2922	1.6641	1.9901	2.3739	2.6387
100	0.6770	0.8452	1.0418	1.2901	1.6602	1.9840	2.3642	2.6259
∞	0.6745	0.8417	1.0365	1.2816	1.6450	1.9602	2.3267	2.5763

B. The Use of the t-Formula for an Individual X Value

The t-formula is used to standardize a normal variable X when the population variance or the population standard deviation of X is UNKNOWN and thus, must be estimated by the sample variance or the sample standard deviation[85]. However, the use of the t-formula is the same as that of the Z-formula. That is, we use the following t-formula:

$$t = \frac{X - \mu}{S}$$

The use of this formula is best demonstrated by the following examples:

Examples >

Given a population mean of 100 and a sample standard deviation of 2, you took a sample of 25 observations. (Please note that the sample standard deviation of 2 was calculated with a divisor of 24 (= n – 1 = 25 – 1) and thus, the correct degrees of freedom to use is 24.)

1. What is the probability that **an individual X value** will be greater than 103.4218?

In finding an answer to this problem, we must remind ourselves of the case with the Z-formula and transformation. That is, we first standardize X into t and find its probability as follows:

$$P(X > 103.4218) = P(\frac{X - \mu}{S} > \frac{103.4218 - \mu}{S})$$

$$= P(t > \frac{103.4218 - 100}{2}) = P(t > 1.7109) = 0.05$$

Alternatively, we can visualize the problem and the solution via the following diagram:

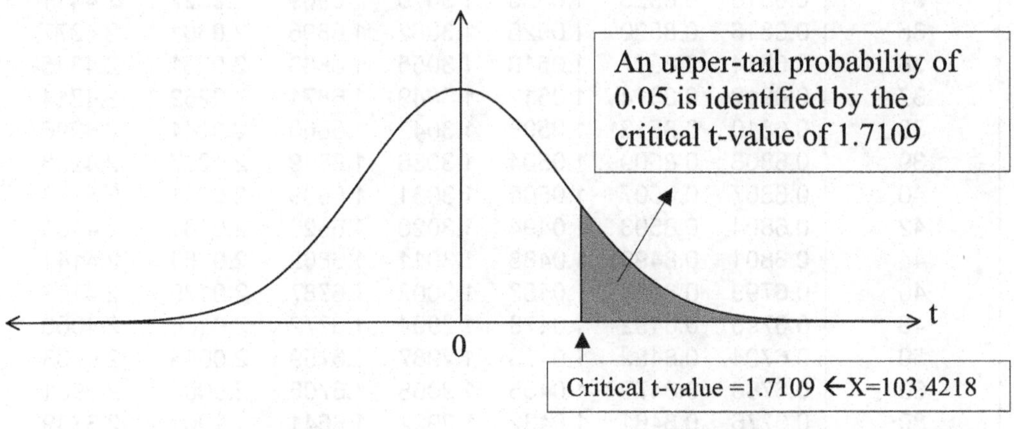

[85] This means, then, whenever the population variance or the population standard deviation or the population standard error is known, the Z-formula must be used. Otherwise, use the t-formula.

Note: In using the t-table, the degrees of freedom (df=24) must be identified first and then, t-value of 1.7109 in that row. Now, move up to the top row where the corresponding probability of 0.05 is found

Note: Instead of using the t-table above, one can use the Excel Spreadsheet command of "=TDIST(X,df,tails)" to calculate the cumulative probability of t at X and (n – 1) degrees of freedom. In this case, because X=1.7109; df=24; and tails=1, the corresponding probability is calculated by Excel as follows[86]:

"=TDIST(1.7109,24,1)" → 0.04999832

2. What is the probability that **an individual X value** will be less than 98.6304?

$$P(X < 98.6304) = P(t < \frac{98.6304 - 100}{2}) = P(t < -0.6848) = 0.25$$

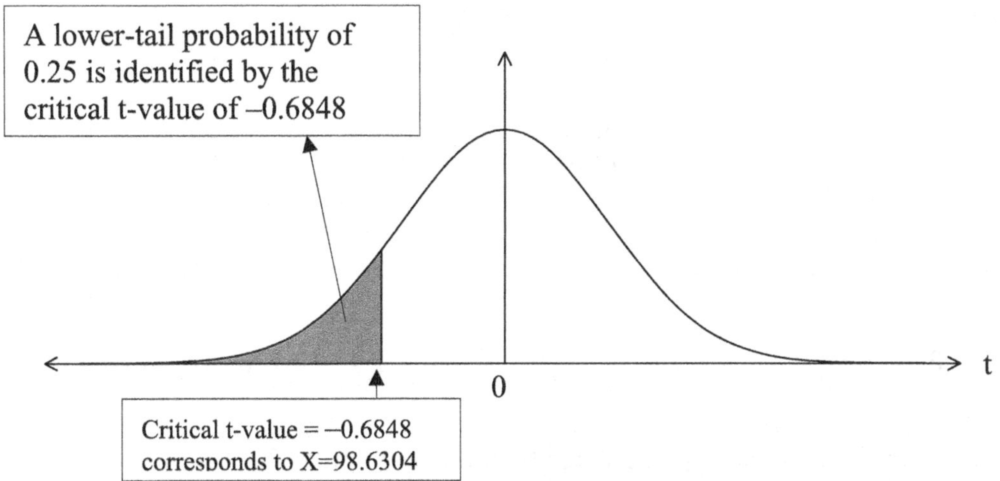

Note: In using the t-table, the degrees of freedom (df=24) must be identified first and then, t-value of 0.6848 in that row. Now, move up to the top row where the corresponding probability of 0.25 is found

Note: Because the t-distribution is symmetrical about the mean, $t_{critical}$ = 0.6848 is used instead of –0.6848. In fact, the Excel Spreadsheet allows only a positive t value as an input and asks for the user to know its +/- sign. Therefore, given the Excel Spreadsheet command of "=TDIST(X,df,tails)," the input of X=0.6848; df=24; and tails=1 yields the following result:

"=TDIST(0.6848,24,1)" → 0.250015383

[86] The Excel Spreadsheet command always gives the cumulative probability of t for ($t_0 > k$), the upper-tail value.

3. What is the probability that **an individual X value** will be greater than 98.6304?

$$P(X > 98.6304) = P(t > \frac{98.6304 - 100}{2}) = P(t > -0.6848) = 1 - 0.25 = 0.75$$

Note: The probability calculated by the Excel Spreadsheet for P(t > 0.6848) is:

"=TDIST(0.6848,24,1)" → 0.250015383

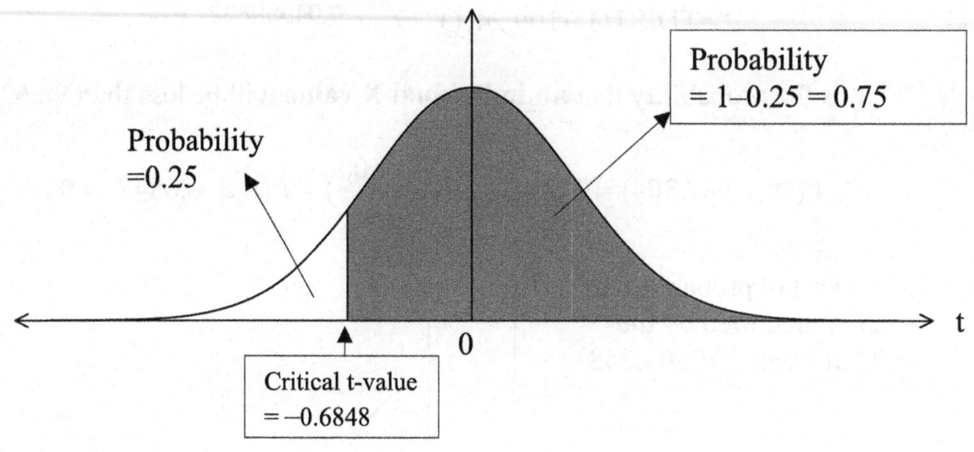

4. What is the probability that **an individual X value** will be greater than 97.3644?

$$P(X > 97.3644) = P(t > \frac{97.3644 - 100}{2}) = P(t > -1.3178) = 1 - 0.10 = 0.90$$

Note: The probability calculated by the Excel Spreadsheet for P(t > 1.3178) is:

"=TDIST(1.3187,24,1)" → 0.100005924

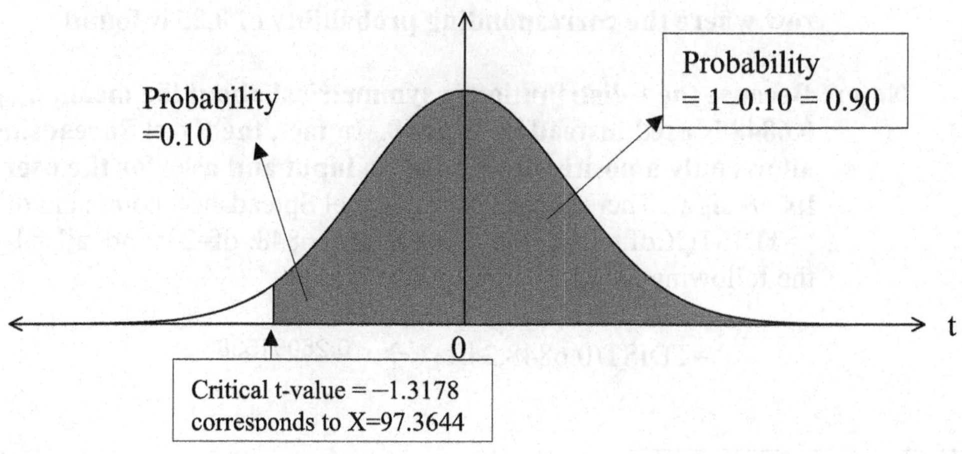

5. What is the probability that **an individual X value** will be between 97.3644 and 104.9844?

$$P(97.3644 \le X \le 104.9844) = P(-1.3178 \le t \le 2.4922) = 1 - 0.1 - 0.01 = 0.89$$

Note 1:

In using the t-table, the degrees of freedom (df=24) is identified first and then, respective t-values of 1.3178 and 2.4922 in that row. Now, move up to the top row where the corresponding probabilities of 0.1 and 0.01 are found

Note 2:

The probability calculated by the Excel Spreadsheet for P(t > 2.4922) is:

"=TDIST(2.4922,24,1)" → 0.009999099 → 0.01

Note 3:

We can graphically visualize this problem and the solution as follows:

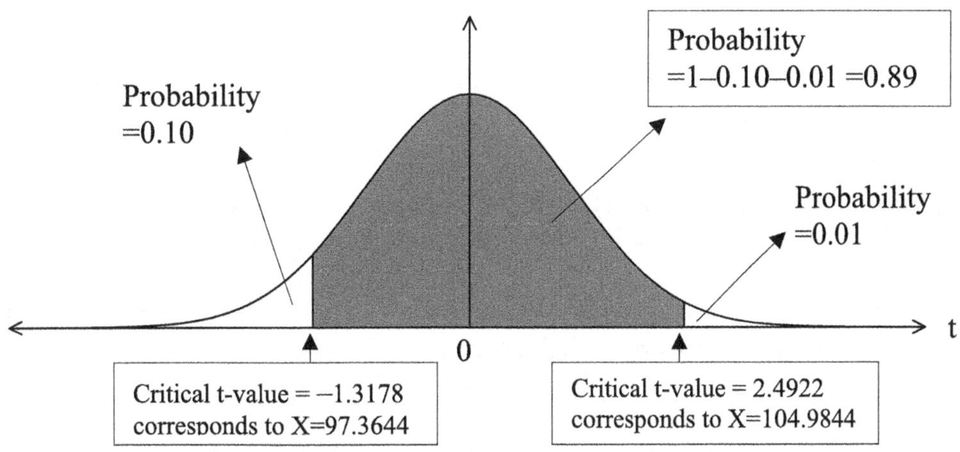

Exercise Problems on
Chapter 8. The t-Probability Distribution

This set of exercise problems has 8 problems, worth a total of 10 points.

1. The difference between a Z-distribution and a t-distribution includes:

    a. a Z-distribution is always bigger than a t-distribution
    b. a Z-distribution is always smaller than a t-distribution
    c. a Z-distribution is always flatter than a t-distribution
    d. none of the above
    e. only (b) and (c) of the above

2. The characteristics of a t-probability distribution curve include _____.

    a. it is somewhat bell-shaped
    b. it is symmetrical from the middle
    c. its mean, median, and mode are all equal and same.
    d. all of the above
    e. only (a) and (b) of the above

3. If there are 20 observations in a sample, its corresponding degrees of freedom for a t-distribution are _____.

    a. 19
    b. 20
    c. 21
    d. not identifiable because the sample mean is not given
    e. not identifiable because the sample variance is not given

4. If the sample size, n, increases to an infinity, the t-value will be _____ the Z-value at the same cumulative probability.

    a. larger than
    b. smaller than
    c. almost equal to
    d. closer to 0 than
    e. only (b) and (c) of the above

5. The transformation of any normal distribution into a standardized t-distribution, t, forces the transformed (or standardized) mean value to be _____ and the transformed (or standardized) standard deviation value to be _____.

    a. 1; 0
    b. 0; 1
    c. 0; 0

d. 1; 1
e. none of the above

6. Assume that a population mean is 10 and a sample standard deviation is 3. What is the t-value of a sample observation, X, that has a value of 4?

   a. 2
   b. -2
   c. 3
   d. -3
   e. not calculable due to lack of information

The following problems are worth 2 points each.

7. Assume that the true average number of 12-oz Coke bottles each person drinks a day is 4. Given a sample with 20 people, you found that a sample standard deviation is 1.4543. What is the probability of finding a person who drinks more than 5 bottles of Coke a day?

   a. 0.05
   b. 0.10
   c. 0.25
   d. 0.30
   e. not calculable due to lack of information

8. Assume that the true average number of 12-oz Coke bottles each person drinks a day is 4. Given a sample with 20 people, you found that a sample standard deviation is 1.4543. What is the probability of finding a person who drinks between 3 and 5 bottles of Coke a day?

   a. 0.20
   b. 0.50
   c. 0.60
   d. 0.68
   e. not calculable due to lack of information

Chapter 9. Sampling Distributions

A sampling distribution refers to the distribution of sample observations and the distribution of their means. Even though we normally do not know the true underlying distribution of the sample observations, statistical theories such as the Central Limit Theorem show that at least the mean of sample observations – the sample mean – follows either a normal Z-distribution or a t-distribution. **If a population standard deviation is known, the sampling distribution of the mean is assumed to follow a normal Z-distribution. If a population standard deviation is unknown, the sampling distribution of the mean is assumed to follow a t-distribution[87].** We shall use this statistical theory extensively without proof throughout the remainder of this book.

This very concept of sampling distributions is the foundation for inferential statistics. **It enables us to examine the likelihood (=probability) of a sample occurring from a known population. It also enables us to infer the likelihood of a unknown population mean occurring from a known sample.** In other words, these two slightly different tasks, arising from two different situations, are the center of inferential statistics.

Because we had studied various sampling methods for collecting primary data in Chapter 2, we jump right in to the issues of understanding the difference between the distribution of a single sample observation and that of the sample mean. Afterwards, **we will study the methods of using Z- and t-distributions to access probability of a sample mean occurring given a known population mean.**

A. Definitions to Understand the Sampling Distributions

Even though there are cases where a population mean is known, we will assume herein that it is not known. Therefore, we are to estimate it by a sample mean and utilize the following statistical properties without proof:

1. The population mean, μ, vs. the sample mean, $\overline{X}$

   a. The mean of a sample, $\overline{X}$

   The population mean is estimated by the mean of a sample.

$$X_1, X_2, X_3, etc \Rightarrow \overline{X} \Rightarrow \mu_X \Rightarrow \mu$$

   A statistical theory states that a sample mean is an unbiased estimate of a population mean[88] → That is, any sample mean can represent and approximate its population mean → $\overline{X} \Rightarrow \mu_X \Rightarrow \mu$.

---

[87] This conclusion based on statistical theories is so important that you must memorize and never forget it.

[88] Recall from Chapter 1 that a (sample) statistic is an estimator of a (population) parameter.

b. The mean of sample means, $\overline{\overline{X}}$

$$\overline{X}_1, \overline{X}_2, \overline{X}_3, etc \Rightarrow \overline{\overline{X}} \Rightarrow \mu_{\overline{X}} \Rightarrow \mu$$

The mean of many sample means, which is denoted as $\overline{\overline{X}}$ and often referred as the grand mean, is also an unbiased estimate of a population mean, just like any sample mean, $\overline{X}$, is.

$$\text{That is, the mean of } \overline{X}'s \Rightarrow \overline{\overline{X}} \Rightarrow \mu_{\overline{X}} \Rightarrow \mu$$

Because this is true, there is no need to collect sample after sample. The mean of a single sample will suffice to estimate the population mean without any bias.

Intuitive Illustration:

Normally, a single sample is taken such as:

Sample #1 → 9, 10, 10, 10, 11 → $\overline{X}_1 = 10$ and $S_1 = \sqrt{0.5} = 0.707$

However, if we assume that 3 additional samples are taken from the same population, we will most likely get different observations in each sample and thus, different means and standard deviations can be calculated as shown below:

Sample #2 → 8, 9, 10, 11, 12 → $\overline{X}_2 = 10$ and $S_2 = \sqrt{2.5} = 1.581$

Sample #3 → 7, 9, 10, 12, 13 → $\overline{X}_3 = 10.2$ and $S_3 = \sqrt{5.7} = 2.387$

Sample #4 → 7, 8, 9, 11, 14 → $\overline{X}_4 = 9.8$ and $S_4 = \sqrt{10.53} = 3.245$

Given these 4 samples above, we note that the mean of these 4 sample means is 10 as shown below:

$$\overline{\overline{X}} = \frac{\overline{X}_1 + \overline{X}_2 + \overline{X}_3 + \overline{X}_4}{4} = \frac{10 + 10 + 10.2 + 9.8}{4} = 10$$

and

the sample variance of these means, denoted as $S^2_{\overline{X}}$, is calculated as:

$$S^2_{\overline{X}} = \frac{0^2 + 0^2 + (0.2)^2 + (-0.2)^2}{3 - 1} = \frac{0.08}{2} = 0.04$$

And the sample standard deviation of these means, denoted as $S_{\overline{X}}$ and known as the "**sample standard error**," is calculated as:

$$S_{\overline{X}} = \sqrt{S_{\overline{X}}^2} = \sqrt{0.04} = 0.2$$

If we conduct such an experiment as above repeatedly for numerous times, we will find the grand mean of these sample means to be a number closer to the population mean because sample means are a measure of central tendency and thus, they should gravitate toward the population mean. This property is known as the "**unbiasedness**" of a sample mean → i.e, a sample mean estimates the population mean without any bias.

2. The standard deviation and the standard error

The standard deviation (or the variance) is a measure of dispersion for a single random variable such as X in relation to its mean. The standard error, on the other hand, is a measure of dispersion for the sample means in relation to their grand mean. Therefore, both measure dispersion but the former measures the variability of a single value X whereas the latter measures that of the mean, $\overline{X}$.

We will first review the formula for the standard deviation and the standard error of the population as follows:

a. The population standard deviation of X → $\sigma = \sigma_X$.

$$\sigma = \sigma_X = \sqrt{\sum_{j=1}^{k} P_j \cdot (X_j - \mu)^2} = \sqrt{\frac{\sum_{i=1}^{N} (X_i - \mu)^2}{N}}$$

b. The population standard deviation of $\overline{X}$ = population standard error, $\sigma_{\overline{X}}$.

$$\sigma_{\overline{X}} = \frac{\sigma_X}{\sqrt{n}} = \frac{\sigma}{\sqrt{n}}$$

where n is the number of sample size[89].

[89] This formula is for the case of an infinite population. For a finite population, we need to use: $\sigma_{\overline{X}} = \sqrt{\frac{N-n}{N-1}}\left(\frac{\sigma}{\sqrt{n}}\right)$. However, as the population s size of N approaches an infinity, it does not make a big difference.

c. The intuitive relationship between σ and $\sigma_{\overline{X}}$.

Note that from the variances and the standard deviations of the 4 samples shown above, we can easily understand that the standard deviations (S_i's) of Samples #1 through #4 are larger than the standard deviation of 4 sample means, $S_{\overline{X}}$. This is because given the observations in all 4 samples, the individual X values are dispersed between 7 and 14 whereas their sample means are dispersed between 9.8 and 10.2. That is, the sample means are distributed more compactly near the population mean than individual X values.

This observation intuitively convinces us that individual Xs vary more than their sample means, $\overline{X}'s$ →Therefore, $S_X > S_{\overline{X}}$. More precisely, their theoretical relationship[90] is:

$$S^2 = S_X^2 = n \cdot S_{\overline{X}}^2 \qquad \rightarrow \qquad S_{\overline{X}} = \frac{S_X}{\sqrt{n}} = \frac{S}{\sqrt{n}}$$

This relationship specifically shows that S_X is $\sqrt{n}$ times greater than $S_{\overline{X}}$. This observation can be applied directly to the case between a population standard deviation and a population standard error. That is, we should infer and know that $\sigma_X > \sigma_{\overline{X}}$.

More precisely, their theoretical relationship[91] is defined as:

$$\sigma^2 = \sigma_X^2 = n \cdot \sigma_{\overline{X}}^2 \qquad \rightarrow \qquad \sigma_{\overline{X}} = \frac{\sigma_X}{\sqrt{n}} = \frac{\sigma}{\sqrt{n}}$$

Once again, σ_X is $\sqrt{n}$ times greater than $\sigma_{\overline{X}}$. Also, we note that $\sigma_{\overline{X}}$ discussed above is called the population standard error of the mean, or simply, the **population standard error**; whereas σ_X is called the standard deviation of the population or simply, the **population standard deviation**.

Important Implication:

In reality, only one sample is taken. However, the statistical theory states that the above relationship between the standard deviation and the standard error is true. Thus, the need to take additional samples does not exist and we simply use the above relationship as the truth in analyzing the distribution of a sample mean.

[90] We accept this relationship without proof. You must know and memorize this formula.
[91] Likewise, we accept this relationship without proof. You must know and memorize this formula.

B. When the Population Mean and Population Standard Deviation are Known

Recall: If a population standard deviation of a normal variable is known, the sampling distribution of its mean follows a normal Z-distribution.

When the population mean and the population standard deviation are known, we can use the following Z formula to standardize a sample mean, $\overline{X}$, into a Z value and analyze the probability of the sample mean occurring:

$$Z = \frac{\overline{X} - \mu}{\sigma_{\overline{X}}} = \frac{\overline{X} - \mu}{\frac{\sigma}{\sqrt{n}}}$$

Note the difference between this Z formula shown above and the Z formula described in Chapter 7. Both are used for standardizing a normal variable. However, **the Z formula shown above standardizes (or normalizes) a sample mean whereas the Z formula in Chapter 7 standardizes a single value of X.**

Always remember that the above Z formula is used when the population standard error or the population standard deviation with the number of observations, n, is known[92].

C. The Comparison of Sampling Distributions of an Individual Value and their Mean

Because statistics is largely a study of samples, everything we do involves samples. In this process, even though an individual sample observation is important and interesting, the sample mean is often the more important and interesting. For example, in order to evaluate the salary level of economics graduates against finance graduates, it is important to know the salary of a single economics and a single finance graduates. However, often, the average salaries based on a large number of economics and finance graduates provide more important and interesting information to a decision maker. That is, the behavior of a sample mean is often more interesting and important.

In a real business world, however, a population mean is either known or unknown. If a population mean is known, we are interested in knowing how likely is the chance that a sample mean is from the population. If it is not

---

[92] In business statistics, σ_X or $\sigma_{\overline{X}}$ is often unknown whereas in engineering, they can be known. For example, when an engineer designs a chocolate chip cookie machine, she has to know the tolerance level of how many chocolate chips are acceptable per each cookie being produced. It might be an average number of 5 chocolate chips with a variability of ± 1 chocolate chip per cookie. If this variability is measured by one standard deviation, then $\sigma = 1$. However, in business data such as sales, profits, etc., the population standard deviation or standard error is often unknown. Thus, it needs to be estimated based on a sample.

known, we are interested in knowing how likely is the chance that the sample mean will represent the unknown population mean.

In this chapter, we will deal with the case of a population mean being known to us. The case of an unknown population mean will be discussed in following chapters[93].

Example 1>

Let's assume that you had invested in 5 of the 30 stocks that make up the Dow Jones Industrial Average. If so, we can quickly understand that the population of stocks is made up of 30 stocks (i.e., N=30) and your sample is comprised of 5 stocks that you own (i.e., n=5). Therefore, you will know the population mean by summing all rates of return from 30 stocks and dividing it by 30 and the population standard deviation by using the formula shown above. Suppose that the population mean return is 10% and the standard deviation is 0.6%. Also, you will know the sample mean by summing all rates of return from your 5 stocks and dividing it by 5 and the sample standard deviation by using the formula shown in Chapter 3. Suppose that the sample mean is found to be 10.3% and the sample standard deviation is 0.5%.

1. What is the probability that you could pick **a single stock** in the Dow Jones Industrial Average that would have yielded more than 11%?

Steps to the Answer:

a. Recognize the availability of the population standard deviation → Given both the population and the sample standard deviation in this case, **we choose the population standard deviation because it is superior information to the sample standard deviation.**

b. Because this problem involves **a single stock** and a known population standard deviation, we must use[94]

$$Z = \frac{X - \mu}{\sigma}$$

c. We now formulate the problem as follows and solve:

$$\text{P}(\text{X} > 11) = P(\frac{X - \mu}{\sigma} > \frac{11 - \mu}{\sigma}) = P(Z > \frac{11 - 10}{0.6})$$

$$= \text{P}(\text{Z} > 1.67) = 0.0475 \rightarrow 4.75\%$$

[93] Obviously, if it is unknown, its estimate must be obtained on the basis of a sample.
[94] Recall the procedure we learned in Chapter 7.

That is, there is only 4.75% probability that you could have picked **a stock** that would have yielded a 11% rate of return from the 30 Dow Jones Industrial Average stocks that had yielded 10% as a group.

2. What is the probability that you could pick **a single stock** in the Dow Jones Industrial Average that would have yielded between 9% and 11%?

Answer:

Based on the above procedure, we now formulate the problem as follows and solve:

$$P(9 < X < 11) = P(\frac{9-10}{0.6} < Z < \frac{11-10}{0.6}) = P(-1.67 < Z < 1.67)$$

$$= 1 - 2 * (0.0475) = 0.905 \rightarrow 90.5\%$$

That is, there is 90.5% probability that you could have picked a stock that would have yielded between 9% and 11%.

3. What is the probability that you could pick 5 stocks in the Dow Jones Industrial Average that would have yielded **an average return** more than 11%?

Steps to the Answer:

a. Recognize that the population standard deviation of 0.6 is given.

b. Because this problem involves an **average** return more than 11% and a known population standard deviation, we must now use the population standard error as follows:

$$Z = \frac{\overline{X} - \mu}{\sigma_{\overline{X}}} \quad \text{where } \sigma_{\overline{X}} = \frac{\sigma}{\sqrt{n}} = \frac{0.6}{\sqrt{5}} = \frac{0.6}{2.236} = 0.2683$$

c. We now formulate the problem as follows and solve:

$$P(\overline{X} > 11) = P(\frac{\overline{X} - \mu}{\sigma_{\overline{X}}} > \frac{11-10}{0.2683}) = P(Z > 3.73) = 0.0001 \rightarrow 0.01\%$$

That is, there is only 0.01% probability that you could have picked a group of 5 stocks that would have yielded an average of 11% or higher rate of return from the 30 Dow Jones Industrial Average stocks that yielded 10%.

4. What is the probability that you could pick 5 stocks in the Dow Jones Industrial Average that would have yielded **an average return** between 9.8% and 10.5%?

Answer:

Based on the above procedure, we now formulate the problem as follows and solve:

$$P(9.8 < \overline{X} < 10.5) = P(\frac{9.8-10}{\frac{0.6}{\sqrt{5}}} < Z < \frac{10.5-10}{\frac{0.6}{\sqrt{5}}}) = P(-0.75 < Z < 1.86)$$

$$= 1 - (0.2266+0.0314) = 1 - 0.2580 = 0.7420 \rightarrow 74.2\%$$

That is, there is 74.2% probability that you could have picked a group of 5 stocks that would have yielded between 9.8% and 10.5%.

Example 2>

Assume that a national intelligence quotient (IQ) research institute published a data that shows a normal distribution[95] with the mean IQ of 110 for all people tested and a standard deviation of 8.

1. John Doe thinks that **he** is smart, having an IQ of 124. What is the probability that John's claim is correct?

   Answer:

   $$P(X > 124) = P(Z > \frac{124-110}{8}) = P(Z > 1.75) = 0.0401 \rightarrow 4\%$$

   Given that only 4% of the people have a higher IQ than he does (or alternatively, 96% of the people have a lower IQ than he does), he can be considered smart.

2. John Doe has signed up for GSB 420 and his classmates all claim that they are smart, too. To verify this claim, Dr. Choi, the GSB 420 instructor, surveyed all 16 students in the class and found that their **average** IQ is 120 with a standard deviation of 10. Can Dr. Choi conclude that he has a "smart" class?

   Answer:

---

[95] In reality, IQ's are not believed to be normally distributed. This is just an example.

$$P(\overline{X} > 120) = P(Z > \frac{120-110}{\frac{8}{\sqrt{16}}}) = P(Z > 5.00) = 0.00000087 \rightarrow 0\%$$

Note 1: Despite the sample standard deviation of 10, the population standard deviation of 8 should be used because the population parameter is always superior to the sample statistic[96] → Thus, use the Z-, not the t-formula.

Note 2: Even though the sample mean of 120 is lower than John's 124, we can definitely conclude that the class is smart because the probability of a sample mean – based on a sample size of 16 – being larger than 120 is almost 0.

Note 3: The fact that 110 is the population mean of a normal distribution means that it is also its median[97]. Thus, the probability of finding a group of 16 people whose average IQ is far above the median value of 110 is very unlikely → Thus, the class as a whole is smart.

D. Examples of When and How to Use the Standard Error

1. Given a population mean of 100 and a population standard deviation of 10, you took a sample of 25 observations.

   a. What is the probability of picking a **sample mean** that is less than 99?

$$P(\overline{X} < 99) = P(Z < \frac{99-100}{\frac{10}{\sqrt{25}}}) = P(Z < -0.5) = 0.3085$$

Note: Instead of using the Z-table in Chapter 7, one can use the Excel Spreadsheet command of "=normdist(X,μ,σ,True)" where X=99; μ=100; σ=2; and True; to calculate the cumulative probability of Z[98] as follows:

"=NORMDIST(99,100,2,TRUE)" → 0.308538

[96] For this reason, often the validity or consistency of a mechanical process is examined by the Z-formula. That is, when a machine is designed to drill a hole, for example, its average diameter and variability have been specified and known. Thus, it is good to use the Z-formula to examine the sample results against this pre-specified mean and standard deviation.

[97] This is often not the case for IQ test scores. It tends to show a negative or left-skewness.

[98] The Excel Spreadsheet command always gives the cumulative probability of Z for (X < k)

b. What is the probability of picking a **sample mean** that is greater than 100.5?

$P(\overline{X}>100.5) = P(Z>0.25) = 0.4013$ ← the Z table solution
$= 1 - 0.5987$ ← the Excel solution

"=NORMDIST(100.5,100,2,TRUE)" → 0.598706

c. What is the probability of picking a **sample mean** that is between 97 and 104?

$P(97<\overline{X}<104) = P(-1.5<Z<2)$
$= 1 - (0.0668+0.0228) = 0.9104$ ← the Z table solution
$= 0.9772 - 0.0668 = 0.9104$ ← the Excel solution

0.066807 and 0.97725

2. Given a population mean of 20 and a population standard error of the mean of 3, you took a sample of 100 observations.

a. What is the probability that **the sample mean** will be less than 18?

$$P(\overline{X}<18) = P(Z<\frac{18-20}{3}) = P(Z<-0.67) = 0.2514$$

b. What is the probability that **the sample mean** will be greater than 17?

$$P(\overline{X}>17) = P(Z>-1) = 1 - 0.1587 = 0.8413$$

c. What is the probability that **the sample mean** will be between 21 and 24?

$P(21<\overline{X}<24) = P(0.33<Z<1.33)$
$= 0.3707 - 0.0918 = 0.2789$ ← the Z table solution
$= 0.9082 - 0.6293 = 0.2789$ ← the Excel solution

E. Sampling Distribution of the Proportion

When evaluating a probability associated with a proportion, always the Z-distribution is used with the following formulas for the mean and the standard error of a proportion.

1. The Sample Mean Proportion, p

p = X/n = (Number of items of interest)/(sample size)

where $0 \le p \le 1$

2. The Standard Error of the Proportion[99]

$$\sigma_p = \sqrt{\frac{\pi(1-\pi)}{n}}$$

where π is the population proportion and n is the number of observations.

3. The Standardized Normal Value for the Proportion, Z[100]

$$Z = \frac{p-\pi}{\sigma_p}$$

4. Example 1

Suppose that 10% of the 1000 taxpayers were audited last year by the IRS.

a. What is the probability that less than 11% will be audited this year, assuming all other things to be unchanged?

$$P(p<0.11) = P(Z < \frac{0.11-0.1}{\sqrt{\frac{0.1(1-0.1)}{1000}}}) = P(Z < 1.05) = 0.8531$$

b. What is the probability that less than 8% will be audited this year?

$$P(p<0.08) = P(Z < \frac{0.08-0.1}{\sqrt{\frac{0.1(1-0.1)}{1000}}}) = P(Z < -2.11) = 0.0174$$

c. What is the probability that between 8% and 11.5% will be audited this year?

$$P(0.08<p<0.115) = P(-2.11<Z<1.58) = 1 - (0.0174 + 0.0571)$$
$$= 0.9429 - 0.0174 = 0.9255$$

---

[99] When the population proportion, π, is unknown, the sample proportion, p, must be used in place of π to calculate the standard error.

[100] When working with a proportion, only the Z-formula and the Z-probability are used.

5. Example 2

A survey showed that 80% of DePaul MBA students love Dr. Choi. If you ask randomly any 10 students, what is the probability that more than 5 students will love Dr. Choi?

Answer:

Identify $\pi = 0.8$; $p = \frac{5}{10} = 0.5$; and n=10.

Therefore,

$$P(p > 0.5) = P(Z > \frac{0.5 - 0.8}{\sqrt{\frac{0.8(1-0.8)}{10}}}) = P(Z > -2.37) = 1 - 0.0089 = 0.9911$$

There is a 99.11% chance that you will find 5 or more students loving Dr. Choi out of 10 that you ask.

6. Example 3

A survey showed that 65% of households in the U.S.A. own a pet animal. If you are to sell door to door a pet-related product to 200 households,

a. what is the probability that more than 70% of the households that you will visit will have a pet animal?

Answer:

Identify $\pi = 0.65$; $p = 0.7$; and n = 200.

Therefore,

$$P(p > 0.7) = P(Z > \frac{0.7 - 0.65}{\sqrt{\frac{0.65(1-0.65)}{200}}}) = P(Z > 1.48) = 1 - 0.9306 = 0.0694$$

There is a 6.94% chance that you will find 70% of the 200 households – that is, 140 households – that will have a pet animal.

b. what is the probability that between 62% and 70% of the households will have a pet animal?

Answer:

Identify $\pi = 0.65$; $p = 0.62$; $p = 0.7$; and n = 200, we find:

$$P(0.62 < p < 0.7) = P(\frac{0.62 - 0.65}{\sqrt{\frac{0.65(1-0.65)}{200}}} < Z < \frac{0.7 - 0.65}{\sqrt{\frac{0.65(1-0.65)}{200}}}) = P(-0.89 < Z < 1.48)$$

$$= 1 - (0.0694 + 0.1867) = 0.9306 - 0.1867 = 0.7439$$

There is a 74.39% chance that you will find between 62% and 70% of the 200 households – that is, between 124 and 140 households – that will have a pet animal.

F. The sample standard deviation vs. the sample standard error

If the population standard deviation is unknown, it has to be estimated on the basis of a sample data by using the following sample standard deviation formula:

a. The sample standard deviation of X → S = S_X.

$$S = S_X = \sqrt{\frac{\sum_{i=1}^{n}(X_i - \overline{X})^2}{n-1}}$$

Using the same logic that explained the relationship between σ and $\sigma_{\overline{X}}$, we can obtain the relationship between the sample standard deviation, S, and the sample standard error, $S_{\overline{X}}$, from the sample variances as follows:

$$S_{\overline{X}}^2 = \frac{S^2}{n} \quad \rightarrow \quad S_{\overline{X}} = \frac{S}{\sqrt{n}}$$

b. The sample standard error of the mean, $S_{\overline{X}}$

$$S_{\overline{X}} = \frac{S}{\sqrt{n}} = \sqrt{\frac{\sum_{i=1}^{n}(X_i - \overline{X})^2}{n \cdot (n-1)}}$$

In this situation where the sample standard error has to be used in place of a population standard error, we need to use a new t-formula that will standardize (or normalize) the mean of the X variable. It is the same t-formula as studied in Chapter 8 with a small modification on the divisor.

c. The Use of the t-formula[101]

The t-formula is used when the population variance or the population standard deviation is UNKNOWN and thus, must be estimated by the sample variance or the sample standard deviation[102]. The t-formula to be used is:

$$t = \frac{\overline{X} - \mu}{S_{\overline{X}}} = \frac{\overline{X} - \mu}{\frac{S}{\sqrt{n}}}$$

However, the use of the t-formula is the same as that of the Z-formula, except that its distribution is subject to the (n – 1) degrees of freedom. This point is best demonstrated by the following examples:

Example 1>

Given a population mean of 100 and a sample standard deviation of 10, you took a sample of 25 observations.

1. What is the probability that **the sample mean** will be greater than 103.4218?

$$\text{P}(\overline{X} > 103.4218) = P(t > \frac{103.4218 - 100}{\frac{10}{\sqrt{25}}}) = P(t > 1.7109) = 0.05$$

Note: If you are not clear of how the answer was obtained in this example, please revisit the use of the t-table described in Chapter 8.

Note: Instead of using the t-table, one can use the Excel Spreadsheet command of "=tdist(X,df,tails)" to calculate the cumulative probability of t at X and (n – 1) degrees of freedom. In this case,

---

[101] Note that these examples are almost identical to the examples shown in Chapter 8. The only and yet, main difference is that we are now assessing the probability of a mean, not an individual value of X. Because we must use the standard error whenever we are to assess the probability of a mean, all of the examples herein use the standard error, not the standard deviation in the divisor of the t-formula.

[102] This means, then, whenever the population variance or the population standard deviation or the population standard error is known, the Z-formula must be used. Otherwise, use the t-formula.

because X=1.7109; df=24; and tails=1, the corresponding probability is calculated by Excel as follows[103]**:**

"=TDIST(1.7109,24,1)" → 0.04999832

2. less than 98.6304?

$$P(\overline{X} < 98.6304) = P(t < \frac{98.6304 - 100}{\frac{10}{\sqrt{25}}}) = P(t < -0.6848) = 0.25$$

"=TDIST(0.6848,24,1)" → 0.250015383

Note: Because the t-distribution is symmetrical about the mean, X=0.6848 is used instead of –0.6848 for calculation by Excel Spreadsheet. That is, given the Excel Spreadsheet command of "=tdist(X,df,tails)," X=0.6848; df=24; and tails=1 were used.

3. greater than 98.6304?

$$P(\overline{X} > 98.6304) = P(t > \frac{98.6304 - 100}{\frac{10}{\sqrt{25}}}) = P(t > -0.6848) = 1 - 0.25 = 0.75$$

Note: The probability calculated by the Excel Spreadsheet for P(t > 0.6848) is:
"=TDIST(0.6848,24,1)" → 0.250015383

4. greater than 97.3644?

$$P(\overline{X} > 97.3644) = P(t > \frac{97.3644 - 100}{\frac{10}{\sqrt{25}}}) = P(t > -1.3178) = 1 - 0.10 = 0.90$$

Note: The probability calculated by the Excel Spreadsheet for P(t > 1.3178) is:

0.100005924

5. between 97.3644 and 104.9844?

---

[103] The Excel Spreadsheet command always gives the cumulative probability of t for (X > k), which is the opposite of the cumulative probability for Z.

$P(97.3644 < \overline{X} < 104.9844) = P(-1.3178 < t < 2.4922) = 1 - 0.1 - 0.01 = 0.89$

Note 1:

The probability calculated by the Excel Spreadsheet for P(t > 2.4922) is:

0.009999099

Note 2:

We can graphically visualize this problem and the solution as follows:

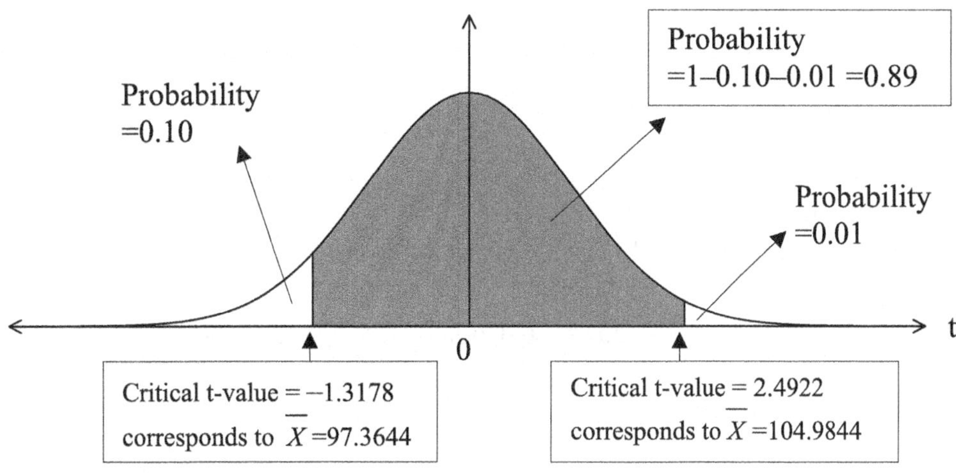

Example 2>

Given a population mean of 20 and a sample **standard error** of the mean of 3, you took a sample of 100 observations.

1. What is the probability that **the sample mean** will be less than 16.1297?

$$P(\overline{X} < 16.1297) = P(t < \frac{16.1294 - 20}{3}) = P(t < -1.2902) = 0.1$$

2. greater than 27.0938?

$$P(\overline{X} > 27.0938) = P(t > 2.3646) = 0.01$$

3. between 17.969 and 24.9812?

$$P(17.969 < \overline{X} < 24.9812) = P(-0.6770 < t < 1.6604) = 1 - 0.25 - 0.05 = 0.7$$

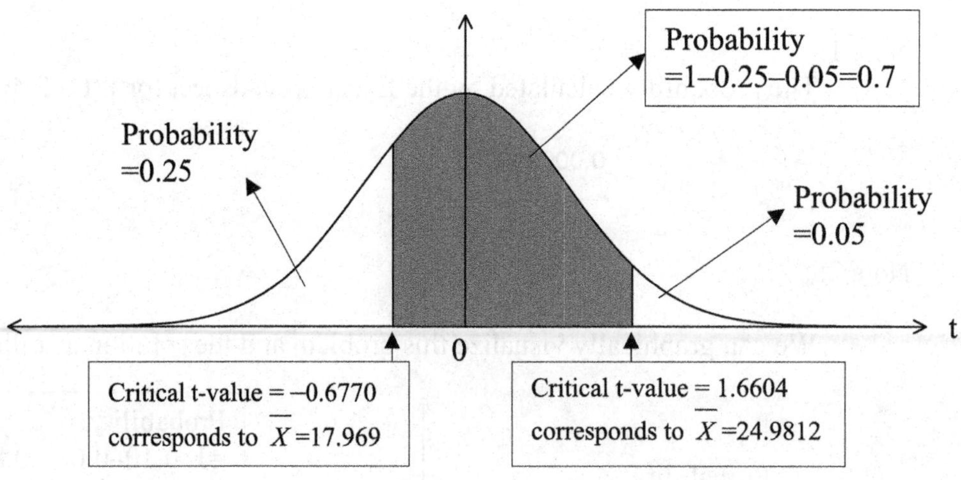

4. between 15 and 23?
   (Hint: when table t-values are not available, use the Excel Spreadsheet.)

   $P(15 < \overline{X} < 23) = P(-1.6667 < t < 1) = 1 - 0.0494 - 0.1599 = 0.7907$

   Note 1:

   $P(t > 1.6667) =$ 0.049367452 and $P(t > 1) =$ 0.159874237

   Note 2:

   We can graphically visualize this problem and the solution as follows:

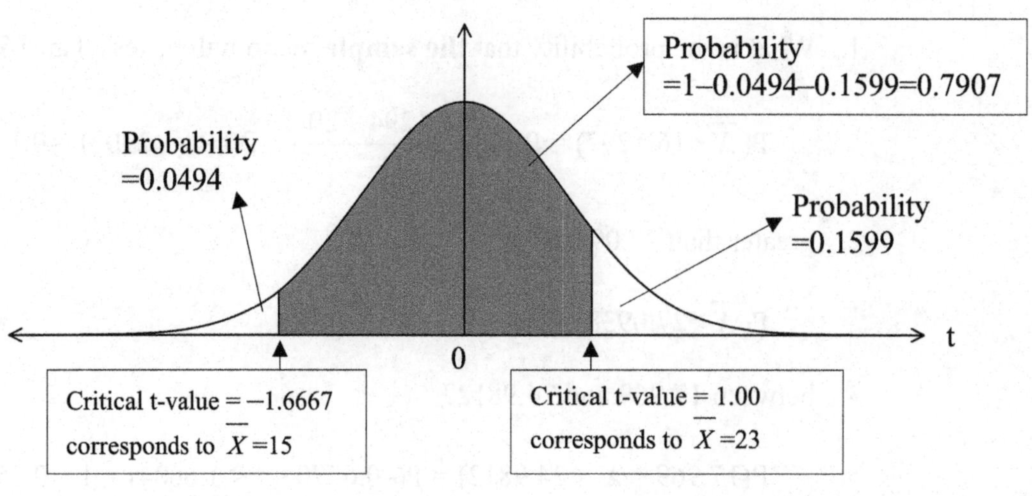

G. The Summary of the Procedures to Identify the Probability

The following steps summarize what we had done so far in assessing a probability associated with a given situation.

Step 1: Identify the nature of the problem by knowing if we are to the probability of an individual X value occurring or a mean of X, $\overline{X}$, occurring?

That is, $P(a \leq X \leq b) = ?$ or $P(a \leq \overline{X} \leq b) = ?$

Or $P(a \leq X) = ?$ or $P(a \leq \overline{X}) = ?$

Step 2: Identify if the problem gives the population standard deviation and asks for a probability of a sample mean occurring or if it deals with a proportion. If so, use the following Z-formula to standardize (or normalize) the $\overline{X}$ value(s):

$$Z = \frac{\overline{X} - \mu_{\overline{X}}}{\sigma_{\overline{X}}} = \frac{\overline{X} - \mu}{\frac{\sigma}{\sqrt{n}}}$$

Step 3: If no population standard deviation is given, use the following t-formula to standardize (or normalize) the $\overline{X}$ value(s):

$$t = \frac{\overline{X} - \mu_{\overline{X}}}{S_{\overline{X}}} = \frac{\overline{X} - \mu}{\frac{S}{\sqrt{n}}}$$

Step 4: Use the Excel Spreadsheet command to identify the probability or look up the probability in the Z- or the t-table.

Summary on the Nature of Problems and Solutions Once More!

Types of Problems concerning an individual X value	Formulas to use		Types of Problems concerning $\overline{X}$	Formulas to use
$P(a < X) = ?$	either $Z=(X-\mu)/\sigma$ or $t=(X-\mu)/S$		$P(a < \overline{X}) = ?$	either $Z=(\overline{X}-\mu)/(\sigma/\sqrt{n})$ or $t=(\overline{X}-\mu)/(S/\sqrt{n})$
$P(X < b) = ?$			$P(\overline{X} < b) = ?$	
$P(a < X < b) = ?$			$P(a < \overline{X} < b) = ?$	

Exercise Problems on
Chapter 9. Sampling Distributions

This set of exercise problems has 13 problems, worth a total of 15 points.

1. The population standard error of the mean, $\sigma_{\overline{X}}$, is defined as _____.

   a. $\dfrac{\sigma_X}{n}$ b. $\dfrac{\sigma_X}{n^2}$ c. $\dfrac{\sigma_X^2}{n}$

   d. $\dfrac{\sigma_X}{\sqrt{n}}$ e. σ_X^2

2. The variance of the sample mean is typically _____ the variance of individual observations in a sample.

   a. larger than b. smaller than c. equal to
   d. less useful than e. only (a) and (d) of the above

3. When calculating the Z value for a sample mean, $\overline{X}$, the denominator used is _____.

   a. the variance of the sample mean
   b. the standard error of the population mean
   c. the variance of the population mean
   d. the standard deviation of the population mean
   e. either (a) or (c) of the above

4. Assume that you took a sample of 9 observations from a population whose mean is 50 and standard deviation is 15. What is the probability that the sample mean will be less than 45?

   a. 0.3707 b. 0.6293 c. 0.1587
   d. 0.8413 e. -1

5. The Central Limit Theorem states that as the sample size _____, the resulting distribution of the sample mean becomes almost _____ regardless of the shape of the population probability distribution.

   a. increases; normal b. increases; abnormal
   c. decreases; normal d. decreases; abnormal
   e. stays the same; normal

6. Even if a population probability distribution is not normal, the standardized normal probability can be used to assess the probability of a mean occurring by invoking ______.

   a. the Chebyshev Rule b. the Bayesian Theorem
   c. the Central Limit Theorem d. the Law of Averages
   e. God

7. The Law of Large Numbers says that as the sample size, n, increases, the difference between the sample mean and the population mean will approach _____.

   a. a positive infinity. b. a negative infinity.
   c. one. d. zero.
   e. a well-defined positive value such as 1 or 2.

8. Given the following sample data,

   20, 24, 28, 30, 22, 26

   the standard deviation is calculated to be _____.

   a. 25 b. 11.6667 c. 3.4155
   d. 14 e. 3.7416

9. Given that a population mean is 20 and a population standard deviation is 5, you collected a sample of 36 observations and found its mean to be 19 and standard deviation to be 4. What is the probability of a single observation having a value greater than 21?

   a. 0.4207 b. 0.4013 c. 0.3446
   d. 0.1151 e. 0.0668

10. Given that a population mean is 20 and a population standard deviation is 5, you collected a sample of 36 observations and found its mean to be 19 and standard deviation to be 4. What is the probability of observing a sample mean smaller than 19?

   a. 0.4207 b. 0.4013 c. 0.3446
   d. 0.1151 e. 0.0668

11. On the basis of 50 people surveyed, a newspaper reported that 30% of them supported the Republican Party. Given this information, what is the probability that the real proportion of Republican Party supporters would be greater than 40%?

a.	0.25	b.	0.5	c.	0.1124
d.	0.0897	e.	0.0618		

The following problems are worth 2 points each.

12. On the basis of 50 people surveyed, a newspaper reported that 30% of them supported the Republican Party. Given this information, what is the probability that the real proportion of Republican Party supporters would be between 25% and 38%?

a.	0.8907	b.	0.6701	c.	0.5
d.	0.2206	e.	0.1001		

13. Given that a population mean is 20 but a population standard deviation is unknown, you collected a sample of 36 observations and found its mean to be 19 and standard deviation to be 4. What is the probability of observing a sample mean between 18.8736 and 21.6251?

a.	0.97	b.	0.94	c.	0.90
d.	0.85	e.	0.55		

Chapter 10. Confidence Interval Construction

Given that the population mean is known, we studied in previous Chapters 7, 8, and 9 how to evaluate a probability of a sampled event occurring, based on a Z-distribution (if the population standard deviation is known) or a t-distribution (if the population standard deviation is unknown). That is, these three chapters answered the basic question, **"Given the population mean, what is the likelihood (=probability) of observing a single value or a sample mean as large or as small as it is, provided that either a population standard error or a sample standard error is available?"**

From this chapter and on, however, we will study the case of an unknown population mean. In fact, the true population mean of a sampled data is often unknown in reality. Therefore, by collecting and analyzing a sample, we often guess the population mean by using the sample mean. Therefore, the sample mean is called the **"point estimate"** of the population mean. However, the probability of the point estimate being exactly the same as the population mean is very small[104]. Therefore, to be safe, we wish to identify a range of values that will capture or contain the true population mean, using the sample mean as a reference point – i.e., a point estimate. This process is known as estimating or constructing a confidence interval for the population mean.

In order to construct a confidence interval, we need the following information: (1) the availability of a population standard deviation; (2) the information about the sample characteristics such as its mean, standard deviation, and number of observations; and (3) the level of confidence that we wish to install in the statement about capturing the population mean within a confidence interval. As before, **if a population standard deviation is known, we use the Z-distribution. If it is unknown, we use the t-distribution in constructing a confidence interval.**

Before actually constructing a confidence interval, let's understand for the last time the big picture behind the reason for constructing a confidence interval by a case scenario as described below.

A Case Scenario:

Suppose that you went on a vacation to a Montana Trout Ranch that advertised in a national newspaper that its average trout size is 20 centimeters (cm). After catching a 10 cm trout, can you sue the ranch owner for a false advertisement? How about if you caught a 19 cm? What can you do to verify the claim that an average trout size is 20 cm?

In order to answer these questions, you must understand that:

(1) An average must be based on more than 1 fish → then how many fish do you need to catch before you can make a judgment? All in the pond or just 10?

[104] Note that this is the very nature of a continuous numerical data. For example, if the population mean is 100.1234, then a point estimate being exactly the same to the last digit is highly unlikely.

(2) You will most likely never catch only 20 cm trouts → If so, what degree of varying sizes will you accept?

(3) The probability of catching a trout that is exactly 20 cm long is almost zero because if you catch a fish that is either 19.999 cm long or 20.001 cm long, you can still claim that it is not 20 cm long and thus, may win an argument.

One way of handling these issues is to construct or identify a range of trout sizes that can be accepted by rational people. Often, if one can say that 95% of the time a trout caught will be between 15 cm and 25 cm (thus, satisfying an average of 20 cm), it may be more accurate and useful, and less confusing about the nature of a trout size. For this reason, the construction (or estimation) of a confidence interval to capture a population mean can yield very valuable information to a decision maker.

Once again, it is worth being reminded that the construction of a confidence interval is critically dependent upon the knowledge about the population standard deviation. **If the population standard deviation is known, a Z-statistic is used. If not, then a t-statistic is used.**

A. Confidence Interval Estimation For the Mean When σ is Known

Because we start with the assumption that the population mean, μ, is unknown, we wish to make an inference about it by collecting and analyzing a sample data. That is, after obtaining $\overline{X}$, on the basis of a sample size of n, we wish to know how closely the sample mean observed can be compared to the population mean that is unknown.

Of course, we can use the estimated sample mean as a point estimate of the population mean. In this case, because a single numerical value of the sample mean is used as an estimate for the population mean, the probability of it matching exactly the population mean is very low. In order to avoid this inherent dilemma, we concede by allowing a range of values – known as a confidence interval – to capture the population mean.

The formula for a confidence interval, when the population standard deviation is known, is derived from the Z formula that we are already familiar with. That is,

Given $Z = \dfrac{\overline{X} - \mu}{\sigma_{\overline{X}}}$, we can solve for μ as follows:

$$\mu = \overline{X} - Z \cdot \sigma_{\overline{X}} = \overline{X} - Z \cdot \frac{\sigma}{\sqrt{n}}$$

Because we are interested in a range of values, not a single value, centered at the point estimate, we must consider the positive and negative sides of the Z distribution, centered at the mean of 0. The revised and final formula for a confidence interval is thus found as:

$$\mu = \overline{X} \pm Z \cdot \sigma_{\overline{X}} = \overline{X} \pm Z \cdot \frac{\sigma}{\sqrt{n}}$$

Recall that in order to construct a confidence interval, we must know the level of confidence we wish to have in our estimation. The typical confidence level chosen is 95%, which means that we will be correct 19 out of 20 estimations, or alternatively, we will be wrong only 1 out of 20 estimations.

The confidence level of 95% means that there is a 5% chance of not capturing the population mean. This 5% chance (or probability) is called a **significance level** and often denoted by a Greek letter, alpha (α). More formally, the relationship between a confidence level (CL) and a significance level, α, is defined as follows:

$$\text{CL} + \alpha = 1$$

→ Therefore, $\text{CL} = 1 - \alpha$ and $\alpha = 1 - \text{CL}$

Alternatively, we can express this in terms of a percentage as:

$$\text{CL\%} + \alpha\% = 100\%$$

→ $\text{CL\%} = 100\% - \alpha\%$ and $\alpha\% = 100\% - \text{CL\%}$

It is very important to note that the significance level of 5% means that we will leave out the top 2.5% of values and the bottom 2.5% of values when constructing a confidence interval. Because we are leaving out a total of 5% extreme values, we are only 95% confident.

Therefore, if you are to find the value of Z corresponding to a 95% confidence level (or a 5% significance level), you must find the Z value that corresponds to the probability of 0.0250 (= 1– 0.9750) in the upper and the lower tails of the Z-probability distribution. As studied in Chapter 7, the corresponding Z values are identified as +1.96 for the top 2.5% and –1.96 for the bottom 2.5%. That is, **Z values of ±1.96 capture 95% of all observations and leave out 5% of them.**[105] **→ Know this!**

For this reason, the above confidence interval formula must be modified as follows to correctly reflect that a confidence interval is bounded by two ends (or tails) of extreme values:

---

[105] For simplicity, the Z-value of 2 (or ±2 standard deviations from the mean) is often used as an approximation in constructing a 95% confidence interval.

$$\mu = \overline{X} \pm Z_{\alpha/2} \cdot \sigma_{\overline{X}} = \overline{X} \pm Z_{\alpha/2} \cdot \frac{\sigma}{\sqrt{n}}$$

where α = a significance level = 1 – Confidence Level. That is, if a significance level, α, is 5%, then the confidence level is 1 – 0.05 = 0.95 = 95%. Of course, if a confidence level is 99%, then the corresponding significance level, α, is 1%.

Also note that the notation of α/2 in $Z_{\alpha/2}$ shows that the confidence interval excludes values in an upper end (or tail) and a lower end (or tail) of the Z-distribution scale[106]. Given a confidence level of 95%, for example, the probability in the upper tail is 5%/2 or 2.5% and that in the lower tail is 5%/2 or 2.5% because the Z distribution is symmetrical about the mean.

Example 1>

If a sample of size 100 yielded a mean of 25, given its population standard deviation of 5, determine the 95% confidence interval for the population mean.

$$\mu = \overline{X} \pm Z_{\alpha/2} \cdot \frac{\sigma}{\sqrt{n}} = 25 \pm 1.96 \cdot \frac{5}{\sqrt{100}} = 25 \pm 0.98 = [24.02, 25.98]$$

Example 2>

If a sample of size 30 yielded a mean of 20, given its population standard deviation of 3, determine the 99% confidence interval for the population mean.

$$\mu = \overline{X} \pm Z_{\alpha/2} \cdot \frac{\sigma}{\sqrt{n}} = 20 \pm 2.575 \cdot \frac{3}{\sqrt{30}} = 20 \pm 1.41 = [18.59, 21.41]$$

Please note here that the closest Z-value corresponding to a probability of 1%/2 = 0.5% = 0.0050 in the upper and lower tails is between 2.57 and 2.58. Thus, we find the mid-value of 2.575 as the Z-value[107].

B. The Meaning of a (100 – α)% Confidence Interval

What does it mean to say that you have constructed a 95% confidence interval? It means that you are 95% confident that the population mean will be captured by the confidence interval that you have just constructed. This also means that this claim of yours will be right 95% of the time and wrong 5% of the time. Therefore, the

[106] An upper end (or tail) corresponds to overestimation and a lower end (or tail) corresponds to underestimation of the population mean.

[107] In doing homework or exam problems, you can use either 2.57, 2.58, or 2.575 to choose the closest value among answer choices.

significance level of 5% (= α) means that there is a 5% chance that you will be wrong but a 95% chance of being right.

In a layman's term, the probability of being correct 95% of the time means being correct 19 out of 20 claims because 19/20 = 95%. Alternatively, then, you will be wrong 5% (=1/20) of the time. By convention, a 95% confidence level is most often used along with a 99% confidence level. However, note that a 99% confidence interval means that you will be right 99 out of 100 times, which demands a high degree of accuracy. **To meet this high demand of accuracy, the size of the 99% confidence interval must be larger than that of the 95% confidence interval.** In a similar vein, note that a 100% confidence interval, if constructed, will have no usefulness for it will most likely to cover too large an interval, stretching from a negative infinity to a positive infinity. This is because only this extreme interval size will cover all possibilities with a certainty of 100%.

C. Confidence Interval Estimation For the Mean When σ is Unknown

When a population standard deviation, σ, or a population standard error, $\sigma_{\overline{X}}$, is not known, it must be estimated via the sample standard error formula as follows:

$$S_{\overline{X}} = \frac{S}{\sqrt{n}} = \sqrt{\frac{\sum (X_i - \overline{X})^2}{n \cdot (n-1)}}$$

Now, the confidence interval formula undergoes the following transformation:

1. $S_{\overline{X}}$ is used in place of $\sigma_{\overline{X}}$
2. $t_{\alpha/2,(n-1)}$ is used in place of $Z_{\alpha/2}$

That is, the 95% confidence interval formula, when σ and $\sigma_{\overline{X}}$ are unknown, is:

$$\mu = \overline{X} \pm t_{\alpha/2,(n-1)} \cdot S_{\overline{X}} = \overline{X} \pm t_{\alpha/2,(n-1)} \cdot \frac{S}{\sqrt{n}}$$

where α = significance level = 1 – Confidence Level; (n–1) = degrees of freedom; and n = sample size.

The value of $t_{\alpha/2,(n-1)}$ is obtained from the t-table shown in Chapter 8 or from the Excel Spreadsheet command of "=TINV(α, df)".

Note: **The Excel Spreadsheet command of "=TINV(α, df)" can be used to find a critical or table value of t for a two-tail test given a significance level, α, and (n – 1) degrees of freedom**[108]**.**

D. Confidence Interval Estimation For the Proportion, p.

When constructing a confidence interval for a proportion, the following transformation needs to be made:

1. The sample proportion, p, is used in place of $\overline{X}$.

$$p = \frac{X}{n}$$

2. The population standard error of proportion, σ_p, is used in place of $\sigma_{\overline{X}}$.

$$\sigma_p = \sqrt{\frac{p(1-p)}{n}}$$

3. The 95% confidence interval for the population proportion, π, is:

$$\pi = p \pm Z_{\alpha/2} \cdot \sigma_p = p \pm Z_{\alpha/2} \cdot \sqrt{\frac{p(1-p)}{n}}$$

Note: Because the population standard error, σ_p, is assumed to be known by the equation shown above, **always Z-value is used in constructing the confidence interval for the proportion.**

Example 1>

Assume that 600 out of 1,000 people attending the White Sox game are male.

a. what is the best expected proportion of today's sox game attendees that would be male?

$$p = \frac{X}{n} = \frac{600}{1000} = 0.6$$

[108] This means that given α, the Excel calculates the t-value for a upper and a lower tails corresponding to α/2.

b. Construct a 95% confidence interval for the population proportion of male attendees.

Answer:

Because the population proportion is unknown, its estimate can be based on the sample proportion of 0.6 and thus, the confidence interval can be constructed as:

$$\pi = p \pm Z_{\alpha/2} \cdot \sqrt{\frac{p(1-p)}{n}} = 0.6 \pm 1.96 \cdot \sqrt{\frac{0.6(1-0.6)}{1000}}$$
$$= 0.6 \pm 0.0304 = [0.5696,\ 0.6304]$$

E. Sample Size Determination

Often before a survey is conducted, the issue of how many sample observations should be collected can be a major issue. Among many methods, the simple one is presented herein.

Assuming that the population standard error is known, the (100 – α)% confidence interval for μ is given as follows:

$$\mu = \overline{X} \pm Z_{\alpha/2} \cdot \sigma_{\overline{X}} = \overline{X} \pm Z_{\alpha/2} \cdot \frac{\sigma}{\sqrt{n}}$$

We note that the last term in the above equation, $\pm Z_{\alpha/2} \cdot \frac{\sigma}{\sqrt{n}}$, is the amount of an estimation error, or the margin of error, around the sample mean. **Because this is related to the sample size, n, it is also known as a sampling error.** If we state that this sampling error as E, we note:

$$\text{Sampling Error} = \text{Margin of Error} = \text{E} = \pm Z_{\alpha/2} \cdot \frac{\sigma}{\sqrt{n}}$$

Solving for n in the above equation yields the sample size, n:

$$n = \left(\frac{Z_{\alpha/2} \cdot \sigma}{E}\right)^2 = \frac{Z_{\alpha/2}^2 \cdot \sigma^2}{E^2}$$

Note 1: **It is not true that the larger the population size, the larger the sample size should be.** As shown in the above equation, the sample size, n, is NOT determined by the population size, N. The sample size is determined by the level of confidence chosen ($Z_{\alpha/2}$), the amount of tolerable error (E), and the population standard deviation (σ).

Note 2: The population standard deviation is often unknown in determining a sample size. Therefore, it is often estimated by taking a small sample or based on a prior experience and knowledge.

Example 1>

Suppose that you are to conduct a survey to collect primary data on the average speed of cars on a highway. From a previous research, the population variance is known to be 36. You wish to have a 95% confidence about your statement by allowing only ±2 m.p.h. of a sampling error. How many cars should you sample?

Answer: Given α=5%, $\sigma = \sqrt{36} = 6$, and E=±2→2,

$$n = \frac{(1.96)^2(6)^2}{(\pm 2)^2} = 34.57 \approx 35$$

Note that the sample size should always be rounded up to the next largest integer in order to ensure the adequate sample size. Otherwise, it will be less than adequate.

Example 2>

Given the situation as above, if you wish to make an inference about the population with a 99% confidence, what should be the minimum sample size?

Answer: Given α=1%, σ=6, and E=±2→2,

$$n = \frac{(2.575)^2(6)^2}{(2)^2} = 59.67 \approx 60$$

Example 3>

Suppose that your competitor had provided the court with the information that 10% of your products are defective, based on their sample analysis. Thus, they claimed that your products are not safe for normal consumption. In order to refute this claim, you decided to conduct your own survey that will give you 95% confidence to be correct with an allowable error of ±3%.

Answer:

Because only a sample proportion of 10% is available, we can assume that the population proportion, π, is equal to a sample proportion of p = 10% = 0.1.

That is, given α =1 – 0.95 = 0.05 = 5%, p = 0.10, and E=±3%→0.03, we note that for the case or proportion, the sample size is determined by:

$$E = \pm Z_{\alpha/2} \cdot \sqrt{\frac{p(1-p)}{n}} \rightarrow E^2 = (Z_{\alpha/2})^2 \cdot \frac{p(1-p)}{n}$$

Therefore, solving for n yields

$$n = (Z_{\alpha/2})^2 \cdot \frac{p(1-p)}{E^2} = \left(\frac{Z_{\alpha/2}}{E}\right)^2 \cdot p \cdot (1-p) \rightarrow$$

$$n = (Z_{0.05/2})^2 \cdot \frac{0.1(1-0.1)}{0.03^2} = 1.96^2 \cdot 100 = 384.16 \approx 385$$

Example 4>

Suppose that your competitor had provided the court with the information that 10% of your products are defective. Thus, they claimed that your products are not safe for normal consumption in light of an industry standard of 7%. In order to refute this claim, you decided to conduct your own survey that will give you 95% confidence to be correct with an allowable error of ±3%. Determine the appropriate sample size that you need to use.

Answer:

When both a population proportion and a sample proportion are available, it is always better to use the population proportion because by definition, a population parameter reflects the true state of the nature unless there is a doubt about the validity of the population proportion.

Therefore, given α =100% – 95% = 5%, π = 0.07, and E=±3%→0.03, we note that the sample size is determined by:

$$E = \pm Z_{\alpha/2} \cdot \sqrt{\frac{\pi(1-\pi)}{n}} \rightarrow E^2 = (Z_{\alpha/2})^2 \cdot \frac{\pi(1-\pi)}{n} \rightarrow n = (Z_{\alpha/2})^2 \cdot \frac{\pi(1-\pi)}{E^2}$$

Therefore,

$$n = (Z_{\alpha/2})^2 \cdot \frac{\pi(1-\pi)}{E^2} \rightarrow$$

$$n = (Z_{0.05/2})^2 \cdot \frac{0.07(1-0.07)}{0.03^2} = 1.96^2 \cdot 72.3333 = 277.87 \approx 278$$

Example 5>

Suppose that you wish to know the proportion (or percentage) of your employees who are smokers with a margin of error of $\pm$ 10%. What sample size would you use if you wish to have a 95% confidence in your survey result?

Answer:

When there is no known proportion – neither population nor sample – to work with, the sample size is determined by assuming the proportion being equal to 0.5. When the proportion is equal to 0.5, the sample size is the largest[109] and thus, provides a robust result.

Therefore,

$$n = (Z_{\alpha/2})^2 \cdot \frac{\pi(1-\pi)}{E^2} \rightarrow$$

$$n = (Z_{0.05/2})^2 \cdot \frac{0.5(1-0.5)}{0.1^2} = 1.96^2 \cdot 25 = 96.04 \approx 97$$

Example 6>

Suppose that you wish to know the proportion (or percentage) of your employees who will respond to e-mails within an hour with a margin of error of $\pm$ 4%. What sample size would you use if you wish to have a 95% confidence in your survey result?

Answer:

Once again, because there is no known proportion to work with, the sample size is once again determined by assuming the proportion being equal to 0.5.

Therefore, $$n = (Z_{\alpha/2})^2 \cdot \frac{\pi(1-\pi)}{E^2} \rightarrow$$

$$n = (Z_{0.05/2})^2 \cdot \frac{0.5(1-0.5)}{0.04^2} = 1.96^2 \cdot 156.25 = 600.25 \approx 601$$

---

[109] Possibly larger than necessary. However, this is preferred because more sample data is better than less. Furthermore, it can be easily proven that when $\pi = 0.5$, the function $n = (Z_{\alpha/2})^2 \cdot \frac{\pi(1-\pi)}{E^2}$ is maximized.

Example 7>

Suppose that you wish to know the proportion (or percentage) of eligible voters in the U.S. who will vote for a Democratic Party in the coming election with a margin of error of $\pm$ 3%, what sample size would you use if you wish to have a 90% confidence in your survey result?

Answer:

Once again, because there is no known proportion to work with, the sample size is once again determined by assuming the proportion being equal to 0.5. However, we can keep the percentages as they are in our calculation as shown below:

That is, convert the proportion of 0.5 to 50% and E = 3%, not 0.03. Also, note that $Z_{0.1/2} = Z_{0.05} = 1.645$.

Therefore, $n = (Z_{\alpha/2})^2 \cdot \frac{\pi(1-\pi)}{E^2}$ →

$$n = (Z_{0.1/2})^2 \cdot \frac{50(100-50)}{3^2} = 1.645^2 \cdot \frac{50 \cdot 50}{9} = 751.67 \approx 752$$

Exercise Problem on
Chapter 10. Confidence Interval Estimation

This set of exercise problems has 17 problems, worth a total of 20 points.

1. A confidence interval is constructed _____.

   a. to show the analyst's confidence in his predicting ability.
   b. to capture the population mean with a certainty.
   c. to capture the population mean with a certain probability.
   d. to capture a sample mean with a certain probability.
   e. to discredit non-confident opponents.

2. A confidence interval utilizes the _____ of a sample as a _____ estimate of the population mean.

   a. standard deviation; point
   b. mean; point
   c. standard deviation; interval
   d. mean; interval
   e. average; confidence

3. A confidence interval constructed with a confidence level of 95% means that _____ of possible mean values will not be included in the estimated interval.

   a. top 2.5% and bottom 2.5%
   b. only top 5%
   c. only bottom 5%
   d. top 5% and bottom 5%
   e. 95%

4. When constructing a confidence interval, if the population standard deviation is known, one would use ____.

   a. a Z value
   b. a t value
   c. an F value
   d. a chi-square value
   e. a 95% confidence level

5. When constructing a confidence interval, if the population standard deviation is NOT known, one would use ____.

   a. a Z value
   b. a t value
   c. an F value
   d. a chi-square value
   e. a 5% significance level

6. The _____ a confidence level and a significance level should be _____.

   a. sum of; zero
   b. difference between; zero
   c. sum of; one
   d. difference between; one
   e. product of; one

7. Approximately 68% of observations in a normal distribution fall within an interval between ____ above and below the mean.

a. one (1) standard deviation
b. two (2) standard deviations
c. three (3) standard deviations
d. one (1) inter-quartile
e. two (2) inter-quartiles

8. As a new Human Resources (HR) director, you found from a random sample of 16 people who left your firm that they had an average stay of 5 years at your firm with a population standard deviation of 2 years. If you are to construct a 95% confidence interval for an average job length for your employees, it is best for you to use _____ and _____ years.

a. a Z value; a standard deviation of 2
b. a t value; a standard deviation of 2
c. a Z value; a standard error of 0.125
d. a t value; a standard error of 0.125
e. a Z value; a standard error of 0.5

9. As a new Human Resources (HR) director, you found from a random sample of 25 people who left your firm that they had an average stay of 5 years at your firm with a sample standard deviation of 3.3 years. If you are to construct a 95% confidence interval for an average job length for your employees, it is best for you to use _____ and _____ years.

a. a Z value; a standard deviation of 3.3
b. a t value; a standard deviation of 3.3
c. a Z value; a standard error of 0.66
d. a t value; a standard error of 0.66
e. a Z value; a standard error of 0.132

10. What is the Z value that needs to be used for construction of a 95% confidence interval?

a. 0.675
b. 0.95
c. 1.96
d. 0.1711
e. 0.05

11. What is the t value that needs to be used for construction of a 95% confidence interval if the number of observations, n, is 21?

a. 0.6870
b. 0.6864
c. 1.7207
d. 2.0796
e. 2.0860

12. In constructing a confidence interval for a population proportion, you can use _____.

a. a Z value only
b. a t value only
c. either a Z value or a t value
d. an F value
e. a chi-square value

13. In conducting a survey, you wish to tolerate only a margin of error of $\pm$ 3% at a confidence level of 95%. Assuming that you know the population standard deviation of your survey is 10%, calculate the minimum sample size, n, that will satisfy your need.

a. 27
b. 36
c. 43
d. 128
e. over 200

14. In conducting a survey, you wish to tolerate only a margin of error of $\pm$ 4% at a confidence level of 99%. Assuming that you know the population standard deviation of your survey is 8%, calculate the minimum sample size, n, that will satisfy your need.

a. 27
b. 36
c. 43
d. 128
e. over 200

The following is worth 2 points each.

15. As a new Human Resources (HR) director, you found from a random sample of 16 people who left your firm that they had an average stay of 5 years at your firm with a population standard deviation of 5 years. Because your boss wishes to know with a 95% confidence that your employees are staying the industry average, you construct a 95% confidence interval for an average job length for your employees as _____.

a. between 2 and 8 years
b. between 2.55 and 7.45 years
c. between 3.1 and 7.3 years
d. between 4.3875 and 5.6125 years
e. between 0 and 10 years

16. As a new Human Resources (HR) director, you found from a random sample of 25 people who left your firm that they had an average stay of 5 years at your firm with a sample standard deviation of 3.3 years. Because your boss wishes to know with a 95% confidence that your employees are staying the industry average, you construct a 95% confidence interval for an average job length for your employees as _____.

a. between 1.7 and 8.3 years
b. between 2.55 and 7.45 years
c. between -1.6 and 11.6 years
d. between 3.64 and 6.36 years
e. between 0 and 8.3 years

17. As a new Human Resources (HR) director, you found from a random sample of 36 people who left your firm that they had an average stay of 4 years at your firm with a sample standard deviation of 3 years. Because your boss wishes to know with a 99% confidence that your employees are staying the industry average, you construct a 99% confidence interval for an average job length for your employees as _____.

a. between 1 and 7 years
b. between 2.55 and 7.45 years
c. between 3.5 and 4.5 years
d. between 2.64 and 5.36 years
e. between 3.77 and 4.23 years

Chapter 11. One-Sample Hypothesis Testing

In this chapter, we are yet asking another question about the unknown population mean. That is, while the previous Chapter 10 showed how a confidence interval can be constructed to capture an unknown population mean, this Chapter 11 takes it one step further by showing us how to accept or reject a hypothesized value of an unknown population mean. We will first hypothesize (or postulate or claim) a specific value for an unknown population mean and then, using a sample data, will test (or verify) its probabilistic (or statistical) occurrence by using statistical testing methods. A set of testing procedures will be described herein.

A. What are we trying to do?

Once again, because we do not know the true nature of a population of our interest, we wish to make an inference about it based on a sample data set. This process is known as "inductive reasoning" because the conclusion is induced from a sample analyzed. Also, because a conclusion or an implication is inferred on the basis of the data, we call this "inferential" statistics.

The inductive reasoning allows us to verify with the sample data what we hypothesize or believe is true or not.[110] That is, it provides a statistical, probability-based method with which we can claim with some confidence a hypothesis - what we believe - is correct or not. Thus, it is important for us to have an idea or a hunch about the nature of the population before we collect and analyze a data set. This raw idea or hunch is called the original hypothesis or the null hypothesis – denoted by either H_o or H_0. Now, based on a sample data, we will draw a conclusion whether our hunch as stated in the null hypothesis can be supported or not. If the data does not support our hunch – that is, we are to reject the null hypothesis – then, by default, we must accept the alternative case of the null hypothesis not being correct. This is known as "accepting the alternative hypothesis" – denoted by either H_a or H_1. Note that rejecting (accepting) the null hypothesis means accepting (rejecting) the alternative hypothesis because these two hypotheses are mutually exclusive. Furthermore, both the null and the alternative hypotheses are "hypotheses" because no one knows with a 100% certainty what the real truth is.

B. Hypothesis Construction Methods

How do we construct a null hypothesis and an alternative hypothesis? Once again, the null hypothesis – H_o or H_0 – is the original hunch or idea that we have. The alternative hypothesis – H_a or H_1 – is just the opposite case of the null hypothesis. Therefore, **if we can construct either one of the two hypotheses, the other remaining hypothesis can be easily identified because they represent mutually exclusive ideas.**

---

[110] This belief or hunch is formally called a "hypothesis" in statistics.

There are the following two approaches to constructing hypotheses:

1. The Theoretical Approach

   There is no rule about which one – the null or the alternative hypothesis – should be constructed first. However, the following conventional procedure is often recommended:

   a. An alternative hypothesis states what we wish to know. It can contain only the following three types of mathematical relations: less than (<), greater than (>), or not equal (≠).

   b. After an alternative hypothesis is established and stated, we can now infer the null hypothesis as an opposite case to the alternative hypothesis.

   c. Note that we can consider the null hypothesis to represent a status quo or an old, established idea or hunch whereas the alternative hypothesis to represent a new or changed idea. Therefore, the purpose of a hypothesis testing is to disprove (or reject) H_0, the status quo or the old idea, in order to prove (or accept) H_1, the new idea. Note that if H_0 is disproved or rejected, then it only means that H_1 is possible. Likewise, if H_0 is proved or accepted, H_1 is not possible within a confidence level. That is, a hypothesis testing will never prove with a 100% certainty if the null or the alternative hypothesis is correct. **It only means that since one is not likely, the other is more likely**.

   d. **A null hypothesis, H_0, always contains the equality sign**[111]. Therefore, it must be of the following form: ≤, ≥, or =.

2. The Practical Approach

   When constructing hypotheses to be tested, **just know that a null hypothesis contains an equality sign whereas an alternative hypothesis does not.**

Examples> State the null and the alternative hypotheses for the following cases.

   a. You wish to know if the average age of DePaul MBA students is still 29.

   Answer:

   Identify which is an old idea and which is a new idea → "Still 29" or "the same as 29" is an old idea and "not 29 or different from 29" is a new idea. → **Alternatively, we concentrate on the fact that the equality sign should belong in the null hypothesis.** That is,

[111] For a two-tail test, the null hypothesis must contain the equality sign. However, for one-tail tests, many variations are possible but for simplicity, we will follow the convention that the null hypothesis should contain the equality sign.

$H_0: \mu = 29$ vs. $H_1: \mu \neq 29$

b. You wish to know if the average age of DePaul MBA students is greater than 29.

Answer:

Identify which is an old idea and which is a new idea → " ... the average age ... is (now) greater than 29" is a new idea; Otherwise, this question would not have been asked → "not greater than or equal to 29" or equivalently, "less than or equal to 29" is an old idea. → **Alternatively, we concentrate on the fact that the equality sign should belong in the null hypothesis.** That is,

$H_0: \mu \leq 29$ vs. $H_1: \mu > 29$

c. You wish to know if the average age of DePaul MBA students is less than 29.

Answer:

Identify which is an old idea and which is a new idea → " ... the average age ... is (now) less than 29" is a new idea; Otherwise, this question would not have been asked → "not less than or equal to 29" or equivalently, "greater than or equal to 29" is an old idea. → **Alternatively, we concentrate on the fact that the equality sign should belong in the null hypothesis.** That is,

$H_0: \mu \geq 29$ vs. $H_1: \mu < 29$

d. You wish to know if the average age of DePaul MBA students has changed from 30.

Answer: The same as #1 above except the age of 30.

That is, $H_0: \mu = 30$ vs. $H_1: \mu \neq 30$

e. You wish to know if the average age of DePaul MBA students has increased from 30.

Answer: The same as #2 above except the age of 30. → **Note that the equality sign belongs in the null hypothesis.** That is,

That is, $H_0: \mu \leq 30$ vs. $H_1: \mu > 30$

f. You wish to know if the average age of DePaul MBA students has decreased from 30.

Answer: The same as #3 above, except the age of 30. → **Note that the equality sign belongs in the null hypothesis.** That is,

That is, $H_0: \mu \geq 30$ vs. $H_1: \mu < 30$

C. Types of Hypothesis Testing

1. Two-Tail vs. One-Tail Tests

a. Two-Tail Tests

$$H_0: \mu = k \qquad \text{vs.} \qquad H_1: \mu \neq k$$

Note that $H_1: \mu \neq k$ means either $\mu > k$ or $\mu < k$. Thus, this type of hypothesis construction is named as a two-tail test.

b. One-Tail Tests → Directional Tests[112]

i. Upper-Tail Test = Right-Sided Test

$$H_0: \mu \leq k \qquad \text{vs.} \qquad H_1: \mu > k$$

ii. Lower-Tail Test = Left-Sided Test

$$H_0: \mu \geq k \qquad \text{vs.} \qquad H_1: \mu < k$$

2. Single vs. Joint (or Composite or Global) Hypotheses Testing

a. Single Hypothesis Testing

→ Any of the above 3 cases of the two-, upper-, and lower-tail tests is called a single hypothesis testing or simply, a hypothesis testing. A variant of this type is shown below:

---

[112] One-tail, one-tailed, or one-sided tests are commonly used to describe a directional test. Furthermore, the name given to each of the directional tests is based on the location of a rejection area which is represented by the alternative hypothesis. That is, an upper-tail or right-sided test identifies the area for rejecting the null hypothesis – thus, accepting the alternative hypothesis – to be at the upper tail or right side of the probability distribution. On the other hand, a lower-tail or left-sided test identifies the location of the rejection area for the null hypothesis to be on the lower-tail or left-side of the probability distribution.

$H_0: \mu_1 = \mu_2$ vs. $H_1: \mu_1 \neq \mu_2$

Note: $H_0: \mu_1 = \mu_2$ → $H_0: \mu_1 - \mu_2 = 0$ → $H_0: \mu_d = 0$
and
$H_1: \mu_1 \neq \mu_2$ → $H_1: \mu_1 - \mu_2 \neq 0$ → $H_1: \mu_d \neq 0$

b. Joint or Composite or Global Hypothesis Testing

When multiple hypotheses are embedded in one, it is called a joint, composite, or global hypothesis. For example,

i. $H_0: \mu_1 = \mu_2 = k$ vs. H_1: not H_0

Note 1:
$H_0: \mu_1 = \mu_2 = k$ → $\mu_1 = k$ and $\mu_2 = k$ → two hypotheses in one.

Note 2:
H_1: not H_0 → either $\mu_1 \neq k$; or $\mu_2 \neq k$; or both μ_1 and $\mu_2 \neq k$
→ three hypotheses in one.

ii. $H_0: \mu_1 = \mu_2 = \mu_3 = k$ vs. H_1: not H_0

Note 1:
$H_0: \mu_1 = \mu_2 = \mu_3 = k$ → $\mu_1 = k$; $\mu_2 = k$; and $\mu_3 = k$ → three hypotheses in one.

Note 2:
H_1: not H_0 → either $\mu_1 \neq k$; or $\mu_2 \neq k$; or $\mu_3 \neq k$; or all μ_i's $\neq k$.

4. Significance Level (α) vs. Confidence Level (1-α)

a. A significance level = the probability of making an error by rejecting H_0 when it is in fact true → Also, known as a Type I error.

b. A confidence level = the probability of being correct by accepting H_0 when it is in fact true.

Note: Confidence Level = 1 – Significance Level

5. Confidence level (1-α) vs. The Power of a Test (1-β)

→ The power of a test = the probability of being correct by rejecting H_0 when it is in fact false.

6. Type I Error (= significance level = α) vs. Type II Error (β)

→ A Type II Error = often denoted as β = the probability of making an error by accepting H_0 when it is in fact false.

Note: The Power of a Test = 1 – Type II Error = $1-\beta$

The following table summarizes the above information:

	Actual Situation	
Decision Given:	H_0 is True	H_0 is False
Fail to Reject H_0 = Accept H_0	Correct Decision: No Error of a $(1-\alpha)$ probability	Type II Error of a (β) probability
Reject H_0	Type I Error of an (α) probability	Correct Decision: No Error of a $(1-\beta)$ probability

Note that (1) there is a trade-off between a reduction in Type I error with that in Type II error. That is, if Type I error is reduced, then Type II error will be increased and vise versa; and (2) only if the sample size, n, increases the reliability of the hypothesis test result will increase, reducing the both types of errors.

7. Interpretation of Hypothesis Testing Conclusions via an Example

In the U.S. court of law, a defendant is innocent until proven guilty. This can be stated as:

H_0: The defendant is innocent → not guilty
H_1: The defendant is not innocent → guilty

Note 1:

If a jury rejects H_0, it means that the jury accepts H_1 → the defendant is guilty. On the other hand, if the jury fails to reject H_0, it means that the jury accepts H_0 and at the same, rejects H_1 → the defendant is not guilty.

However, when you accept H_0, it does not prove that the defendant is truly innocent. It only means that the defendant is not guilty, given the evidence presented to the jury and to the extent of what the jury believes. That is why the expression, "fail to reject H_0" is more appropriate to use in hypothesis testing than "do not reject" or "accept" H_0. However, for convenience and simplicity, the term, "accept H_0" will be used herein, instead of "fail to reject H_0" or "do not reject H_0."

Note 2:

If God knows that the defendant is innocent and the jury fail to reject H_0 at a given probability, $(1 - \alpha)$, then there is no error in the decision. The probability of making this correct decision is called the confidence level.

If God knows that the defendant is innocent and the jury rejects H_0 at a given probability, α, then there is an error in the decision. This error probability of α is called the significance level or often, the Type I error.

If God knows that the defendant is NOT innocent and the jury fails to reject H_0 at a given probability, β, then there is an error in the decision. This error probability, β, is called the Type II error or β risk.

If God knows that the defendant is NOT innocent and the jury rejects H_0 at a given probability, $(1 - \beta)$, then there is no error in the decision. The probability of making this correct decision is called the power of a test.

Note 3:

If a large confidence level is chosen, then there is a greater chance of making Type II error. That is, while the chance of making a correct decision to accept a true H_0 increases, it is also true that the chance of making a wrong decision to accept a false H_0 increases when the confidence level increases.

Note 4:

The 95% and 99% confidence levels are most popularly used. Furthermore, the 95% confidence level is MORE frequently used than 99% because it identifies reasonable outliers whereas the 99% confidence level identifies the most extreme outliers[113].

D. Decision Criteria for Two-tail vs. One-tail (or directional) Tests

1. Two-tail Tests

→ There are two rejections areas - one upper and one lower tails.

→ Given a significance level of α, it has to be split into two – upper and lower tails – because each of the two rejection areas should have a probability of $\alpha/2$. Therefore, when the probabilities of both tails are summed, it will be equal to α.

→ When the calculated Z-value falls within the upper and the lower critical Z values, we fail to reject H_0 and thus, accept H_0.

---

[113] Of course, this generalization is also subjective in that what should constitute an outlier is in the eyes of the analyst. Choosing a 95% confidence level, however, allows more frequent rejection of H_0 than a 99% confidence level.

Graphically, the decision criteria for a two-tail test is depicted as follows:

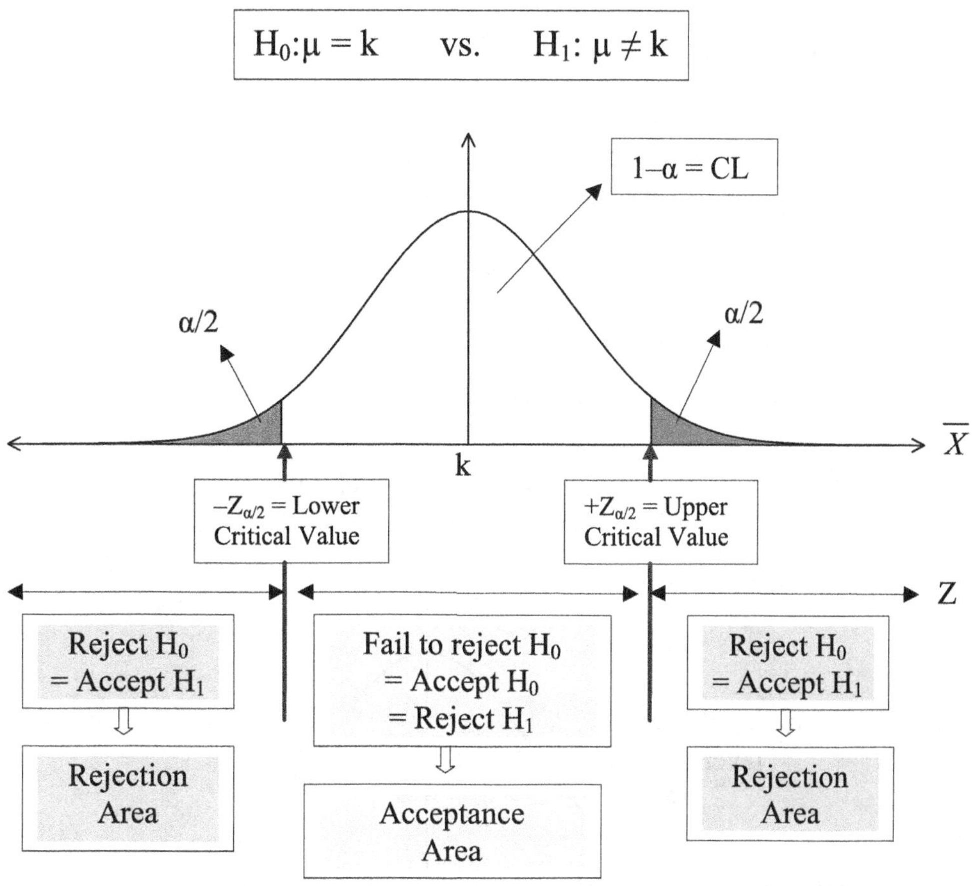

Note 1:

If any observation or a mean falls in either of the two rejection areas in the above diagram, it means that the observed value can be considered as an unlikely value given the sample data. That is why we reject the null hypothesis, which in turn means that we accept the alternative hypothesis.

Note 2:

When the mean or an observation falls within the acceptance region, we treat the following expressions equivalent and use them interchangeably:

Do not reject H_0 = fail to reject H_0 = accept H_0 = reject H_1.

Alternatively, reject H_0 = accept H_1.

However, statisticians prefer the expression of "fail to reject H_0" because the data did not support the rejection of H_0 and thus, by default, one must assume that the status quo or the old idea is still valid. That is, in absence of any evidence to reject H_0, one has failed to reject H_0.

2. One-Tail Tests

a. Lower-Tail Tests → Left-Sided Tests

→ There is one rejection area that is at the lower-end (or lower-tail or left-side) of the Z probability distribution → Note that the rejection area corresponds with the acceptance of the alternative hypothesis.[114]

→ Also note that the rejection area in this lower-tail test is larger than that in the two-tail test due to the difference between α used for a one-tail test and α/2 used for a two-tail test.

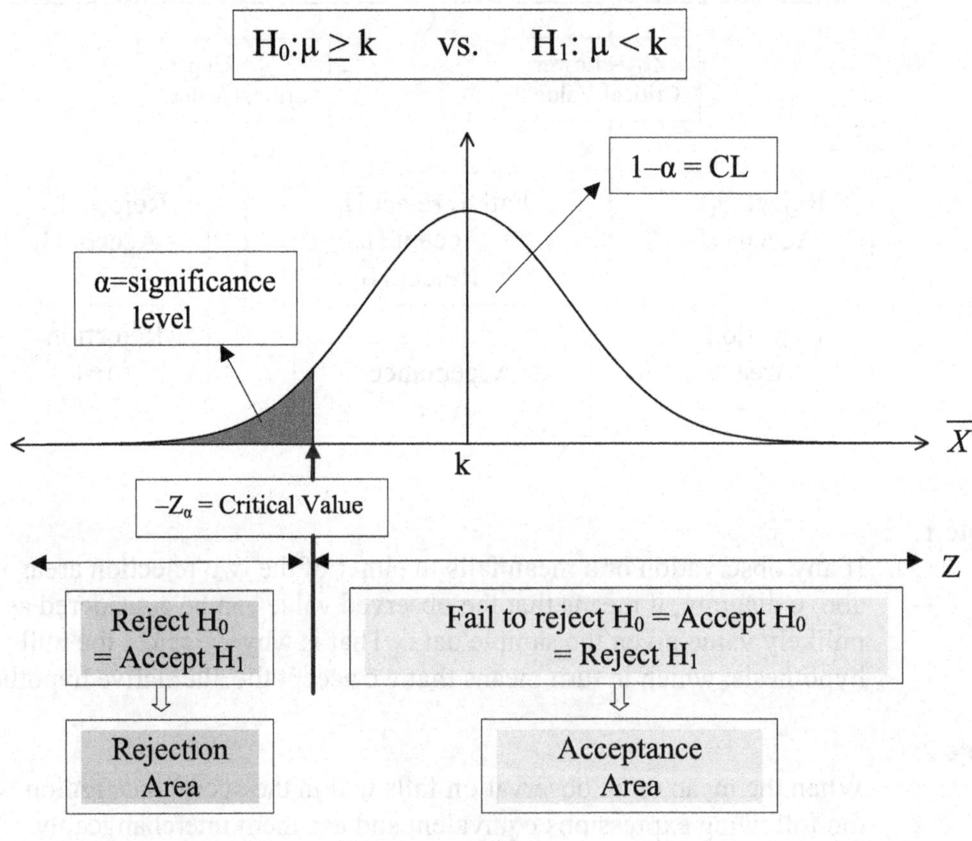

[114] This observation makes a perfect sense because when one rejects the null hypothesis, an alternative hypothesis must be accepted. Therefore, the acceptance region for the alternative hypothesis is labeled as the rejection area of the null hypothesis.

b. Upper-Tail Tests → Right-Sided Tests

→ There is one rejection area that is at the upper-end (or upper-tail or right-side) of the Z probability distribution → Note that the rejection area corresponds with the alternative hypothesis.

→ Also note that the rejection area in this upper-tail test is larger than that in the two-tail test due to the difference between α used for a one-tail test and $\alpha/2$ used for a two-tail test.

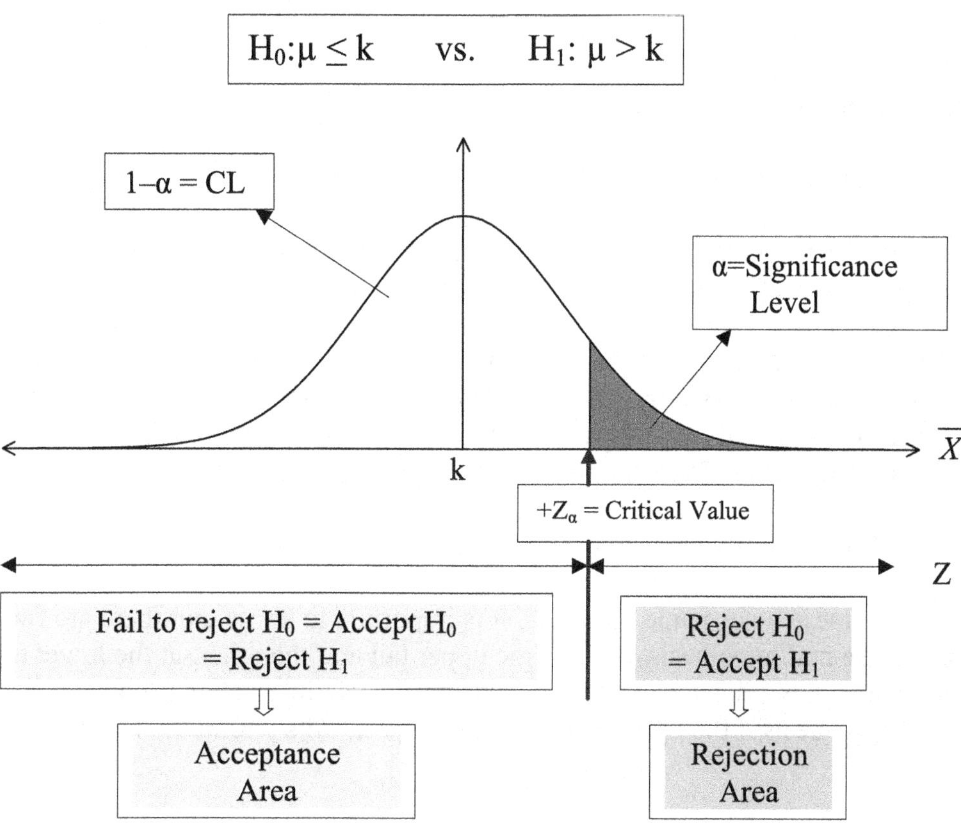

3. Rejection vs. No-rejection (=Acceptance) Regions

The rejection and the acceptance regions for the 3 types of hypothesis testing are as shown in the above 3 graphs → **Memorize these three graphs.**

Note once again that the following expressions are considered synonymous:

(1) Do not reject H_0 = Fail to reject H_0 = Accept H_0 = Reject H_1

(2) Reject H_0 = Accept H_1

Section 1.
The Z-Test for the Mean When σ is Known

When conducting a hypothesis test for the mean, the presence of the population variance or the population standard deviation is very critical. If either one of them is known, then the Z-test as described hereafter must be used. On the other hand, if both of them are unknown, then the t-test as described in Section 2 below must be used.

Regardless of a Z-test or a t-test being used, there are three approaches to testing the hypothesis for its significance. They are the critical value approach, the p-value approach, and the confidence interval approach. **All three approaches must yield the same identical conclusion.** Which approach to use depends on the preference and need of the analyst given the information and the convenience of its use.

1. The Confidence Interval Approach

This approach is an extension of Chapter 10 that discussed the construction of a confidence interval. The procedure for constructing a $(1-\alpha)$ confidence interval is exactly the same. Once a confidence interval is so constructed, then we examine if the hypothesized value, k, falls within that interval. If it does, we accept H_0. If not, we reject H_0 and accept H_1. **Therefore, the key idea behind using this confidence interval approach is to identify the upper and the lower limits of the confidence interval before anything else.** Let's study this more in detail.

a. Two-Tail Tests of $H_0{:}\mu = k$ vs. $H_1{:}\mu \neq k$

In a two-tail test, as its name indicates, it is important to know that there are two areas of rejecting the null hypothesis – one at the upper tail and the other, at the lower tail. Keeping this in mind, we can construct the confidence interval as follows[115] and accept the null hypothesis if the hypothesized value of μ, k, falls within the interval:

That is,

If $\overline{X} - Z_{\alpha/2} \cdot \frac{\sigma}{\sqrt{n}} \leq k \leq \overline{X} + Z_{\alpha/2} \cdot \frac{\sigma}{\sqrt{n}}$, then accept H_0: $\mu = k$.

Otherwise, reject H_0: $\mu = k$ → accept H_1: $\mu \neq k$. That is, if the confidence interval does not capture the hypothesized mean of k, then reject the null hypothesis.

The following graphical presentation re-enforces what is concluded above.

---

[115] In theory, however, the confidence-interval approach for hypothesis testing calls for the point estimate to be the hypothesized value, k, instead of the sample mean, $\overline{X}$. That is, the correct equation for a confidence interval should be: $k - Z_{\alpha/2} \cdot \frac{\sigma}{\sqrt{n}} \leq \overline{X} \leq k + Z_{\alpha/2} \cdot \frac{\sigma}{\sqrt{n}}$. However, for convenience and clear intuitive appeal, this alternative confidence interval equation is used throughout this book.

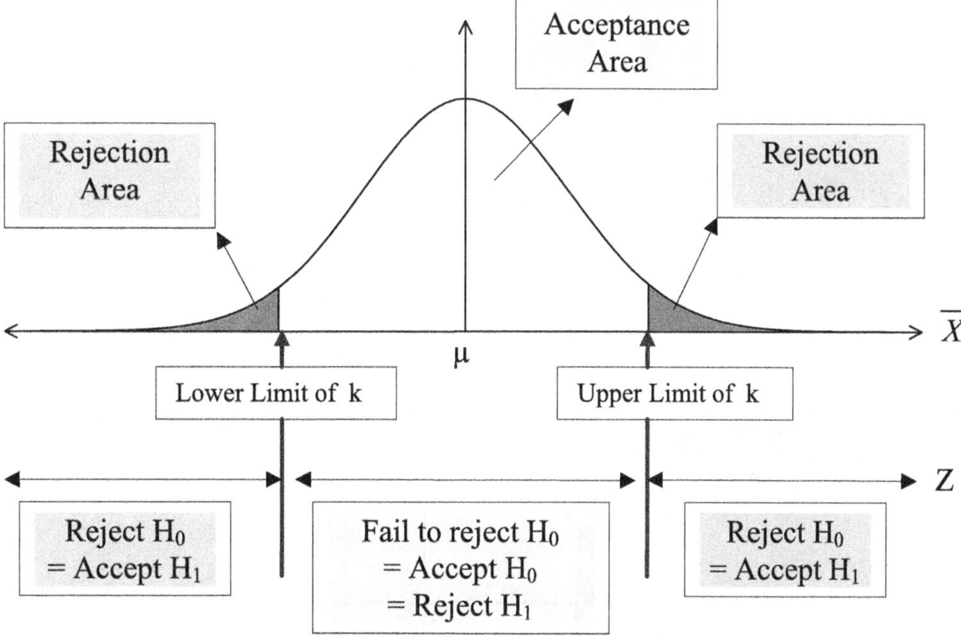

Note: Lower Limit of k = Lower Boundary = $\overline{X} - Z_{\alpha/2} \cdot \sigma_{\overline{X}} = \overline{X} - Z_{\alpha/2} \cdot \dfrac{\sigma}{\sqrt{n}}$

Upper Limit of k = Upper Boundary = $\overline{X} + Z_{\alpha/2} \cdot \sigma_{\overline{X}} = \overline{X} + Z_{\alpha/2} \cdot \dfrac{\sigma}{\sqrt{n}}$

Once these upper and lower limits of k are established, check where the hypothesized value of k lies. If it falls within these limits, accept H_0: μ = k. If it falls outside of these limits, reject H_0: μ = k and accept H_1:μ ≠ k.

b. One-Tail Tests

In a one-tail test, as its name indicates, it is important to know that there is only one area of rejecting the null hypothesis – either at the upper tail or at the lower tail. Therefore, in a strict sense, a confidence interval approach does not define a closed interval, but an open interval of either greater than or less than a specific value. Keeping this in mind, we can construct the confidence interval as follows and accept the null hypothesis if the hypothesized value of μ, k, falls within a specific region:

i. **Lower-Tail Tests = Left-Sided Tests of** H_0: $\mu \geq k$ vs. H_1: $\mu < k$

The decision criteria is:

If $k - Z_\alpha \cdot \dfrac{\sigma}{\sqrt{n}} \leq \overline{X}$, then accept H_0: $\mu \geq k$ for a lower-tail test.

Otherwise, reject H_0: $\mu \geq k$ → accept H_1: $\mu < k$.

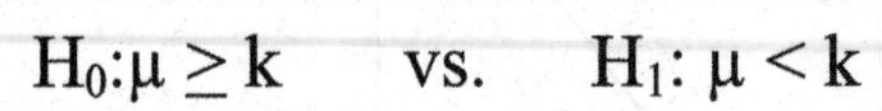

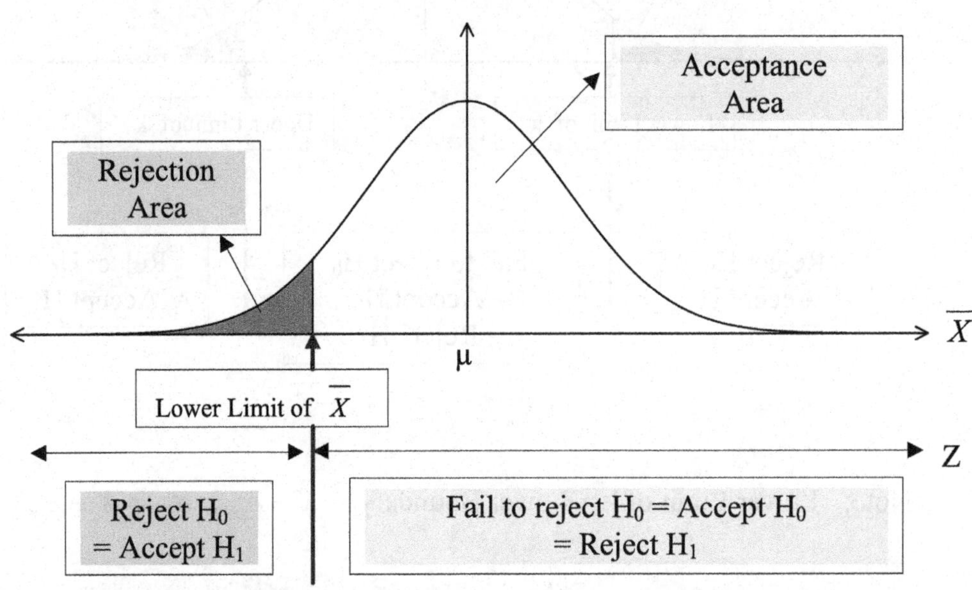

Note: Lower Limit of $\overline{X}$ = Lower Boundary = $k - Z_\alpha \cdot \sigma_{\overline{X}} = k - Z_\alpha \cdot \dfrac{\sigma}{\sqrt{n}}$

Once this lower limit of $\overline{X}$ is established, check where the sample mean, $\overline{X}$, lies. If it falls above it, accept H_0. If it falls below it, reject H_0.

ii. **Upper-Tail Tests = Right-Sided Tests of** H_0: $\mu \leq k$ vs. H_1: $\mu > k$.

The decision criteria is:

If $\overline{X} \leq k + Z_\alpha \cdot \dfrac{\sigma}{\sqrt{n}}$, then accept H_0: $\mu \leq k$ for an upper-tail test.

Otherwise, reject H_0: $\mu \leq k$ → accept H_1: $\mu > k$.

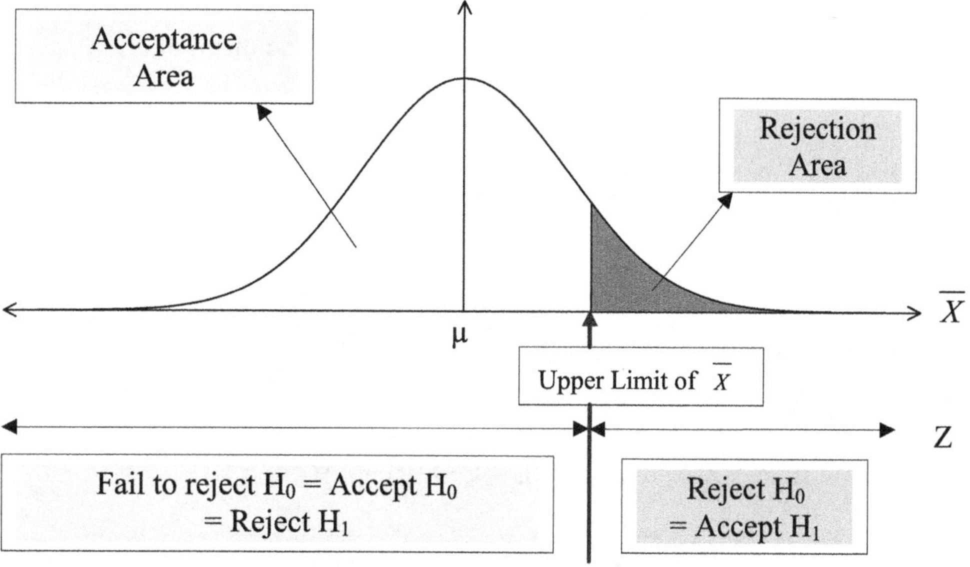

Note: Upper Limit of $\overline{X}$ = Upper Boundary = $k + Z_\alpha \cdot \sigma_{\overline{X}} = k + Z_\alpha \cdot \frac{\sigma}{\sqrt{n}}$

Once this upper limit of $\overline{X}$ is established, check where the sample mean, $\overline{X}$, lies. If it falls below it, accept H_0. If it falls above it, reject H_0.

2. The Critical Z-Value Approach[116]

The key idea behind using this critical Z-value approach is to identify the upper and the lower critical (or table) values of Z before anything else.

The following shows the steps used in the critical Z-value approach:

Step 1. Check if a Z-test is valid → that is, if $\sigma^2, \sigma,$ or $\sigma_{\overline{X}}$ is known, continue with the following steps. Otherwise, go to the t-test.

Step 2. **Choose the significance level and find the critical Z-value from the Z-table** → Note that the critical value of Z for a two-tail test is $Z_{\alpha/2}$ and Z_α for a one-tail test.

[116] This approach is called a "critical value" approach because the Z value found in the table serves as the "critical" value that determines the acceptance vs. the rejection regions.

Note: Instead of using the Z-table in Chapter 7, one can use the Excel Spreadsheet command of "=normsinv(α)" to find a table Z value given a probability, α. → A note of caution: You must still remember that for a two-tail test, α/2 must be input in place of α when using the Excel.

Step 3. Calculate $Z_c = \dfrac{\overline{X}-\mu}{\sigma_{\overline{X}}} = \dfrac{\overline{X}-\mu}{\frac{\sigma}{\sqrt{n}}}$ for the mean of X, $\overline{X}$.[117]

Step 4. Apply the decision criteria

i. For a two-tail test → H_0: $\mu = k$ vs. H_1: $\mu \neq k$

If $|Z_c| \leq Z_{\alpha/2}$, or $-Z_{\alpha/2} \leq Z_c \leq Z_{\alpha/2}$, then fail to reject H_0: $\mu = k$ → accept H_0: $\mu = k$.

If $|Z_c| > Z_{\alpha/2}$, reject H_0: $\mu = k$ → accept H_1: $\mu \neq k$.

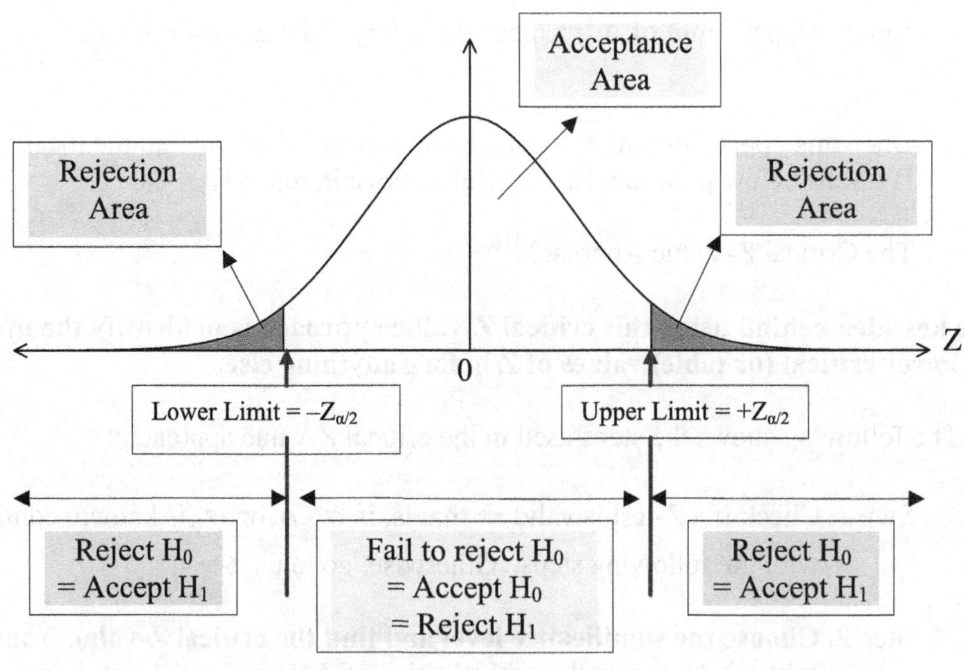

117 For an individual value of X, the calculated Z value would be: $Z_c = \dfrac{X-\mu}{\sigma}$. However, because hypothesis testing involves the test of the mean, this formula based on $\sigma_{\overline{X}}$ is used hereafter.

ii. For a one-tail test

(a) Lower-Tail Test → H_0: $\mu \geq k$ vs. H_1: $\mu < k$

Step 1. **Choose the significance level, α, and find the Z-value from the Z-Table →** Look for $-Z_\alpha$ **→ Remember that a lower-tail test uses a negative Z value its critical value.**

Note: Instead, one can use the Excel Spreadsheet command of "=normsinv(α)" to find a table Z value given a probability, α.

Step 2. Calculate $Z_c = \dfrac{\overline{X}-\mu}{\sigma_{\overline{X}}} = \dfrac{\overline{X}-\mu}{\dfrac{\sigma}{\sqrt{n}}}$ for the mean of X.

Step 3. Decision Criteria:

If $Z_c \geq -Z_\alpha$, then fail to reject H_0: $\mu \geq k$ → accept H_0

If $Z_c < -Z_\alpha$, reject H_0: $\mu \geq k$ → Accept H_1: $\mu < k$

H_0: $\mu \geq k$ vs. H_1: $\mu < k$

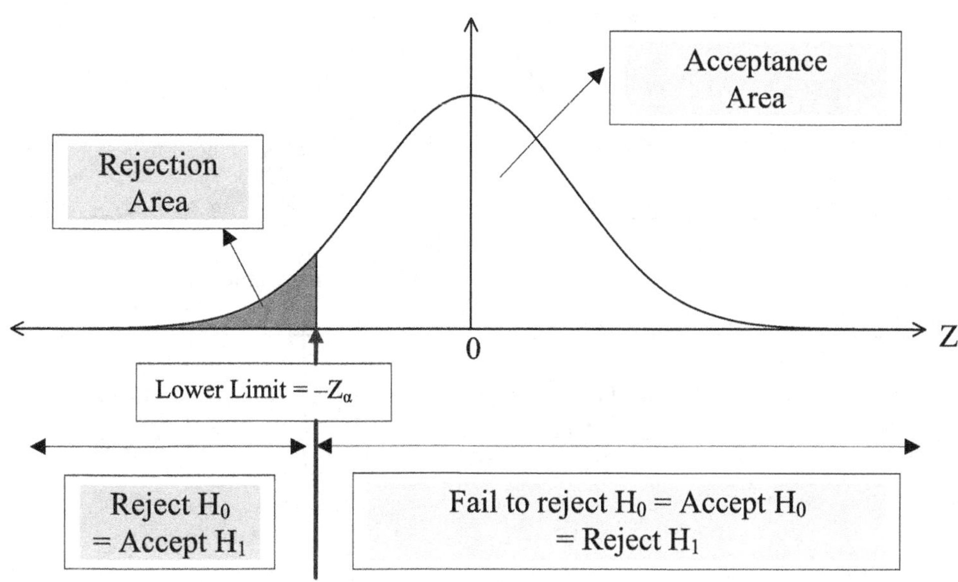

(b) Upper-Tail Test → H_0: $\mu \leq k$ vs. H_1: $\mu > k$

Step 1. **Choose the significance level, α, and find the Z-value from the Z-table →** Look for $+Z_{\alpha}$ **→ Remember that an upper-tail test uses a positive Z value as its critical value.**

Note: Instead of using the Z-table in the textbook, one can use the Excel Spreadsheet command of "=normsinv(α)" to find a table Z value given a probability, α.

Step 2. Calculate $Z_c = \dfrac{\overline{X}-\mu}{\sigma_{\overline{X}}} = \dfrac{\overline{X}-\mu}{\frac{\sigma}{\sqrt{n}}}$ for the mean of X.

Step 3. Decision Criteria:

If $Z_c \leq +Z_{\alpha}$, then fail to reject H_0: $\mu \leq k$ → accept H_0

If $Z_c > +Z_{\alpha}$, reject H_0: $\mu \leq k$ → Accept H_1: $\mu > k$

$H_0: \mu \leq k$ vs. $H_1: \mu > k$

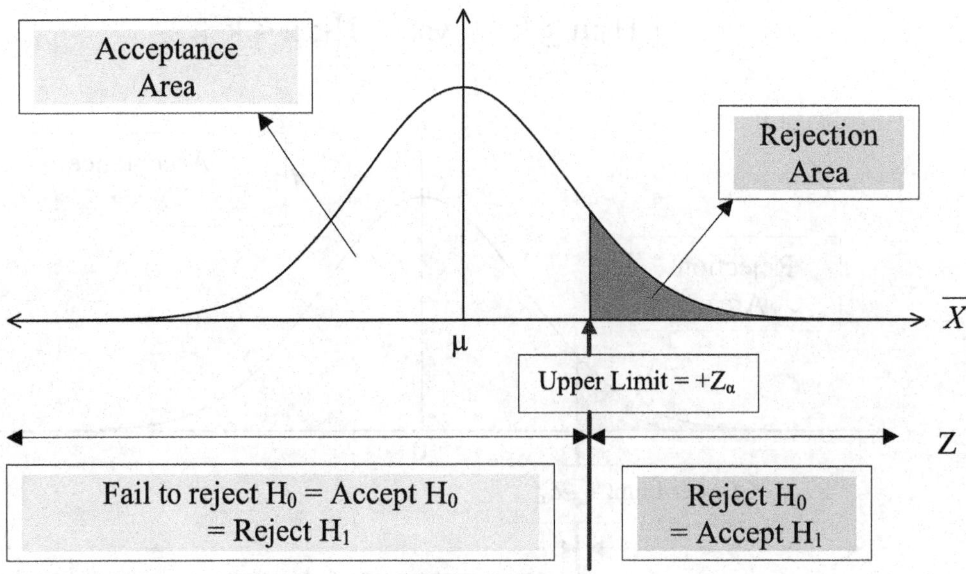

3. The p-Value Approach[118] → Compare p-value vs. chosen α

The key idea behind using this p-value approach is to identify the probability associated with the calculated Z-value, Z_c, AND the significance level, α, before anything else.

Step 1. Calculate $Z_c = \dfrac{\overline{X} - \mu}{\sigma_{\overline{X}}} = \dfrac{\overline{X} - \mu}{\frac{\sigma}{\sqrt{n}}}$ for the mean of X.

For an illustration purpose, let's assume that we identify it as 2.00. That is, $Z_c = 2.00$.

Step 2. Given the value of Z_c, look for the probability of rejection in the Z-table.

That is, given $Z_c = 2.00$, we find $P(Z > Z_c = 2.00) = 0.0228$.

Step 3. Identify the p-value associated with Z_c.

i. If a two-tail test, the probability associated with each tail must be calculated and then, summed to get the p-value.

This summed probability-value is called the p-value. → Alternatively, the p-value can be obtained by multiplying the probability of one-tail by 2 because there are two equal tails.

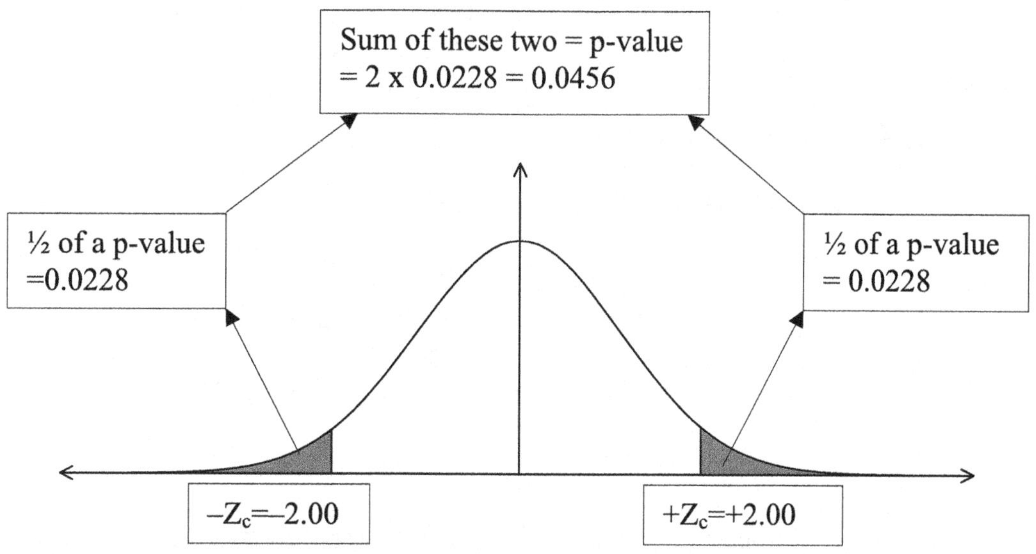

Therefore, the p-value in the above case is 0.0228 x 2 = 0.0456.

---

[118] This approach should be called a "critical" p-value approach because the probability obtained on the basis of the calculated Z-value defines the "critical" level of probability that determines the acceptance and the rejection regions.

Suppose that we chose a significance level, $\alpha = 0.05$. If so, we know that it has to be split into two pieces for a two-tail test → $\alpha/2 = 0.05/2 = 0.0250$ → This means that the corresponding Z-value from the table is ± 1.96, which defines the critical value for accepting or rejecting H_0.

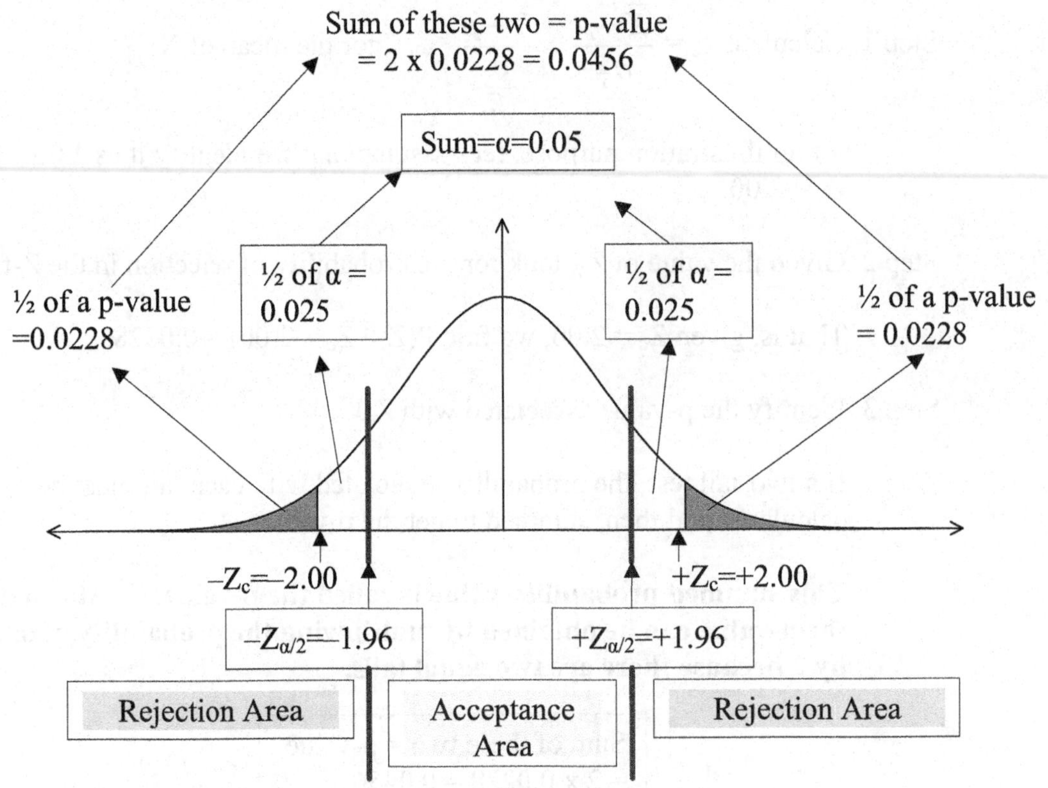

Step 4. Decision Criteria:

The decision criteria for a two-tail test are:

If p-value $\geq$ a chosen α,
then fail to reject H_0: $\mu = k$ → accept H_0: $\mu = k$

If p-value < a chosen α, reject H_0 → accept H_1.

Therefore, in the above case,

because (p-value of 0.0456) < (a chosen α of 0.05),
we reject H_0 → accept H_1.

It is often much simpler to memorize the above decision criteria and apply them, rather than try to logically deduct the above diagram to make a decision.

ii. If a one-tail test, only the probability of one side is calculated. This one-sided probability-value is called the p-value.

The Case of a Lower-Tail Test:

Step 1. Calculate $Z_c = \dfrac{\overline{X}-\mu}{\sigma_{\overline{X}}} = \dfrac{\overline{X}-\mu}{\dfrac{\sigma}{\sqrt{n}}}$ for the mean of X.

For an illustration purpose, let's assume that we identify it as –1.23. That is, $Z_c = -1.23$.

Step 2. Given the value of Z_c, look for the probability of rejection in the Z-table.

That is, given $Z_c = -1.23$, we find $P(Z < Z_c = -1.23) = 0.1093$.

Step 3. Identify the p-value associated with Z_c and a significance level, α, such as 0.05, for example.

$H_0: \mu \geq k$ vs. $H_1: \mu < k$

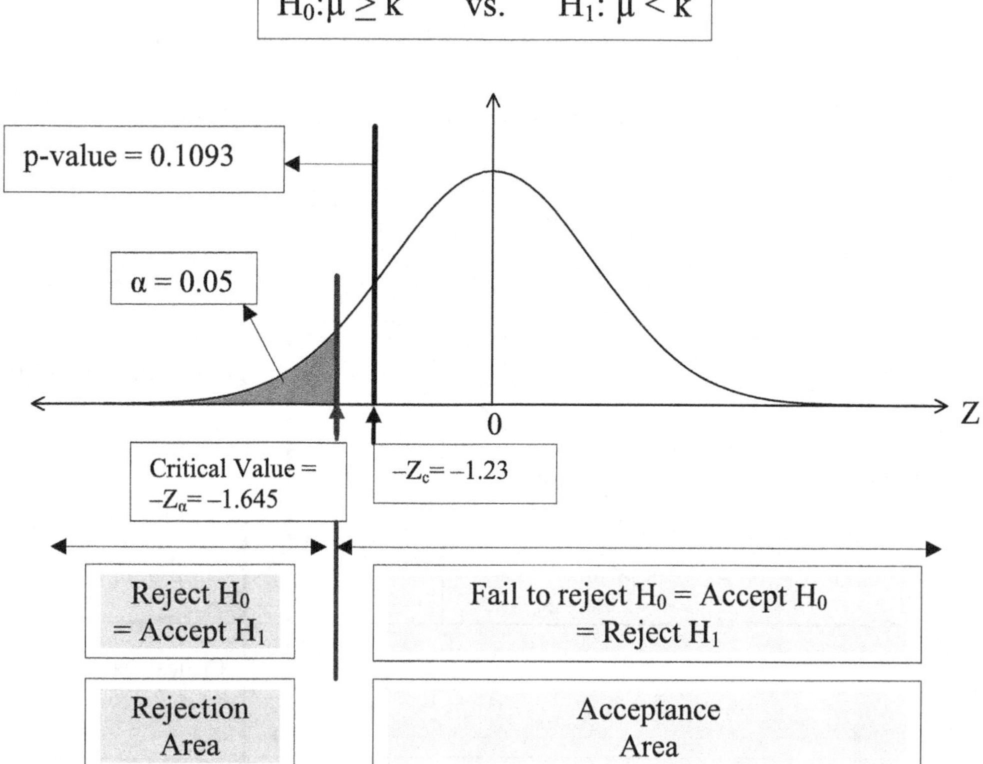

Step 4. Decision Criteria:

The decision criteria for a lower-tail test are:

If p-value $\geq$ a chosen α,
then fail to reject H_0: $\mu = k \rightarrow$ accept H_0: $\mu = k$

If p-value < a chosen α, reject $H_0 \rightarrow$ accept H_1.

Therefore, in the above case,

because (p-value of 0.1093) > (a chosen α of 0.05),
we fail to reject $H_0 \rightarrow$ accept H_0.

The Case of an Upper-Tail Test:

Step 1. Calculate $Z_c = \dfrac{\overline{X} - \mu}{\sigma_{\overline{X}}} = \dfrac{\overline{X} - \mu}{\frac{\sigma}{\sqrt{n}}}$ for the mean of X.

For an illustration purpose, let's assume that we identify it as +1.55. That is, $Z_c = 1.55$.

Step 2. Given the value of Z_c, look for the probability of rejection in the Z-table.

That is, given $Z_c = 1.55$, we find $P(Z > Z_c = 1.55) = 0.0606$.

Step 3. Identify the p-value associated with Z_c and a significance level, α, such as 0.01, for example.

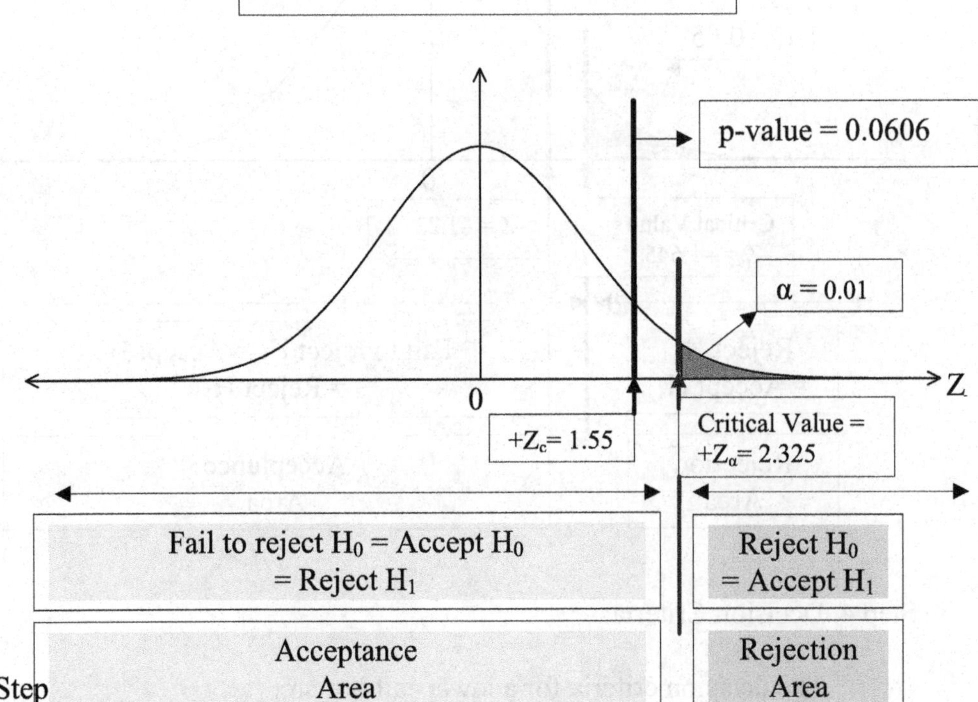

Step

The decision criteria for an upper-tail test are:

If p-value $\geq$ a chosen α,
then fail to reject H_0: $\mu \leq k$ → accept H_0: $\mu \leq k$

If p-value < a chosen α, reject H_0 → accept H_1.

Therefore, in the above case,

because (p-value of 0.0606) > (a chosen α of 0.01),
we accept H_0: $\mu \leq k$ at a 1% significance level[119].

Summary Note:

Decision Criteria for all 3 types of hypotheses – two-tail, upper-tail, or lower-tail – are exactly the same as summarized below:

That is,

If p-value $\geq$ a chosen α,
then fail to reject all null hypotheses → accept the null hypothesis.

That is,

Accept H_0: $\mu = k$ or H_0: $\mu \geq k$ or H_0: $\mu \leq k$

If p-value < a chosen α, reject H_0 → accept H_1.

Memorize this decision criteria!!!

Hint: Instead of using the Z-table for a probability associated with a Z_c value, one can use the Excel Spreadsheet command of "=normdist(X,μ,σ,True)" to calculate the cumulative probability of Z.

---

[119] If you were to use the critical value approach for this case, you would make the same conclusion of accepting H_0 because ($Z_c = 1.55$) < ($Z_{0.01} = 2.325$).

The Summary of Hypothesis Testing Methods Using the Z-statistic

Hypothesis Testing Method	Two-Tail Test	Lower-Tail Test	Upper-Tail Test
	H_0: μ=k vs. H_1: μ≠k	H_0: μ ≥ k vs. H_1: μ < k	H_0: μ ≤ k vs. H_1: μ > k
Rejection Area(s) Shaded	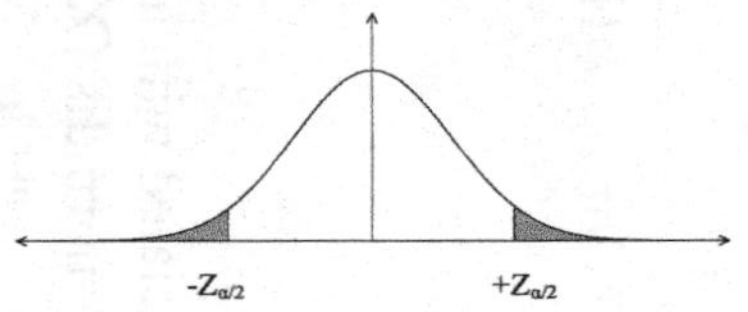 $-Z_{\alpha/2}$ $+Z_{\alpha/2}$	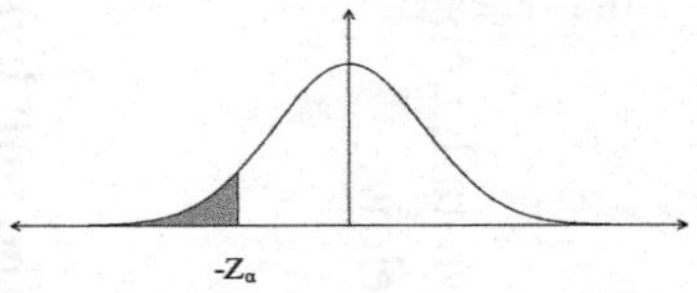 $-Z_\alpha$	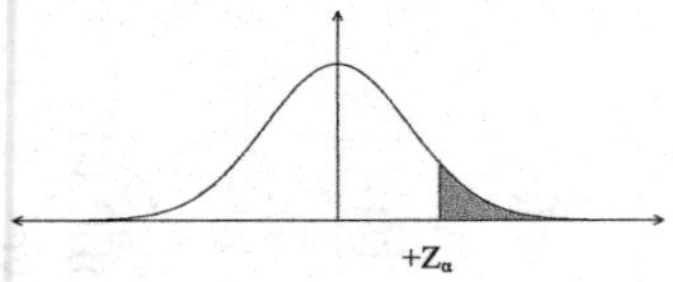 $+Z_\alpha$
Critical Value	If $-Z_{\alpha/2} \le Z_c \le +Z_{\alpha/2}$, accept H_0: μ = k. Otherwise, reject H_0 → accept H_1	If $Z_c \ge -Z_\alpha$, accept H_0: μ ≥ k. Otherwise, reject H_0 → accept H_1	If $Z_c \le +Z_\alpha$, accept H_0: μ ≤ k. Otherwise, reject H_0 → accept H_1
p-Value	If p-value calculated ≥ α chosen, accept H_0: μ = k. Otherwise, reject H_0 → accept H_1	If p-value calculated ≥ α chosen, accept H_0: μ ≥ k. Otherwise, reject H_0 → accept H_1	If p-value calculated ≥ α chosen, accept H_0: μ ≤ k. Otherwise, reject H_0 → accept H_1
Confidence Interval	If $\overline{X} - Z_{\alpha/2} \cdot \sigma_{\overline{X}} \le k \le \overline{X} + Z_{\alpha/2} \cdot \sigma_{\overline{X}}$, accept H_0: μ = k. Otherwise, reject H_0 → accept H_1	If $k - Z_\alpha \cdot \sigma_{\overline{X}} \le \overline{X}$, accept H_0: μ ≥ k. Otherwise, reject H_0 → accept H_1	If $\overline{X} \le k + Z_\alpha \cdot \sigma_{\overline{X}}$, accept H_0: μ ≤ k. Otherwise, reject H_0 → accept H_1

Section 2.
The t-Test for the Mean When σ is Unknown

For the t-test, everything is the same as the Z-test shown in the previous section except that instead of Z, now a t-value needs to be calculated and its corresponding probability value.

Note 1: $t_c = \dfrac{X-\mu}{S}$ for an individual value of X or

$$t_c = \frac{\overline{X}-\mu}{S_{\overline{X}}} = \frac{\overline{X}-\mu}{\frac{S}{\sqrt{n}}}$$ for the mean of X, $\overline{X}$.

Note 2: The degrees of freedom[120], (n-1), must be taken into consideration.

Because the concept for the 3 approaches of the confidence interval, the critical table-value, and the p-value approaches are the same for Z and t, we highlight only the decision criteria here below.

1. The Confidence Interval Approach

Construct the confidence interval as follows and accept the null hypothesis if the hypothesized value of μ, k, falls within the following interval:

That is,

If $\overline{X} - t_{\alpha/2,n-1} \cdot \dfrac{S}{\sqrt{n}} \leq k \leq \overline{X} + t_{\alpha/2,n-1} \cdot \dfrac{S}{\sqrt{n}}$,
then accept H_0: $\mu = k$ for a two-tail test.

If $k - t_{\alpha,n-1} \cdot \dfrac{S}{\sqrt{n}} \leq \overline{X}$, then accept H_0: $\mu \geq k$ for a lower-tail test.

If $\overline{X} \leq k + t_{\alpha,n-1} \cdot \dfrac{S}{\sqrt{n}}$, then accept H_0: $\mu \leq k$ for an upper-tail test.

Note that the decision criteria described above for testing all three types of hypotheses are exactly the same whether we use the Z-statistic or t-statistic.

---

[120] Note that the degree of freedom is (n-1) because the mean of the sample, $\overline{X}$, is calculated and known in this case. This forces the loss of 1 degree of freedom. That is, if $\overline{X}$ is known, we can forgo 1 data point out of n.

2. The Critical t-Value Approach[121]

Step1. Check if a t-test is valid → that is, if $\sigma^2, \sigma,$ or $\sigma_{\overline{X}}$ is unknown, continue with the following steps. Otherwise, go to the Z-test.

Step 2. **Choose the significance level and find the t-value from the t-Table →** call this the table value or critical value of t or $t_{\alpha/2,n-1}$ for a two-tail test or $t_{\alpha,n-1}$, for a one-tail test.

Note: Instead of using the t-table in Chapter 8, one can use the Excel Spreadsheet command of "=tdist(X,df,tails)" to calculate the cumulative probability of t and "=tinv(α, df)" to find a table t value given a probability, α.

Step 3. Calculate $t_c = \dfrac{\overline{X}-\mu}{S_{\overline{X}}} = \dfrac{\overline{X}-\mu}{\frac{S}{\sqrt{n}}}$ for the mean of X, $\overline{X}$.

Step 4. Decision Criteria

i. Two-tail tests: H_0: $\mu = k$ vs. H_1: $\mu \neq k$

If $|t_c| \le t_{\alpha/2,n-1}$, or $-t_{\alpha/2,n-1} \le t_c \le t_{\alpha/2,n-1}$,

then fail to reject H_0: $\mu=k$ → accept H_0: $\mu=k$.

Otherwise, reject H_0 → accept H_1: $\mu \neq k$.

ii. One-tail tests

(a) Lower-Tail Test: H_0: $\mu \geq k$ vs. H_1: $\mu < k$

Decision Criteria:

If $t_c \geq -t_{\alpha,n-1}$, then fail to reject H_0: $\mu \geq k$ → accept H_0: $\mu \geq k$.

Otherwise, reject H_0 → accept H_1: $\mu < k$.

(b) Upper-Tail Test: H_0: $\mu \leq k$ vs. H_1: $\mu > k$

[121] As studies in Chapter 8, the Excel Spreadsheet command of "=TINV(significance level, degrees of freedom)" yields the critical or table t-value for a two-tail test. Therefore, if one is to calculate the table t-value for a two-tail test via Excel, do NOT divide the significance level by 2. On the other hand, when calculating table t-value for a one-tail test via Excel, we must multiply the significance level by 2.

Decision Criteria:

If $t_c \leq +t_{\alpha,n-1}$, then fail to reject H_0: $\mu \leq k$ → accept H_0

Otherwise, reject H_0 → accept H_1: $\mu > k$

Note once again that the decision criteria described above for testing all three types of hypotheses are exactly the same whether we use the Z-statistic or t-statistic.

3. The p-Value Approach

Step 1. Calculate $t_c = \dfrac{X - \mu}{S}$ for an individual value of X or

Calculate $t_c = \dfrac{\overline{X} - \mu}{S_{\overline{X}}} = \dfrac{\overline{X} - \mu}{\frac{S}{\sqrt{n}}}$ for the mean of X.

Step 2. Given the value of t_c, look for the probability of rejection in the t-table with (n-1) degrees of freedom.

If a two-tail test, the sum of both ends needs to be calculated.
If a one-tail test, only one side is calculated.
These probability-values are called p-values.

Step 3. For the p-value approach, the decision criteria for all 3 types of hypotheses tests – two-tail, lower-tail, or upper-tail – are once again the same whether we use the Z-statistic or t-statistic.

That is,

If p-value $\geq$ a chosen α,

then fail to reject all null hypotheses
→ accept H_0: $\mu = k$ or H_0: $\mu \geq k$ or H_0: $\mu \leq k$

Otherwise, reject H_0 → accept H_1.

The Summary of Hypothesis Testing Methods Using the t-statistic

Hypothesis Testing Method	Two-Tail Test	Lower-Tail Test	Upper-Tail Test
	H_0: $\mu = k$ vs. H_1: $\mu \neq k$	H_0: $\mu \geq k$ vs. H_1: $\mu < k$	H_0: $\mu \leq k$ vs. H_1: $\mu > k$
Rejection Area(s) Shaded	$-t_{\alpha/2,n-1}$ $+t_{\alpha/2,n-1}$	$-t_{\alpha,n-1}$	$+t_{\alpha,n-1}$
Critical Value	If $-t_{\alpha/2,n-1} \leq t_c \leq +t_{\alpha/2,n-1}$, accept H_0: $\mu = k$. Otherwise, reject $H_0 \rightarrow$ accept H_1	If $t_c \geq -t_\alpha$, accept H_0: $\mu \geq k$. Otherwise, reject $H_0 \rightarrow$ accept H_1	If $t_c \leq +t_\alpha$, accept H_0: $\mu \leq k$. Otherwise, reject $H_0 \rightarrow$ accept H_1
p-Value	If p-value calculated $\geq \alpha$ chosen, accept H_0: $\mu = k$. Otherwise, reject $H_0 \rightarrow$ accept H_1	If p-value calculated $\geq \alpha$ chosen, accept H_0: $\mu \geq k$. Otherwise, reject $H_0 \rightarrow$ accept H_1	If p-value calculated $\geq \alpha$ chosen, accept H_0: $\mu \leq k$. Otherwise, reject $H_0 \rightarrow$ accept H_1
Confidence Interval	If $\overline{X} - t_{\alpha/2,n-1} \cdot S_{\overline{X}} \leq k \leq \overline{X} + t_{\alpha/2,n-1} \cdot S_{\overline{X}}$, accept H_0: $\mu = k$. Otherwise, reject $H_0 \rightarrow$ accept H_1	If $k - t_{\alpha,n-1} \cdot S_{\overline{X}} \leq \overline{X}$, accept H_0: $\mu \geq k$. Otherwise, reject $H_0 \rightarrow$ accept H_1	If $\overline{X} \leq k + t_{\alpha,n-1} \cdot S_{\overline{X}}$, accept H_0: $\mu \leq k$. Otherwise, reject $H_0 \rightarrow$ accept H_1

Exercise Problems for Hypothesis Testing

A. Given the following information,

Suppose that you know from previous experience that the average speed of a car on the highway is 60 m.p.h.[122] The population standard deviation is 8 m.p.h. In order to find out whether the average speed of a car on the same highway has remained the same or not, you conducted a survey and found that the average speed of 25 cars surveyed was 64 m.p.h.

1. Verify at a 5% significance level if the average car speed had changed.

H_0: $\mu = 60$ vs. H_1: $\mu \neq 60$

a. by using the critical value approach.

Because $(Z_c = \dfrac{64-60}{\dfrac{8}{\sqrt{25}}} = 2.50) > (Z_{0.05/2}=1.96)$,
we reject H_0: μ=60 → accept H_1: $\mu \neq 60$ at a 5% significance level.

b. by using the p-value approach.

Because $P(Z_c>2.50) = 0.0062$ from the Z-table in Chapter 7,
p-value = 2 x 0.0062 = 0.0124 = 1.24%

Because (p-value=1.24%) < (α = 5%),
we reject H_0: μ=60 → accept H_1: $\mu \neq 60$ at a 5% significance level.

c. by using the confidence-interval approach.

Because $\overline{X} - Z_{\alpha/2} \cdot \dfrac{\sigma}{\sqrt{n}} \leq k \leq \overline{X} + Z_{\alpha/2} \cdot \dfrac{\sigma}{\sqrt{n}}$ defines the region of accepting H_0, we find

$$64 - 1.96 \cdot \frac{8}{\sqrt{25}} \leq (k = 60) \leq 64 + 1.96 \cdot \frac{8}{\sqrt{25}}$$

$$60.864 \nleq (k=60) < 67.136$$

We note that this 95% confidence interval does not capture μ=60=k. Thus, we reject H_0: μ=60 → accept H_1: $\mu \neq 60$ at a 5% significance level.

---

[122] This description implies that 60 m.p.h. can be considered as the population mean.

2. Verify at a 1% significance level if the average car speed had changed.

H_0: μ = 60 vs. H_1: μ ≠ 60

a. by using the critical value approach.

Because (Z_c=2.50) < ($Z_{0.01/2}$=2.575),
we fail to reject H_0: μ=60 → accept H_0:μ=60→ reject H_1: μ≠60
at a 1% significance level.

b. by using the p-value approach.

Because P(Z_c>2.50) = 0.0062 from the Z-table in Chapter 7,
p-value = 2 x 0.0062 = 0.0124 = 1.24%

Because (p-value=1.24%) > (α = 1%),
we fail to reject H_0: μ=60 → accept H_0:μ=60→ reject H_1: μ≠60
at a 1% significance level.

c. by using the confidence-interval approach.

Because $\overline{X} - Z_{\alpha/2} \cdot \frac{\sigma}{\sqrt{n}} \le k \le \overline{X} + Z_{\alpha/2} \cdot \frac{\sigma}{\sqrt{n}}$ defines the region of accepting H_0, we find

$$64 - 2.575 \cdot \frac{8}{\sqrt{25}} \le (k = 60) \le 64 + 2.575 \cdot \frac{8}{\sqrt{25}}$$

$$59.88 < (k=60) < 68.12$$

We note that this 99% confidence interval does capture μ=k=60. Thus, we fail to reject H_0: μ=60 → accept H_0:μ=60→ reject H_1: μ≠60 at a 1% significance level.

B. Given the following information,

Suppose that you know from previous experience that the average speed of a car on the highway is 60 m.p.h. In order to find out whether the average speed of a car on the same highway has remained the same or not, you conducted a survey and found that the average speed of 16 cars surveyed was 63.90375 m.p.h. and the sample standard deviation was 6 m.p.h.

1. Verify at a 5% significance level if the average car speed had changed

H_0: μ = 60   vs. H_1: μ ≠ 60

a. by using the critical value approach.

Because $\left(t_c = \dfrac{63.90375 - 60}{\dfrac{6}{\sqrt{16}}} = 2.6025\right) > (t_{0.05/2,(16-1)} = 2.1315)$,

we reject H_0: μ=60 → accept H_1: μ≠60 at a 5% significance level.

b. by using the p-value approach.

Because $P(t_c > 2.6025) = 0.01$ from the t-table in Chapter 8 with (16-1) d.f.,
p-value = 2 x 0.01 = 0.02 = 2%

Because (p-value=2%) < (α = 5%),
we reject H_0: μ=60 → accept H_1: μ≠60 at a 5% significance level.

c. by using the confidence-interval approach.

Because $\overline{X} - t_{\alpha/2,(16-1)} \cdot \dfrac{S}{\sqrt{n}} \le k \le \overline{X} + t_{\alpha/2,15} \cdot \dfrac{S}{\sqrt{n}}$ defines the region of accepting H_0, we find

$$64 - 2.1315 \cdot \frac{6}{\sqrt{16}} \le (k = 60) \le 64 + 2.1315 \cdot \frac{6}{\sqrt{16}}$$

$$60.8028 \not< (k=60) < 67.1973$$

We note that this 95% confidence interval does not capture μ=k=60. Thus, we reject H_0: μ=60 → accept H_1:μ≠60 at a 5% significance level.

2. Verify at a 1% significance level if the average car speed had changed

H_0: μ = 60   vs. H_1: μ ≠ 60

a. by using the critical value approach.

Because $(t_c = 2.6025) < (t_{0.01/2,(16-1)} = 2.9467)$,
we fail to reject H_0: μ=60 → accept H_0: μ=60 at a 1% significance level.

b. by using the p-value approach.

Because $P(t_c > 2.6025) = 0.01$ from the t-table in Chapter 8 with 15 d.f., p-value = 2 x 0.01 = 0.02 = 2%

Because (p-value=2%) > (α = 1%),
we fail to reject H₀: μ=60 → accept H₀: μ=60 at a 1% significance level.

c. by using the confidence-interval approach.

Because $\overline{X} - t_{\alpha/2,(n-1)} \cdot \frac{S}{\sqrt{n}} \le k \le \overline{X} + t_{\alpha/2,(n-1)} \cdot \frac{S}{\sqrt{n}}$ defines the region of accepting H_0, we find

$$64 - 2.9467 \cdot \frac{6}{\sqrt{16}} \le (k = 60) \le 64 + 2.9467 \cdot \frac{6}{\sqrt{16}}$$

59.4837 < (k=60) < 68.3238

We note that this 99% confidence interval does capture μ=k=60. Thus, we fail to reject H_0: μ=60 → accept H_0: μ=60 at a 1% significance level.

C. Given the following information,

Suppose that you know from previous experience that the average speed of a car on the highway is 60 m.p.h. The population standard deviation is 8 m.p.h. In order to find out whether the average speed of a car on the same highway has increased or not, you conducted a survey and found that the average speed of 25 cars surveyed was 64 m.p.h.

1. Verify at a 5% significance level if the average car speed had increased

a. State the null and the alternative hypotheses.

H_0: μ ≤ 60 vs. H_1: μ > 60

b. by using the critical value approach.

Because this is a upper-tail test, we must use the positive value of Z by looking it up in the Z table. That is, identify the critical or table value of Z at a significance level of α, Z_α. Therefore, we find $Z_\alpha = Z_{0.05} = 1.645$.

We now calculate Z value, Z_c, via the Z-formula as follows:

$$Z_c = \frac{\overline{X} - \mu}{\sigma_{\overline{X}}} = \frac{64 - 60}{\frac{8}{\sqrt{25}}} = 2.50$$

Finally, we compare Z_c against Z_α as follows and make a decision about accepting or rejecting H_0:

Because $(Z_c=2.50) > (Z_{0.05}=1.645)$,
we reject H_0: $\mu \leq 60$ → accept H_1: $\mu > 60$ at a 5% significance level.

c. by using the p-value approach.

Because $P(Z_c>2.50) = 0.0062$ from the Z-table,
p-value = 0.0062 = 0.62% ← No need to double this p-value because this is a one-tail test.

Because (p-value=0.62%) < (α = 5%),
we reject H_0: $\mu \leq 60$ → accept H_1: $\mu > 60$ at a 5% significance level.

d. by using the confidence-interval approach.

Because $\overline{X} \leq k + Z_\alpha \cdot \sigma_{\overline{X}}$ defines the region of accepting H_0, we find[123]

$$64 \not\leq 60 + 1.645 \cdot \frac{8}{\sqrt{25}} = 62.632$$

This shows that the sample mean, $\overline{X}$, of 64 does not support the claim that the population mean (μ) can be hypothesized to be less than or equal to 60 at a 95% confidence level. Therefore, we reject H_0: $\mu \leq 60$ → accept H_1: $\mu > 60$ at a 5% significance level.

2. Verify at a 1% significance level if the average car speed had increased.

a. State the null and the alternative hypotheses.

H_0: $\mu \leq 60$ vs. H_1: $\mu > 60$

b. by using the critical value approach.

Because $(Z_c=2.50) > (Z_{0.01}=2.33)$,
we reject H_0: $\mu \leq 60$ → accept H_1: $\mu > 60$ at a 1% significance level.

c. by using the p-value approach.

Because $P(Z_c>2.50) = 0.0062$ from the Z-table,
p-value = 0.0062 = 0.62%

[123] Alternatively, we can use the following confidence interval: $\overline{X} - Z_\alpha \cdot \frac{\sigma}{\sqrt{n}} \leq k$.

Because (p-value=0.62%) < (α = 1%),
we reject H_0: $\mu \leq 60$ → accept H_1: $\mu > 60$ at a 1% significance level.

d. by using the confidence-interval approach.

Because $\overline{X} \leq k + Z_\alpha \cdot \frac{\sigma}{\sqrt{n}}$ defines the region of accepting H_0,

we find

$$64 \nleq (60 + 2.33 \cdot \frac{8}{\sqrt{25}} = 63.728)$$

This shows that the sample mean, $\overline{x}$, of 64 does not support the claim that the population mean (μ) can be hypothesized to be less than or equal to 60 at a 99% confidence level. Therefore, we reject H_0: $\mu \leq 60$ → accept H_1: $\mu > 60$ at a 1% significance level.

D. Given the following information,

You know from previous experience that the average speed of a car on the highway is 60 m.p.h. In order to find out whether the average speed of a car on the same highway has increased or not, you conducted a survey and found that the average speed of 16 cars surveyed was 63.90375 m.p.h. and the sample variance was 36 m.p.h.

1. Verify at a 5% significance level if the average car speed had increased

a. State the null and the alternative hypotheses.

H_0: $\mu \leq 60$ vs. H_1: $\mu > 60$

b. by using the critical value approach.

Because $(t_c = \frac{63.90375 - 60}{\frac{\sqrt{36}}{\sqrt{16}}} = 2.6025) > (t_{0.05,(16-1)} = 1.7531)$,

Reject H_0: $\mu \leq 60$ → Accept H_1: $\mu > 60$.

c. by using the p-value approach.

Because $P(t_c > 2.6025) = 0.01$ from the t-table with 15 d.f.,
p-value = 0.01 = 1%

Because (p-value=1%) < (α = 5%),
we reject H_0: $\mu \leq 60$ → accept H_1: $\mu > 60$

d. by using the confidence-interval approach.

Because $\overline{X} \leq k + t_{\alpha,n-1} \cdot \frac{S}{\sqrt{n}}$ defines the acceptance region for H_0, we find

$$63.90375 \nleq (60 + 1.7531 \cdot \frac{\sqrt{36}}{\sqrt{16}} = 62.62965)$$

This shows that the sample mean, $\overline{x}$, of 63.90375 does not support the claim that the population mean (μ) can be hypothesized to be less than or equal to 60 at a 95% confidence level. Therefore, we reject H_0: $\mu \leq 60$ → accept H_1: $\mu > 60$ at a 5% significance level.

2. Verify at a 1% significance level if the average car speed had increased.

a. State the null and the alternative hypotheses.

H_0: $\mu \leq 60$ vs. H_1: $\mu > 60$

b. by using the critical value approach.

Because $(t_c = \frac{63.90375 - 60}{\frac{6}{\sqrt{16}}} = 2.6025) = (t_{0.01,(16-1)} = 2.6025)$,

we are at a borderline for either accepting or rejecting H_0: $\mu \leq 60$ → the convention in such a case is to accept H_0: $\mu \leq 60$ because the equality sign belongs in H_0.

c. by using the p-value approach.

Because $P(t_c > 2.6025) = 0.01$ from the t-table with 15 d.f.,
p-value = 0.01 = 1%

Because (p-value=1%) = (α = 1%),
we accept H_0: $\mu \leq 60$ because the equality sign belongs in H_0.

d. by using the confidence-interval approach.

Because $\overline{X} \leq k + t_{\alpha,n-1} \cdot \frac{S}{\sqrt{n}}$ defines the region of accepting H_0,

we find

$$63.90375 \leq (60 + 2.6025 \cdot \frac{\sqrt{36}}{\sqrt{16}} = 63.90375)$$

Because the equality holds, we accept H_0: $\mu \leq 60$ at a 1% significance level.

E. Given the following information,

You know from previous experience that the average speed of a car on the highway is 60 m.p.h. The population variance is 64 m.p.h. In order to find out whether the average speed of a car on the same highway has decreased or not, you conducted a survey and found that the average speed of 25 cars surveyed was 56 m.p.h.

1. Verify at a 5% significance level if the average car speed had decreased

   a. State the null and the alternative hypotheses.

   H_0: $\mu \geq 60$ vs. H_1: $\mu < 60$

   b. by using the critical value approach.

   Because ($Z_c = \dfrac{56-60}{\dfrac{\sqrt{64}}{\sqrt{25}}} = -2.5$) < ($Z_{0.05} = -1.645$),

   we reject H_0: $\mu \geq 60$ → Accept H_1: $\mu < 60$.

   c. by using the p-value approach.

   Because $P(Z_c < -2.5) = 0.0062$ from the Z-table,
   p-value = 0.0062 = 0.62%

   Because (p-value=0.62%) < (α = 5%),
   we reject H_0: $\mu \geq 60$ → Accept H_1: $\mu < 60$.

   d. by using the confidence-interval approach.

   Because $k - Z_\alpha \cdot \dfrac{\sigma}{\sqrt{n}} \leq \overline{X}$ defines the acceptance region for H_0,

   we find

   $$(60 - 1.645 \cdot \frac{\sqrt{64}}{\sqrt{25}} = 57.368) \nleq 56$$

   This shows that the sample mean, $\overline{x}$, of 56 does not support the claim that the population mean (μ) can be hypothesized to be greater than or equal to 60 at a 95% confidence level. Therefore, we reject H_0: $\mu \geq 60$ → Accept H_1: $\mu < 60$ at a 5% significance level.

2. Verify at a 1% significance level if the average car speed had decreased

   a. State the null and the alternative hypotheses.

   H_0: $\mu \geq 60$ vs. H_1: $\mu < 60$

b. by using the critical value approach.

Because ($Z_c = -2.5$) < ($Z_{0.01} = -2.33$),
we reject H_0: $\mu \geq 60$ → Accept H_1: $\mu < 60$.

c. by using the p-value approach.

Because $P(Z_c < -2.5) = 0.0062$ from the Z-table,
p-value = 0.0062 = 0.62%

Because (p-value=0.62%) < (α = 1%),
we reject H_0: $\mu \geq 60$ → Accept H_1: $\mu < 60$.

d. by using the confidence-interval approach.

Because $k - Z_{\alpha} \cdot \frac{\sigma}{\sqrt{n}} \leq \overline{X}$ defines the acceptance region for H_0,

we find

$$(60 - 2.33 \cdot \frac{\sqrt{64}}{\sqrt{25}} = 56.272) \nleq 56$$

This shows that the sample mean, $\overline{X}$, of 56 does not support the claim that the population mean (μ) can be hypothesized to be greater than or equal to 60 at a 99% confidence level. Therefore, we reject H_0: $\mu \geq 60$ → Accept H_1: $\mu < 60$ at a 1% significance level.

F. Given the following information,

You know from previous experience that the average speed of a car on the highway is 60 m.p.h. In order to find out whether the average speed of a car on the same highway has decreased or not, you conducted a survey and found that the average speed of 16 cars surveyed was 56.09625 m.p.h. and the sample standard deviation was 6 m.p.h.

1. Verify at a 5% significance level if the average car speed had decreased

a. State the null and the alternative hypotheses.

H_0: $\mu \geq 60$ vs. H_1: $\mu < 60$

b. by using the critical value approach.

Because ($t_c = \frac{56.09625 - 60}{\frac{6}{\sqrt{16}}} = -2.6025$) < ($t_{0.05,(16-1)} = -1.7531$),

Reject H_0: $\mu \geq 60$ → Accept H_1: $\mu < 60$.

c. by using the p-value approach.

Because $P(t_c < -2.6025) = 0.01$ from the t-table with 15 d.f.,
p-value = 0.01 = 1%

Because (p-value=1%) < (α = 5%),
we reject H_0: $\mu \geq 60$ → accept H_1: $\mu < 60$

d. by using the confidence-interval approach.

Because $k - t_{\alpha,n-1} \cdot \frac{S}{\sqrt{n}} \leq \overline{X}$ defines the acceptance region for H_0,

we find $(60 - 1.7531 \cdot \frac{6}{\sqrt{16}} = 57.37035) \nleq 56.09625$

Therefore, we reject H_0: $\mu \geq 60$ → accept H_1: $\mu < 60$.

2. Verify at a 1% significance level if the average car speed had decreased

a. State the null and the alternative hypotheses.

H_0: $\mu \geq 60$ vs. H_1: $\mu < 60$

b. by using the critical value approach.

Because $(t_c = -2.6025) \geq (t_{0.01,(16-1)} = -2.6025)$,
we accept H_0: $\mu \geq 60$ because the equality sign belongs in H_0.

c. by using the p-value approach.

Because $P(t_c < -2.6025) = 0.01$ from the t-table with 15 d.f.,
p-value = 0.01 = 1%

Because (p-value=1%) ≥ (α = 1%),
we accept H_0: $\mu \geq 60$ because the equality sign belongs in H_0.

d. by using the confidence-interval approach.

Because $k - t_{\alpha,n-1} \cdot \frac{S}{\sqrt{n}} \leq \overline{X}$ defines the acceptance region for H_0,

we find $(60 - 2.6025 \cdot \frac{6}{\sqrt{16}} = 56.09625) \leq 56.09625$

Therefore, we accept H_0: $\mu \geq 60$ because the equality sign belongs in H_0.

Exercise Problems on
Chapter 11. One-Sample Hypothesis Testing

This set of exercise problems has 50 Questions, worth a total of 50 points.

1. Inductive reasoning allows one to draw a conclusion about _____ from ______.

   a. a sample; a population   b. a sample; another sample
   c. a population; a sample   d. a population; another population
   e. the degree of dispersion; a standard normal distribution

2. The null hypothesis contains by convention ______.

   a. an equality sign   b. an inequality sign
   c. a plus and/or minus sign   d. a multiplication and/or division sign
   e. a value of zero or larger

3. An alternative hypothesis contains by convention _____.

   a. no equality sign   b. no inequality sign
   c. a plus and/or minus sign   d. a multiplication and/or division sign
   e. a value of zero or larger

4. The null hypothesis is also called the ____ hypothesis and expressed as _____.

   a. original; H_0 or H_o b. original; H_1 or H_a
   c. important; H_0 or H_o d. important; H_1 or H_a
   e. new; not H_0

5. The purpose of a hypothesis testing is to disapprove the _____ hypothesis in order to prove the _____ hypothesis.

   a. alternative; null   b. null; alternative
   c. opposite; alternative   d. null; another null
   e. new; old

6. An example of a null hypothesis in single hypothesis testing is _____ and that of composite or joint hypothesis testing is _____.

   a. H_0: $\mu_1 = \mu_2 = 0$; H_0: $\mu = 0$
   b. H_0: $\mu_1 = \mu_2 = 0$; H_0: $\mu \neq 0$
   c. H_0: $\mu \neq 0$; H_0: $\mu_1 = \mu_2 = 0$
   d. H_0: $\mu = 0$; H_0: $\mu_1 = \mu_2 = 0$
   e. H_0: $\mu \neq 0$; H_0: $\mu_1 \neq \mu_2 \neq 0$

7. When an alternative hypothesis is stated as Ha: $\mu \neq 0$, it means a _____ test.

a. one-tailed
b. two-tailed
c. composite
d. only (a) and (c) of the above
e. only (b) and (c) of the above

8. When an alternative hypothesis is states as Ha: $\mu > 0$, it means a _____ test.

a. one-tailed or directional
b. directional or upper-tail
c. directional or lower-tail
d. only (a) and (b) of the above
e. only (a) and (c) of the above

9. A two-tail test means that there is/are ______ rejection region(s) and a one-tail test means that there is/are _____ rejection region(s).

a. two; one
b. two; two
c. one; one
d. one; two
e. more than 2; only one

10. A significance level, α, is

a. a measure of Type I error.
b. a measure of Type II error.
c. equal to 1 minus a confidence level.
d. only (a) and (c) of the above.
e. only (b) and (c) of the above.

11. Rejecting the null hypothesis when it is in fact true involves an error called _____.

a. Type I error measured by a significance level.
b. Type II error measured by a significance level.
c. Type I error measured by a power of the test.
d. Type II error measured by a power of the test.
e. Type I error measured by a confidence level.

12. Accepting the null hypothesis when it is in fact false involves an error called _____.

a. Type II error often represented as $(1- \beta)$.
b. Type I error often represented as α.
c. Type II error often represented as α.
d. Type I error often represented as β.
e. Type II error often represented as β.

13. The power of the test is measured as _____.

a. α
b. $(1- \alpha)$
c. β
d. $(1- \beta)$
c. $(\alpha + \beta)$

14 The best way to minimize both Type I and Type II errors is _____.

a. to use experts of error minimization
b. to decrease the sample size
c. to increase the sample size
d. to choose a significance level of zero
e. to choose a power of the test to be zero

15. Which of the following is an approach (or a method) that can be used to conduct hypothesis testing?

a. the critical Z-value approach
b. the p-value approach
c. the confidence interval approach
d. all of the above
e. none of the above

16. If one sees a set of hypotheses as: H_0: $\mu \leq k$ vs. H_1: $\mu > k$, it is called a/an _____ test and if one sees a set of hypotheses as: H_0: $\mu \geq k$ vs. H_1: $\mu < k$, it is called a/an _____ test.

a. one-tail; upper-tail
b. lower-tail; upper-tail
c. regular two-tail; reverse two-tail
d. upper-tail; lower-tail
e. lower-tail; one-tail

17. The decision criterion for a two-tail test, using the critical Z-value approach and a significance level of α, includes _____.

a. If $|Z_c| \leq Z_{\alpha/2}$, or $-Z_{\alpha/2} \leq Z_c \leq +Z_{\alpha/2}$, then fail to reject H_0: μ=k → accept H_0
b. If $|Z_c| \leq Z_{\alpha}$, or $-Z_{\alpha} \leq Z_c \leq +Z_{\alpha}$, then fail to reject H_0: μ=k → accept H_0
c. If $|Z_c| \leq Z_{\alpha/2}$, or $-Z_{\alpha/2} \leq Z_c \leq +Z_{\alpha/2}$, then reject H_0: μ=k → accept H_a
d. If $|Z_c| \leq Z_{\alpha}$, or $-Z_{\alpha} \leq Z_c \leq +Z_{\alpha}$, then reject H_0: μ=k → accept H_a
e. none of the above.

18. The decision criterion for an upper-tail test, using the critical Z-value approach and a significance level of α, includes _____.

a. If $|Z_c| \leq Z_{\alpha/2}$, or $-Z_{\alpha/2} \leq Z_c \leq +Z_{\alpha/2}$, then fail to reject H_0 → accept H_0
b. If $Z_c \leq Z_\alpha$, then fail to reject H_0 → accept H_0
c. If $Z_c \geq Z_\alpha$, then fail to reject H_0 → accept H_0
d. If $Z_c > Z_\alpha$, then reject H_0 → accept H_a
e. only (b) and (d) of the above

19. The decision criterion for a lower-tail test, using the critical Z-value approach and a significance level of α, includes _____.

a. If $Z_c \leq -Z_{\alpha/2}$, then fail to reject H_0 → accept H_0
b. If $Z_c \geq -Z_{\alpha}$, then fail to reject H_0 → accept H_0
c. If $Z_c < -Z_{\alpha}$, then reject H_0 → accept H_a
d. only (b) and (c) of the above
e. none of the above

20. The decision criterion for a two-tail test, using the p-value or the prob-value approach and a significance level of α, includes _____.

a. If p-value $\geq \alpha$, then fail to reject the null hypothesis → accept H_0
b. If p-value $< \alpha$, then fail to reject the null hypothesis → accept H_0
c. If p-value $\geq \alpha/2$, then fail to reject the null hypothesis → accept H_0
d. If p-value $< \alpha/2$, then fail to reject the null hypothesis → accept H_0
e. none of the above

21. The decision criterion for a two-tail test, using the p-value or the prob-value approach and a significance level of α, includes _____.

a. If p-value $\geq \alpha$, then reject the null hypothesis → accept H_a
b. If p-value $< \alpha$, then reject the null hypothesis → accept H_a
c. If p-value $\geq \alpha/2$, then reject the null hypothesis → accept H_a
d. If p-value $< \alpha/2$, then reject the null hypothesis → accept H_a
e. none of the above

22. The decision criterion for an upper-tail test, using the p-value or the prob-value approach and a significance level of α, includes _____.

a. If p-value $> \alpha$, then fail to reject the null hypothesis → accept H_0
b. If p-value $< \alpha$, then fail to reject the null hypothesis → accept H_0
c. If p-value $> \alpha/2$, then fail to reject the null hypothesis → accept H_0
d. If p-value $< \alpha/2$, then fail to reject the null hypothesis → accept H_0
e. none of the above

23. The decision criterion for a lower-tail test, using the p-value or the prob-value approach and a significance level of α, includes _____.

a. If p-value $> \alpha$, then reject the null hypothesis → accept H_a
b. If p-value $< \alpha$, then reject the null hypothesis → accept H_a
c. If p-value $> \alpha/2$, then reject the null hypothesis → accept H_a
d. If p-value $< \alpha/2$, then reject the null hypothesis → accept H_a
e. none of the above

24. The decision criterion for a two-tail test, using the confidence interval approach and a significance level of α, includes _____.

a. if a hypothesized value such as $\mu = k$ falls within a confidence interval, fail to reject the null hypothesis → accept H_0

b. if a hypothesized value such as $\mu = k$ falls outside a confidence interval, fail to reject the null hypothesis → accept H_0

c. if a hypothesized value such as $\mu = k$ falls within a confidence interval, reject the null hypothesis → accept H_a

d. only (b) and (c) of the above.

e. none of the above

25. For hypothesis testing, a Z-test is used when the _____ standard deviation is _____ whereas a t-test is used when it is _____.

a. population; known; unknown b. population; unknown; known

c. sample; known; unknown d. sample; unknown; known

e. population; unknown; unknown

26. The decision criteria based on a critical (Z- or t-) value approach for accepting or rejecting a null hypothesis are ______.

a. more complicated for a Z-test because it relies on the sample standard deviation which is difficult to calculate.

b. more complicated for a t-test than a Z-test due to the degrees-of-freedom issue.

c. more complicated for a Z-test than a t-test due to the degrees-of-freedom issue.

d. more complicated for a t-test because it relies on the standardized normal probability distribution.

e. only (a) and (c) of the above

Use the following general situation to answer Questions 27 through 37.

Suppose that you are hired as a consultant to examine the production process of sugar-glazed donuts. The donut machine used is designed to put on 10 grams of sugar with an allowable variability, as measured by the standard deviation, of 1.5 grams for each donut made.

In order to see if the donut machine is doing working properly, you took a random sample of 25 donuts and found that the average amount of sugar on each donut was 11 grams.

27. If you wish to examine if the observed average of 11 grams per donut is statistically the same as the machine-designed weight of 10 grams, which of the following accurately states the null hypothesis that you wish to test?

a. H_1: $\mu = 11$ b. H_0: $\mu = 11$ c. H_1: $\mu = 10$

d. H_0: $\mu = 10$ e. none of the above

28. Which test statistic would you use?

a. Z-value b. t-value c. F-value
d. Chi-square value e. none of the above

29. Assuming that you are to test the following hypothesis using the above information and a critical Z-value approach at a 5% significance level,

H_0: $\mu = 10.5$ vs. H_1: $\mu \neq 10.5$

you would find the calculated Z-value, Z_c, to be _____ and the critical (or table) Z-value to be _____.

a. 0.33; 1.96 b. 0.33; 1.645 c. 1.67; 1.96
d. 1.67; 1.645 e. 1.96; 1.96

30. If the absolute value of the calculated Z-value is greater than the critical (or table) Z-value, your conclusion would be _____.

a. accept the null hypothesis. b. reject the null hypothesis.
c. fail to reject the null hypothesis d. only (a) and (c) of the above
e. none of the above

31. If the absolute value of the calculated Z-value is less than the critical (or table) Z-value, your conclusion would be _____.

a. accept the null hypothesis. b. reject the null hypothesis.
c. fail to reject the null hypothesis d. only (a) and (c) of the above
e. none of the above

32. Using the p-value approach, if the calculated Z-value is 2.27 for a two-tail test, you conclusion would be to _______ at a 5% significance level.

a. accept the null hypothesis. b. reject the null hypothesis.
c. fail to reject the null hypothesis d. only (a) and (c) of the above
e. none of the above

33. Using the p-value approach, if the calculated Z-value is 2.27 for a two-tail test, you conclusion would be to _______ at a 1% significance level.

a. accept the null hypothesis. b. reject the null hypothesis.
c. fail to reject the null hypothesis d. only (a) and (c) of the above
e. none of the above

34. Using the confidence-interval approach, if the sample mean is 11 grams and a population standard deviation of 1.5 grams, the 95% confidence interval for a two-tail test is found to be _____.

a. between 9.5 and 12.5 b. between 8 and 14
c. between 10.41 and 11.59 d. between 8.06 and 13.94
e. between 9.41 and 10.59

35. Using the confidence-interval approach for a two-tail test, if the hypothesized value falls within the constructed confidence interval, your conclusion would be to _____.

a. accept the null hypothesis b. reject the null hypothesis
c. fail to reject the null hypothesis d. only (a) and (c) of the above
e. none of the above

36. Assuming that you are to test the following hypothesis using the above information and a critical Z-value approach at a 5% significance level,

H_0: $\mu \geq 10.5$ vs. H_1: $\mu < 10.5$

you would find the calculated Z-value, Z_c, to be _____ and the critical (or table) Z-value to be _____.

a. 0.33; –1.96 b. 0.33; –1.645 c. 1.67; –1.96
d. 1.67; –1.645 e. 1.96; –1.645

37. If the calculated Z-value is greater than the critical (or table) Z-value for the above one-tail test, your conclusion would be _____.

a. accept the null hypothesis that $\mu \geq 10.5$.
b. reject the null hypothesis that $\mu \geq 10.5$.
c. fail to reject the null hypothesis that $\mu \geq 10.5$.
d. only (a) and (c) of the above
e. none of the above

Use the following general situation to answer Questions 38 through 46.

Suppose that you are hired as a marketing consultant to examine the average age of shoppers in a major shopping mall. After asking randomly chosen 100 people of their age, you received 49 unbiased truthful answers. Upon tabulation, you found that the average age was 35 with a variance of 1.96.

38. If you wish to examine if the observed average age of 35 is statistically the same as the average age of 38 that the president of the shopping mall strongly believes on the basis of

his daily observations, which of the following accurately states the null hypothesis that you wish to test?

a. H_0: $\mu = 35$ (b.) H_0: $\mu = 38$ c. H_0: $\mu \geq 35$
d. H_0: $\mu \leq 38$ e. H_0: $\mu > 35$

39. Which test statistic would you use?

a. Z-value b. t-value c. F-value
d. Chi-square value e. none of the above

40. Assuming that you are to test the following hypothesis using the above information and a critical t-value approach at a 5% significance level,

H_0: $\mu = 36.5$ vs. H_1: $\mu \neq 36.5$

you would find the calculated t-value, t_c, to be _____ and the critical (or table) t-value to be _____ and the corresponding degrees of freedom to be _____.

a. –7.5; $\pm$2.0106; 48 b. –7.5; $\pm$2.0096; 49
c. 5.357; $\pm$2.0106; 48 d. 5.357; $\pm$2.0096; 49
e. –7.65; $\pm$2.0106; 48

41. If the absolute value of the calculated t-value is greater than the critical (or table) t-value, your conclusion would be _____.

a. accept the null hypothesis. b. reject the null hypothesis.
c. fail to reject the null hypothesis d. only (a) and (c) of the above
e. none of the above

42. If the absolute value of the calculated t-value is less than the critical (or table) t-value, your conclusion would be _____.

a. accept the null hypothesis. b. reject the null hypothesis.
c. fail to reject the null hypothesis d. only (a) and (c) of the above
e. none of the above

43. Using the p-value approach, if the calculated t-value is 2.0423 with 30 degrees of freedom for a two-tail test, you conclusion would be to _______ at a 1% significance level.

a. accept the null hypothesis. b. reject the null hypothesis.
c. fail to reject the null hypothesis d. only (a) and (c) of the above
e. none of the above

44. Using the confidence-interval approach, if the sample mean age is 35 and the sample variance is 1.96 on the basis of 49 observations, the 95% confidence interval for a two-tail test is found to be _____.

a. between 33.04 and 36.96
b. between 33.6 and 36.4
c. between 34.6 and 35.4
d. between 34.72 and 35.28
e. between 31.06 and 38.94

45. Assuming that you are to test the following hypothesis using the above information and a critical t-value approach at a 5% significance level,

H_0: $\mu \leq 36.5$ vs. H_1: $\mu > 36.5$

you would find the calculated t-value, t_c, to be _____ and the critical (or table) t-value to be _____.

a. –7.5; 2.0106
b. 7.5; 2.0096
c. 5.357; 2.0106
d. 5.357; 1.6772
e. –7.5; 1.6772

46. If the calculated t-value is greater than the critical (or table) t-value for the above one-tail test, your conclusion would be _____.

a. accept the null hypothesis that $\mu \leq 36.5$.
b. reject the null hypothesis that $\mu \leq 36.5$.
c. fail to reject the null hypothesis that $\mu \leq 36.5$.
d. only (a) and (c) of the above
e. none of the above

47. When conducting a two-tail test, if the p-value is ____ than the chosen significance level, you would reject the _____ hypothesis at the _____.

a. smaller; null; p-value
b. large; null; significance level
c. smaller; alternative; significance level
d. larger; alternative; p-value
e. smaller; null; significance level

48. When conducting hypothesis testing, 3 approaches are commonly used. The three approaches include _____.

a. the critical Z- or t-value approach
b. the confidence or significance approach
c. the p-value or the prob-value approach
d. all of the above
e. only (a) and (c) of the above

49. When using the 3 approaches to hypothesis testing, the conclusion drawn _____.

a. should be the same regardless of which approach is used.
b. can be different depending upon what approach is used.
c. the p-value or prob-value approach yields the most accurate and consistent result.
d. only (a) and (c) of the above
e. only (b) and (c) of the above

50. The confidence interval associated with a 1% significance level is _____ that associated with a 5% significance level.

a. narrower than
b. wider than
c. the same as
d. can be narrower and can be wider depending on the size of the hypothesized value
e. is a better measure of accuracy than

Chapter 12. Two-Sample Hypothesis Testing

Up to this point, we have studied the hypothesis testing of a single mean against a specific hypothesized value such k. However, in this chapter, we broaden the analysis into two-sample cases. That is, we wish to examine if the means of two independent samples[124] are the same or not. Similarly, we will examine if the variances of two independent samples are the same or not.

In Section 1, the hypothesis testing of the equality of two independent sample means will be studied by using the Z-test and the t-test. In Section 2, on the other hand, the hypothesis testing of the equality of two independent sample variances will be examined via the F-test.

The most important point in this chapter is to understand whether the two samples are independent from each other or not and accordingly, when to use the Z-test, t-test, and F-test. **The procedures and interpretations of hypothesis testing essentially remain the same as in the previous chapters for the Z- and t-tests.** Furthermore, the critical Z- or t-value approach to hypothesis testing is most dominantly used for two-sample cases[125].

Section 1.
The Tests for the Difference Between Two Independent Means

In this section, we will study the hypothesis testing of the equality[126] of two independent sample means.

A. The Z-test When σ_1^2 and σ_2^2 of Two Independent Populations are Known

1. The Types of Hypotheses to be Tested.

a. For a two-tail test

$H_0 : \mu_1 = \mu_2$ is equivalent to $H_0 : \mu_1 - \mu_2 = 0$
and $H_0 : \mu_D = 0$ where $\mu_D = \mu_1 - \mu_2$

$H_1 : \mu_1 \neq \mu_2$ is equivalent to $H_1 : \mu_1 - \mu_2 \neq 0$
and $H_1 : \mu_D \neq 0$ where $\mu_D = \mu_1 - \mu_2$

b. For an upper-tail test

---

[124] If two samples are believed to be dependent, a method different from what is described herein should be used. The dependent cases will not be discussed herein.

[125] Therefore, only the critical Z- and t-value approach is presented herein.

[126] The "difference" between two means is equivalent to the "equality" of two means because if they are equal, there is no difference. Likewise, if they are not different, they are equal.

$H_0: \mu_1 \geq \mu_2$ is equivalent to $H_0: \mu_1 - \mu_2 \geq 0$
and $H_0: \mu_D \geq 0$ where $\mu_D = \mu_1 - \mu_2$
$H_1: \mu_1 < \mu_2$ is equivalent to $H_1: \mu_1 - \mu_2 < 0$
and $H_1: \mu_D < 0$ where $\mu_D = \mu_1 - \mu_2$

c. For a lower-tail test

$H_0: \mu_1 \leq \mu_2$ is equivalent to $H_0: \mu_1 - \mu_2 \leq 0$
and $H_0: \mu_D \leq 0$ where $\mu_D = \mu_1 - \mu_2$
$H_1: \mu_1 > \mu_2$ is equivalent to $H_1: \mu_1 - \mu_2 > 0$
and $H_1: \mu_D \leq 0$ where $\mu_D = \mu_1 - \mu_2$

2. The Decision Criteria for an Equality Test[127] of

$$H_0: \mu_1 = \mu_2 \qquad \text{vs.} \qquad H_1: \mu_1 \neq \mu_2$$

Alternatively,

$$H_0: \mu_D = 0 \qquad \text{vs.} \qquad H_1: \mu_D \neq 0 \quad \text{where } \mu_D = \mu_1 - \mu_2$$

Step 1. Calculate the Z-test statistic, Z_c, as follows:

$$Z_c = \frac{\overline{X}_D - \mu_D}{\sigma_D} = \frac{(\overline{X}_1 - \overline{X}_2) - (\mu_1 - \mu_2)}{\sqrt{\frac{\sigma_1^2}{n_1} + \frac{\sigma_2^2}{n_2}}}$$

Step 2. Select a significance level, α, and find the corresponding critical table value of Z, $Z_{\alpha/2}$.

Step 3. Apply the decision criteria as follows:

If $|Z_c| \leq Z_{\alpha/2}$, then accept the null hypothesis of $H_0: \mu_1 = \mu_2$ or $H_0: \mu_1 - \mu_2 = 0$ at an α% significance level.

Otherwise, reject the null hypothesis and accept the alternative hypothesis of $H_1: \mu_1 \neq \mu_2$ or $H_1: \mu_1 - \mu_2 \neq 0$ at an α% significance level.

[127] The critical Z- or t-value approach is most commonly used. Therefore, only this method is described herein.

3. An Example:

Suppose that $\overline{X_1}$ =10, $\overline{X_2}$ =15, σ_1^2= 5, and σ_2^2= 6 where n_1 = 20 and n_2 = 30. Test if both sample means are equal to each other at a 5% significance level.

a. The Hypothesis to be tested

$$H_0: \mu_1 = \mu_2 \rightarrow H_0: \mu_1 - \mu_2 = 0 \text{ vs.}$$
$$H_1: \mu_1 \neq \mu_2 \rightarrow H_1: \mu_1 - \mu_2 \neq 0$$

b. $$Z_c = \frac{(\overline{X_1} - \overline{X_2}) - (\mu_1 - \mu_2)}{\sqrt{\frac{\sigma_1^2}{n_1} + \frac{\sigma_2^2}{n_2}}} = \frac{(10-15)-(0)}{\sqrt{\frac{5}{20} + \frac{6}{30}}} = \frac{-5}{\sqrt{0.45}} = -7.45$$

c. $Z_{\alpha/2} = Z_{0.05/2}$ = 1.96 from the Z-table.

d. Because $|Z_c|$=7.45 > $Z_{\alpha/2}$=1.96, reject the null hypothesis and accept the alternative hypothesis of $H_1: \mu_1 \neq \mu_2$ or $H_1: \mu_1 - \mu_2 \neq 0$ at a 5% significance level.→ The two means are statistically NOT equal at a 5% significance level.

B. The t-test When σ_1^2 and σ_2^2 of Two Independent Populations are Unknown But Assumed to Be Equal → Their sample variances, S_1^2 and S_2^2, are known.

1. The Types of Hypotheses to be Tested.

$H_0: \mu_1 = \mu_2$ is equivalent to $H_0: \mu_1 - \mu_2 = 0$ vs.
$H_1: \mu_1 \neq \mu_2$ is equivalent to $H_1: \mu_1 - \mu_2 \neq 0$

$H_0: \mu_1 \geq \mu_2$ is equivalent to $H_0: \mu_1 - \mu_2 \geq 0$ vs.
$H_1: \mu_1 < \mu_2$ is equivalent to $H_1: \mu_1 - \mu_2 < 0$

$H_0: \mu_1 \leq \mu_2$ is equivalent to $H_0: \mu_1 - \mu_2 \leq 0$ vs.
$H_1: \mu_1 > \mu_2$ is equivalent to $H_1: \mu_1 - \mu_2 > 0$

2. The Procedures and Decision Criteria for an Equality Test of

$$H_0: \mu_1 = \mu_2 \quad \text{vs.} \quad H_1: \mu_1 \neq \mu_2$$

Step 1. Calculate the t-test statistic, t_c, as follows:

$$t_c = \frac{\overline{X_D} - \mu_D}{\sqrt{adjusted(S_P^2)}} = \frac{(\overline{X_1} - \overline{X_2}) - (\mu_1 - \mu_2)}{\sqrt{S_p^2 \cdot \left(\frac{1}{n_1} + \frac{1}{n_2}\right)}}$$

where

$$\text{Pooled Sample Variance} = S_p^2 = \frac{(n_1 - 1)S_1^2 + (n_2 - 1)S_2^2}{(n_1 - 1) + (n_2 - 1)}$$

Step 2. Select a significance level, α, and find the corresponding critical table value of t, $t_{\alpha/2, n_1+n_2-2}$.

Step 3. Apply the decision criteria as follows:

If $|t_c| \le t_{\alpha/2, n_1+n_2-2}$, then accept the null hypothesis of $H_0 : \mu_1 = \mu_2$ or $H_0 : \mu_1 - \mu_2 = 0$.

Otherwise, reject the null hypothesis and accept the alternative hypothesis of $H_1 : \mu_1 \ne \mu_2$ or $H_1 : \mu_1 - \mu_2 \ne 0$.

3. An Example:

Suppose that $\overline{X_1} = 10$, $\overline{X_2} = 12.5$, $S_1^2 = 8$, and $S_2^2 = 9$ where $n_1 = 20$ and $n_2 = 15$. Assuming that both sample variances are equal, test if both sample means are equal to each other at a 5% significance level.

a. The Hypothesis to be tested

$$H_0 : \mu_1 = \mu_2 \rightarrow H_0 : \mu_1 - \mu_2 = 0 \text{ vs.}$$
$$H_1 : \mu_1 \ne \mu_2 \rightarrow H_1 : \mu_1 - \mu_2 \ne 0$$

b. Calculate the t-test statistic, t_c, as follows:

$$t_c = \frac{(\overline{X_1} - \overline{X_2}) - (\mu_1 - \mu_2)}{\sqrt{S_p^2 \cdot \left(\frac{1}{n_1} + \frac{1}{n_2}\right)}} = \frac{(10 - 12.5) - 0}{\sqrt{8.42 \cdot \left(\frac{1}{20} + \frac{1}{15}\right)}} = \frac{-2.5}{\sqrt{0.98}} = -2.525$$

where the Pooled Variance =

$$S_p^2 = \frac{(n_1-1)S_1^2+(n_2-1)S_2^2}{(n_1-1)+(n_2-1)} = \frac{19\cdot 8+14\cdot 9}{(20-1)+(15-1)} = \frac{278}{33} = 8.42$$

c. At a significance level, α, of 5%, the corresponding critical table value of t, $t_{\alpha/2,n_1+n_2-2}=t_{0.05/2,20+15-2}=t_{0.025,33}=2.0345$.

d. Because $|t_c|$=2.525 > $t_{\alpha/2,n_1+n_2-2}$=2.0345, reject the null hypothesis and accept the alternative hypothesis of $H_1: \mu_1 \neq \mu_2$ or $H_1: \mu_1 - \mu_2 \neq 0$ at a 5% significance level.→ The two means are statistically NOT equal at a 5% significance level.

An Example of a Two-Sample Test of Temperature

1. Given the following temperature data on the Memorial Day and the Labor Day in the Chicago area, can you verify the claim that Labor Day is as warm as Memorial Day at a 95% confidence level? Before conducting the analysis, identify all assumptions that you need to make to conduct this analysis.

Year	Memorial Day			Labor Day		
	High	Low	Average	High	Low	Average
2000	67	44	55.5	70	61	65.5
2001	67	50	58.5	85	62	73.5
2002	78	49	63.5	82	66	74
2003	71	46	58.5	65	61	63
2004	66	58	62	80	56	68
2005	74	46	60	87	58	72.5
2006	91	68	79.5	71	60	65.5
Average=	73.42857	51.57143	62.5	77.14286	60.57143	68.85714
Variance=	78.95238	73.28571	62.91667	71.14286	9.952381	19.80952

Answer:

Assumptions:
(1) 7 observations are adequate enough for the analysis.
(2) Memorial Day and Labor Day temperatures are independent of each other.
(3) Variances for Memorial and Labor Days are assumed to be equal.

Solution:

a. The Hypothesis to be tested

$$H_0: \mu_M = \mu_L \rightarrow H_0: \mu_M - \mu_L = 0 \text{ vs.}$$
$$H_1: \mu_M \neq \mu_L \rightarrow H_1: \mu_M - \mu_L \neq 0$$

b. Because population variances are unknown, the t-test is used by calculating the t-test statistic, t_c, as follows:

$$t_c = \frac{(\overline{X_M} - \overline{X_L}) - (\mu_M - \mu_L)}{\sqrt{S_p^2 \cdot \left(\frac{1}{n_M} + \frac{1}{n_L}\right)}} = \frac{(62.5 - 68.86) - 0}{\sqrt{41.36 \cdot \left(\frac{1}{7} + \frac{1}{7}\right)}} = \frac{-6.36}{\sqrt{11.82}} = \frac{-6.36}{3.4376} = -1.85$$

where

Pooled Variance =

$$S_p^2 = \frac{(n_M - 1)S_M^2 + (n_L - 1)S_L^2}{(n_M - 1) + (n_L - 1)} = \frac{6 \cdot 62.91 + 6 \cdot 19.81}{(7-1) + (7-1)} = \frac{496.32}{12} = 41.36$$

because $S_M^2 = 62.91$ and $S_L^2 = 19.81$

c. At a significance level, α, of 5%, the corresponding critical table value of t, $t_{\alpha/2, n_1+n_2-2} = t_{0.05/2, 7+7-2} = t_{0.025,12} = 2.1788$.

d. Because $|t_c| = 1.85 < t_{\alpha/2, n_1+n_2-2} = 2.1788$, accept the null hypothesis of $H_0 : \mu_1 = \mu_2$ at a 5% significance level. → The two means are statistically equal at a 5% significance level.

Because we can intuitively conclude that the sample variances are not equal, we should resort to the case of unequal variances by utilizing the Satterthwaite's Separate-Variance t-test, which may be found in an advanced textbook[128].

2. Can you verify the claim that Labor Day is warmer than Memorial Day by 6 degrees?

Answer:

Because the two temperature means were judged to be the same as shown above, we can claim that the Labor Day is not warmer than Memorial Day by 6 degrees. (Of course, we have not done the correct statistical test and thus, there is a room for us to be sued if we insist on this conclusion. Therefore, be careful about what you say and claim.)

[128] This statistical test is not described herein.

3. Do you think that there is a significant difference in temperature variability between Memorial Day and Labor Day at a 5% significance level? Can you tell which day has a more erratic temperature fluctuation? (Hint: One must adopt the procedure described in Section 2 below to test for temperature variability.)

Answer: The following is a preview of Section 2 of this Chapter.

Test $H_0: \sigma_M^2 = \sigma_L^2$ vs $H_1: \sigma_M^2 \neq \sigma_L^2$ at a 5% significance level.

a. Calculate $F_c = \frac{\sigma_M^2}{\sigma_L^2} = \frac{62.92}{19.81} = 3.1762$

b. Find $F_{table,\,0.05/2} = F_{7-1,0.025}^{7-1} = 5.12 = F_U$

c. Since $(F_c = 3.1762) < (F_U = 5.12)$,
accept $H_0: \sigma_M^2 = \sigma_L^2$ at a α% significance level.

Therefore, the temperature variability between Memorial Day and Labor Day is statistically judged the same despite the difference in their numerical magnitude. This (somewhat surprising) result may be due to the small sample size of 7.

Section 2.
The F-Test for the Difference Between Two Independent Variances

In this section, we will study the hypothesis testing of the equality of two independent sample variances. First of all, however, we must know how to use F-probability tables.

A. The F-Probability Tables

Because an F-statistic is a ratio of two variances, it is always a positive number. This means that (1) the value of an F-statistic is non-negative; (2) this characteristic makes an F-probability distribution generally skewed to the right as shown in the diagram below; and (3) it is sensitive to the number of observations used in calculating the two variances which are called the degrees of freedom (DF). If the number of observations for the numerator variance is N_1 and that for the denominator variance is N_2, then n_1 $(=N_1 - 1)$ and n_2 $(=N_2 - 1)$ are the respective degrees of freedom for the numerator and the denominator variances and denoted as $F^{n_1}_{n_2,\alpha}$. Note that α is the significance level chosen for a hypothesis testing.

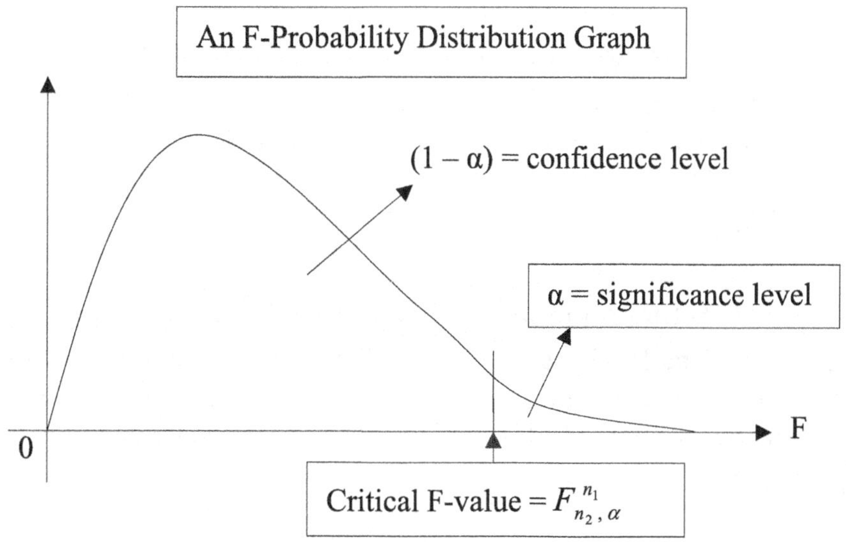

Depending on the level of significance, we can now generate following F-tables via the Excel command of "=FINV(Probability, Numerator DF, Denominator DF)" where the probability is the significance level; the numerator degrees of freedom is denoted as n_1; and the denominator degrees of freedom is denoted as n_2. For each level of significance chosen, a separate F table is generated.

The F-values shown inside the body of the table is the critical value that defines the probability in the upper tail only. F-probability tables are arranged first by the significance level; then, the degrees of freedom associated with the numerator and the denominator. Therefore, in order to find, for example, the critical value of $F^{n_1}_{n_2,\alpha}$ from the table, you need to identify the F-table that has the α value that you are looking for; then, identify the respective degrees of freedom, n_1 and n_2.

More specifically, the following steps can be taken:

Step 1:
Identify the appropriate table corresponding to the significance level, α, chosen. For example, if $\alpha = 0.05=5\%$, find the table that says "Upper-tail Probability of 0.05". If $\alpha = 0.01=1\%$, find the table that says "Upper-tail Probability of 0.01".

Step 2:
Identify the numerator degrees of freedom, n_1, from the first row and the denominator degrees of freedom, n_2, from the first column.

Step 3:
The intersection of the row associated with n_2 and the column associated with n_1 identifies the table value of F, $F_t = F^{n_1}_{n_2,\alpha}$.

Examples:

1. Find the critical (or table) value of $F^{8}_{10,0.05}$, using the partial F-tables shown below.
   (1) Go to the F-table with α =0.05. (2) Find the numerator DF, n_1, of 8 from the top row. (3) Find the denominator DF, n_2, of 10. (4) Identify the value at the intersection of the row and the column. (5) It is 3.07 which is the critical value of F. That is, $F^{8}_{10,0.05}=3.07$.

2. Find the critical (or table) value of $F^{6}_{5,0.025}$, using the partial F-tables shown below.
   (1) Go to the F-table with α =0.025. (2) Find the numerator DF, n_1, of 6 from the top row. (3) Find the denominator DF, n_2, of 5. (4) Identify the value at the intersection of the row and column. (5) It is 6.98 which is the critical value of F. That is, $F^{6}_{5,0.025}=6.98$.

3. Find the critical (or table) value of $F^{3}_{9,0.01}$, using the partial F-tables shown below.
   (1) Go to the F-table with α =0.01. (2) Find the numerator DF, n_1, of 3 from the top row. (3) Find the denominator DF, n_2, of 9. (4) Identify the value at the intersection of the row and column. (5) It is 6.99 which is the critical value of F. That is, $F^{3}_{9,0.01}=6.99$.

Table 12-1. The Critical Values of F for a Upper-Tail Probability of 0.05 (=5%)

DF	Numerator Degrees of Freedom, n_1										
n_2	**1**	**2**	**3**	**4**	**5**	**6**	**7**	**8**	**9**	**10**	**12**
1	161	199	216	225	230	234	237	239	241	242	244
2	18.51	19.00	19.16	19.25	19.30	19.33	19.35	19.37	19.38	19.40	19.41
3	10.13	9.55	9.28	9.12	9.01	8.94	8.89	8.85	8.81	8.79	8.74
4	7.71	6.94	6.59	6.39	6.26	6.16	6.09	6.04	6.00	5.96	5.91
5	6.61	5.79	5.41	5.19	5.05	4.95	4.88	4.82	4.77	4.74	4.68
6	5.99	5.14	4.76	4.53	4.39	4.28	4.21	4.15	4.10	4.06	4.00
7	5.59	4.74	4.35	4.12	3.97	3.87	3.79	3.73	3.68	3.64	3.57
8	5.32	4.46	4.07	3.84	3.69	3.58	3.50	3.44	3.39	3.35	3.28
9	5.12	4.26	3.86	3.63	3.48	3.37	3.29	3.23	3.18	3.14	3.07
10	4.96	4.10	3.71	3.48	3.33	3.22	3.14	3.07	3.02	2.98	2.91

Table 12-2. The Critical Values of F for a Upper-Tail Probability of 0.025 (=2.5%)

DF	Numerator Degrees of Freedom, n_1										
n_2	**1**	**2**	**3**	**4**	**5**	**6**	**7**	**8**	**9**	**10**	**12**
1	648	799	864	900	922	937	948	957	963	969	977
2	38.51	39.00	39.17	39.25	39.30	39.33	39.36	39.37	39.39	39.40	39.41
3	17.44	16.04	15.44	15.10	14.88	14.73	14.62	14.54	14.47	14.42	14.34
4	12.22	10.65	9.98	9.60	9.36	9.20	9.07	8.98	8.90	8.84	8.75
5	10.01	8.43	7.76	7.39	7.15	6.98	6.85	6.76	6.68	6.62	6.52
6	8.81	7.26	6.60	6.23	5.99	5.82	5.70	5.60	5.52	5.46	5.37
7	8.07	6.54	5.89	5.52	5.29	5.12	4.99	4.90	4.82	4.76	4.67
8	7.57	6.06	5.42	5.05	4.82	4.65	4.53	4.43	4.36	4.30	4.20
9	7.21	5.71	5.08	4.72	4.48	4.32	4.20	4.10	4.03	3.96	3.87
10	6.94	5.46	4.83	4.47	4.24	4.07	3.95	3.85	3.78	3.72	3.62

Table 12-3. The Critical Values of F for a Upper-Tail Probability of 0.01 (=1%)

DF	Numerator Degrees of Freedom, n_1										
n_2	**1**	**2**	**3**	**4**	**5**	**6**	**7**	**8**	**9**	**10**	**12**
1	4052	4999	5403	5625	5764	5859	5928	5981	6022	6056	6106
2	98.50	99.00	99.17	99.25	99.30	99.33	99.36	99.37	99.39	99.40	99.42
3	34.12	30.82	29.46	28.71	28.24	27.91	27.67	27.49	27.35	27.23	27.05
4	21.20	18.00	16.69	15.98	15.52	15.21	14.98	14.80	14.66	14.55	14.37
5	16.26	13.27	12.06	11.39	10.97	10.67	10.46	10.29	10.16	10.05	9.89
6	13.75	10.92	9.78	9.15	8.75	8.47	8.26	8.10	7.98	7.87	7.72
7	12.25	9.55	8.45	7.85	7.46	7.19	6.99	6.84	6.72	6.62	6.47
8	11.26	8.65	7.59	7.01	6.63	6.37	6.18	6.03	5.91	5.81	5.67
9	10.56	8.02	6.99	6.42	6.06	5.80	5.61	5.47	5.35	5.26	5.11
10	10.04	7.56	6.55	5.99	5.64	5.39	5.20	5.06	4.94	4.85	4.71

B. The Test for the Equality of Two Variances

Test at a α% significance level $H_0 : \sigma_1^2 = \sigma_2^2$ vs $H_1 : \sigma_1^2 \neq \sigma_2^2$

1. Calculate $F_{calculated} = F_c = \dfrac{S_1^2}{S_2^2}$

   where the numerator is always greater than the denominator.

   That is, $S_1^2 \geq S_2^2$. Therefore, we always find $F_c \geq 1$.

2. Obtain a table value of F – denoted as either $F_{table,\alpha/2}$ or F_t or F_U – from the F-table with n_1 (=N_1–1) for the numerator and n_2 (=N_2–1) degrees of freedom for a denominator at a α/2% significance level. Note that n_1 is the sample size for S_1^2 and n_2 is the sample size for S_2^2.

3. If $\dfrac{1}{F^*_{table,\alpha/2}} \leq F_c \leq F_{table,\alpha/2}$, then accept $H_0 : \sigma_1^2 = \sigma_2^2$ at a α% significance level where $F^*_{table,\alpha/2} = F_U^*$ = the table value of F with n_2 (=N_2–1) for the numerator and n_1 (=N_1–1) degrees of freedom for a denominator at a α/2% significance level → Note that the degrees of freedom are reversed from the case of $F_{table,\alpha/2}$.

   Otherwise, reject $H_0 : \sigma_1^2 = \sigma_2^2$ → Accept $H_1 : \sigma_1^2 \neq \sigma_2^2$ at a α% significance level.

4. Alternatively,

   If $F_c \leq F_{table,\alpha/2}$, then accept $H_0 : \sigma_1^2 = \sigma_2^2$ at a α% significance level.
   If $F_c > F_{table,\alpha/2}$, then reject $H_0 : \sigma_1^2 = \sigma_2^2$ → Accept $H_1 : \sigma_1^2 \neq \sigma_2^2$ at a α% significance level.

C. The Tests for the Inequality of Two Variances: Case 1

Test at a α% significance level $H_0 : \sigma_1^2 \leq \sigma_2^2$ vs $H_1 : \sigma_1^2 > \sigma_2^2$

1. Calculate $F_{calculated} = F_c = \dfrac{S_1^2}{S_2^2}$

   where the numerator is always greater than the denominator.

   That is, $S_1^2 \geq S_2^2$. Therefore, we always find $F_c \geq 1$.

2. Obtain a table value of F – F_{table} or F_t – from the F-table with n_1 (=N_1–1) and n_2 (=N_2–1) degrees of freedom at a α% significance level. Note that N_1 is the sample size for S_1^2 and N_2 is the sample size for S_2^2.

3. If $F_c \leq F_{table,\alpha}$, then accept $H_0 : \sigma_1^2 \leq \sigma_2^2$ at a α% significance level.
   If $F_c > F_{table,\alpha}$, then reject $H_0 : \sigma_1^2 \leq \sigma_2^2$ → Accept $H_1 : \sigma_1^2 > \sigma_2^2$.

D. The Tests for the Inequality of Two Variances: Case 2

Test at a α% significance level $H_0 : \sigma_1^2 \geq \sigma_2^2$ vs $H_1 : \sigma_1^2 < \sigma_2^2$

1. Calculate $F_{calculated} = F_c = \dfrac{S_1^2}{S_2^2}$

   where the numerator is always greater than the denominator.

   That is, $S_1^2 \geq S_2^2$. Therefore, we always find $F_c \geq 1$.

2. Obtain $F^*_{table,\alpha} = F_U^*$ = the table value of F with n_2 (=N_2–1) for the numerator and n_1 (=N_1–1) degrees of freedom for a denominator at a α% significance level → Note that the degrees of freedom are reversed from the case of $F_{table,\alpha}$.

3. If $F_c \geq \dfrac{1}{F^*_{table,\alpha}}$, then accept $H_0 : \sigma_1^2 \geq \sigma_2^2$ at a α% significance level.

   If $F_c < \dfrac{1}{F^*_{table,\alpha}}$, then reject $H_0 : \sigma_1^2 \geq \sigma_2^2$ → Accept $H_1 : \sigma_1^2 < \sigma_2^2$.

Note:

$F_U = F_{table}$ and $F_L = \dfrac{1}{F_U^*} = \dfrac{1}{F^*_{table}}$ with respect to the appropriate α.

Exercise Problems for Section 2
The F-Test for the Difference Between Two Independent Variances

A. Suppose that you own an ice cream shop in Chicago. Over the last 10 years, you found the following monthly revenues and variances of monthly revenues from your operation during the summer and the winter months:

	Summer	Winter
Monthly Revenue	$500	$200
Variances	$400	$100

1. Verify at a 5% and a 1% significance levels if seasonal differences affect revenue variability.

a. Test $H_0 : \sigma_s^2 = \sigma_w^2$ vs $H_1 : \sigma_s^2 \neq \sigma_w^2$ at a 5% significance level.

i. Calculate $F_c = \frac{\sigma_s^2}{\sigma_w^2} = \frac{400}{100} = 4$

ii. Find $F_{table,\ 0.05/2} = F_{10-1\ ,0.025}^{10-1} = 4.03 = F_U$

iii. Since $(F_c = 4) < (F_U = 4.03)$, accept $H_0 : \sigma_s^2 = \sigma_w^2$ at a 5% significance level.

b. Test $H_0 : \sigma_s^2 = \sigma_w^2$ vs $H_1 : \sigma_s^2 \neq \sigma_w^2$ at a 1% significance level.

i. Calculate $F_c = \frac{\sigma_s^2}{\sigma_w^2} = \frac{400}{100} = 4$

ii. Find $F_{table,\ 0.01/2} = F_{10-1\ ,0.005}^{10-1} = 6.54 = F_U$

iii. Since $(F_c = 4) < (F_U = 6.54)$, accept $H_0 : \sigma_s^2 = \sigma_w^2$ at a 1% significance level.

2. Verify at a 5% and a 1% significance levels if summer shows a higher revenue variability than winter.

a. Test $H_0 : \sigma_s^2 \leq \sigma_w^2$ vs $H_1 : \sigma_s^2 > \sigma_w^2$ at a 5% significance level.

i. Calculate $F_c = \frac{\sigma_s^2}{\sigma_w^2} = \frac{400}{100} = 4$

ii. Find $F_{table,\,0.05} = F_{10-1\ ,0.05}^{10-1} = 3.18 = F_U$

iii. Since ($F_c = 4$) > ($F_U = 3.18$), reject $H_0 : \sigma_s^2 \le \sigma_w^2$ → accept $H_1 : \sigma_s^2 > \sigma_w^2$ at a 5% significance level.

b. Test $H_0 : \sigma_s^2 \le \sigma_w^2$ vs $H_1 : \sigma_s^2 > \sigma_w^2$ at a 1% significance level.

i. Calculate $F_c = \frac{\sigma_s^2}{\sigma_w^2} = \frac{400}{100} = 4$

ii. Find $F_{table,\,0.01} = F_{10-1\ ,0.01}^{10-1} = 5.35 = F_U$

iii. Since ($F_c = 4$) < ($F_U = 5.35$), accept $H_0 : \sigma_s^2 \le \sigma_w^2$ at a 1% significance level.

3. Verify at a 5% and a 1% significance levels if summer shows a lower revenue variability than winter.

a. Test $H_0 : \sigma_s^2 \ge \sigma_w^2$ vs $H_1 : \sigma_s^2 < \sigma_w^2$ at a 5% significance level.

i. Calculate $F_c = \frac{\sigma_s^2}{\sigma_w^2} = \frac{400}{100} = 4$

ii. Find $F_L = \frac{1}{F_U^*} = \frac{1}{F_{table}^*} = \frac{1}{3.18} = 0.3145$

Note: In this case, because the numerator and denominator degrees of freedom are the same, $F_{table} = F_{table}^*$

iii. Since ($F_c = 4$) > ($F_L = 0.3145$), accept $H_0 : \sigma_s^2 \ge \sigma_w^2$

b. Test $H_0 : \sigma_s^2 \ge \sigma_w^2$ vs $H_1 : \sigma_s^2 < \sigma_w^2$ at a 1% significance level.

i. Calculate $F_c = \frac{\sigma_s^2}{\sigma_w^2} = \frac{400}{100} = 4$

ii. Find $F_L = \frac{1}{F_U^*} = \frac{1}{F_{table}^*} = \frac{1}{5.35} = 0.1869$

Note: In this case, because the numerator and denominator degrees of freedom are the same, $F_{table} = F_{table}^*$

iii. Since ($F_c = 4$) > ($F_L = 0.1869$), accept $H_0 : \sigma_s^2 \geq \sigma_w^2$

B. Suppose that you own two ice cream shops – one in Chicago and the other in Rolling Meadows. During the summer months over many years, you had observed sales revenues and found the following monthly revenues and variabilities of monthly revenues for these two shops:

	Chicago	Rolling Meadows
Monthly Revenue	\$3000	\$2500
Standard Deviation	\$45	\$25
Number of months (n)	31	41

1. Verify at a 5% and a 1% significance levels if both locations have the same revenue variability

a. Test $H_0 : \sigma_{CH}^2 = \sigma_{RM}^2$ vs $H_1 : \sigma_{CH}^2 \neq \sigma_{RM}^2$ at a 5% significance level.

i. Calculate $F_c = \frac{\sigma_{CH}^2}{\sigma_{RM}^2} = \frac{45^2}{25^2} = \frac{2025}{625} = 3.24$

ii. Find $F_{table,\ 0.05/2} = F_{41-1\ ,0.025}^{31-1} = F_{40,0.025}^{30} = 1.94 = F_U$

iii. Since ($F_c = 3.24$) > ($F_U = 1.94$), reject $H_0 : \sigma_{CH}^2 = \sigma_{RM}^2$ → Accept $H_1 : \sigma_{CH}^2 \neq \sigma_{RM}^2$ at a 5% significance level.

b. Test $H_0 : \sigma_{CH}^2 = \sigma_{RM}^2$ vs $H_1 : \sigma_{CH}^2 \neq \sigma_{RM}^2$ at a 1% significance level.

i. Calculate $F_c = \frac{\sigma_{CH}^2}{\sigma_{RM}^2} = \frac{45^2}{25^2} = \frac{2025}{625} = 3.24$

ii. Find $F_{table,\ 0.01/2} = F_{41-1\ ,0.005}^{31-1} = F_{40,0.005}^{30} = 2.40 = F_U$

iii. Since ($F_c = 3.24$) > ($F_U = 2.40$), reject $H_0 : \sigma_{CH}^2 = \sigma_{RM}^2$ → Accept $H_1 : \sigma_{CH}^2 \neq \sigma_{RM}^2$ at a 1% significance level.

2. Verify at a 5% and a 1% significance levels if the Chicago shop has a higher revenue variability than the Rolling Meadows shop.

   a. Test $H_0: \sigma^2_{CH} \leq \sigma^2_{RM}$ vs $H_1: \sigma^2_{CH} > \sigma^2_{RM}$ at a 5% significance level.

      i. Calculate $F_c = \frac{\sigma^2_{CH}}{\sigma^2_{RM}} = \frac{45^2}{25^2} = \frac{2025}{625} = 3.24$

      ii. Find $F_{table,\ 0.05} = F^{31-1}_{41-1\ ,0.05} = F^{30}_{40,0.05} = 1.74 = F_U$

      iii. Since ($F_c = 3.24$) > ($F_U = 1.74$), reject $H_0: \sigma^2_{CH} \leq \sigma^2_{RM}$ → Accept $H_1: \sigma^2_{CH} > \sigma^2_{RM}$ at a 5% significance level.

   b. Test $H_0: \sigma^2_{CH} \leq \sigma^2_{RM}$ vs $H_1: \sigma^2_{CH} > \sigma^2_{RM}$ at a 1% significance level.

      i. Calculate $F_c = \frac{\sigma^2_{CH}}{\sigma^2_{RM}} = \frac{45^2}{25^2} = \frac{2025}{625} = 3.24$

      ii. Find $F_{table,\ 0.01} = F^{31-1}_{41-1,0.01} = F^{30}_{40,0.01} = 2.20 = F_U$

      iii. Since ($F_c = 3.24$) > ($F_U = 2.20$), reject $H_0: \sigma^2_{CH} \leq \sigma^2_{RM}$ → Accept $H_1: \sigma^2_{CH} > \sigma^2_{RM}$ at a 1% significance level.

3. Verify at a 5% and a 1% significance levels if the Chicago shop has a lower revenue variability than the Rolling Meadows shop.

   a. Test $H_0: \sigma^2_{CH} \geq \sigma^2_{RM}$ vs $H_1: \sigma^2_{CH} < \sigma^2_{RM}$ at a 5% significance level.

      i. Calculate $F_c = \frac{\sigma^2_{CH}}{\sigma^2_{RM}} = \frac{45^2}{25^2} = \frac{2025}{625} = 3.24$

      ii. Find $F_L = \frac{1}{F^*_U} = \frac{1}{F^*_{table}} = \frac{1}{1.79} = 0.5586$

      Note: In this case, because the numerator and denominator degrees of freedom are reversed, $F_{table} \neq F^*_{table}$. In fact, we find $F^*_{table,0.05} = F^{41-1}_{31-1,0.05} = F^{40}_{30,0.05} = 1.79$

iii. Since ($F_c = 3.24$) > ($F_L = 0.5586$), accept $H_0 : \sigma^2_{CH} \geq \sigma^2_{RM}$ at a 5% significance level.

b. Test $H_0 : \sigma^2_{CH} \geq \sigma^2_{RM}$ vs $H_1 : \sigma^2_{CH} < \sigma^2_{RM}$ at a 1% significance level.

i. Calculate $F_c = \dfrac{\sigma^2_{CH}}{\sigma^2_{RM}} = \dfrac{45^2}{25^2} = \dfrac{2025}{625} = 3.24$

ii. Find $F_L = \dfrac{1}{F^*_U} = \dfrac{1}{F^*_{table}} = \dfrac{1}{2.30} = 0.4347$

Note: In this case, because the numerator and denominator degrees of freedom are reversed, $F_{table} \neq F^*_{table}$. In fact, we find $F^*_{table,0.01} = F^{41-1}_{31-1,0.01} = F^{40}_{30,0.01} = 2.30$

iii. Since ($F_c = 3.24$) > ($F_L = 0.4347$), accept $H_0 : \sigma^2_{CH} \geq \sigma^2_{RM}$ at a 1% significance level.

Exercise Problems on
Chapter 12. Two-Sample Hypothesis Testing

This set of exercise problems has 20 problems, worth a total of 20 points.

A. Please answer Questions 1 through 5, using the following scenario.

Suppose your molding machine #1 is to known to produce an average of 10 defects with a variability(=standard deviation) of $\pm$ 5 defects from the mean per day. On the other hand, your molding machine #2 is to known to produce an average of 12 defects with a variability (=standard deviation) of $\pm$ 7 defects from the mean per day.

1. If you are to know the average number of defects from Machine #1 is the same as that from #2, which of the following correctly represents the hypotheses to be tested?

a. $H_0 : \mu_1 = \mu_2$ vs. $H_1 : \mu_1 - \mu_2 = 0$
b. $H_0 : \mu_1 - \mu_2 = 0$ vs. $H_1 : \mu_1 \neq \mu_2$
c. $H_0 : \mu_1 \neq \mu_2$ vs. $H_1 : \mu_1 - \mu_2 = 0$
d. $H_0 : \mu_1 \neq \mu_2$ vs. $H_1 : \mu_1 - \mu_2 \neq 0$
e. $H_0 : \mu_1 - \mu_2 \leq 0$ vs. $H_1 : \mu_1 - \mu_2 > 0$

2. To test the above hypothesis, you need to use the ____ statistic because the population ______ are assumed to be known.

a. Z; means
b. t; means
c. Z; variances
d. t; variances
e. F; variances

3. On the basis of 200 samples from each machine, you found the average number of defects from Machine #1 is 5 and that from Machine #2 is 8. Using the standard deviation information above, the appropriate test statistic for testing the equality of two means is calculated as ______.

a. -3.288
b. 3.288
c. -4.932
d. 4.932
e. -3

4. If the absolute value of the calculated test statistic is ______ than the critical (=table) value at the correct significance level, you would ______ the null hypothesis of equal means.

a. greater; accept
b. less; reject
c. greater; reject
d. less; accept
e. only (c) and (d) of the above

5. If the calculated test statistic is -3, and the critical (=table) value is 2, you would _____ the null hypothesis of testing the equality of two means.

a.	reject	b.	accept	c.	fail to reject
d.	only (b) and (c) of the above	e.	not make a decision about		

B. Please answer Questions 6 through 11, using the following scenario.

Your boss asked you to see if the monthly sales revenue of Store #1 is the same as that of Store #2. You examined the monthly sales for Store #1 for the last 2 years – that is, 24 monthly sales revenue data – and the monthly sales for Store #2 for the 3 years – that is, 36 monthly sales data. The following information is found:

	Store #1	Store #2
Average Monthly Sales	$100,000	$105,000
Monthly Sales Variance	$30,000	$40,000

6. What test statistic is appropriate to use in this case?

a.	Z-statistic	b.	t-statistic	c.	F-statistic
d.	Chi-square	e.	indeterminate		

7. Assuming that you need to use a t-test, what is the value of the pooled variance?

a.	69,000.00	b.	10,983.72	c.	183.37
d.	36,034.48	e.	189.83		

8. Assuming that you need to use a t-test, what is the appropriate degrees of freedom?

a.	23	b.	35	c.	58
d.	59	e.	60		

9. Assuming that you need to use a t-test, what is the calculated t-value?

a.	–99.95	b.	99.95	c.	–199.95
d.	199.95	e.	5000.00		

10. Assume that you need to use a t-test. If the absolute value of the calculated t-statistic is _____ than the critical (=table) value of t at the correct significance level, you would _____ the null hypothesis of equal means.

a.	greater; accept	b.	less; reject	c.	greater; reject
d.	less; accept	e.	only (c) and (d) of the above		

11. If the calculated t-statistic is –1, and the critical (=table) value of t is 2, you would _____ the null hypothesis of testing the equality of two means.

a. reject b. accept c. fail to reject
d. only (b) and (c) of the above e. not make a decision about

C. Please answer Questions 12 through 20, using the following scenario.

Suppose that you own two ice cream shops – one in Chicago and the other in Rolling Meadows. During the summer months over many years, you had observed sales revenues and found the following monthly revenues and variabilities of monthly revenues for these two shops:

	Chicago	Rolling Meadows
Monthly Revenue	$3000	$2500
Standard Deviation	$35	$25
Number of months (n)	21	31

12. The set of hypotheses that can test the equality of revenue variabilities at both locations can be states as:

a. $H_0: \sigma_{CH}^2 = \sigma_{RM}^2$ vs $H_1: \sigma_{CH}^2 \neq \sigma_{RM}^2$
b. $H_0: \sigma_{CH}^2 - \sigma_{RM}^2 = 0$ vs $H_1: \sigma_{CH}^2 - \sigma_{RM}^2 \neq 0$
c. $H_0: \sigma_{CH}^2 \neq \sigma_{RM}^2$ vs $H_1: \sigma_{CH}^2 = \sigma_{RM}^2$
d. any of the above
e. only (a) and (b) of the above

13. At a 5% significance level, if you are to test the equality of revenue variabilities at both locations, you need to conduct an F-test. The calculated F-statistic is ____.

a. 1.96 b. -1.96 c. 1.4
d. -1.4 e. 2.20

14. At a 5% significance level, if you are to test the equality of revenue variabilities at both locations, you need to conduct an F-test. The critical table F-statistic is ____ with the numerator degree of freedom of _____ and the denominator degree of freedom of _____.

a. 1.96; 21; 31 b. 1.96; 20; 30 c. 2.20; 20; 30
d. 1.93; 20; 30 e. 2.04; 30; 20

15. If the calculated F-statistic is _____ than the critical (=table) value of F at the correct significance level, you would _____ the null hypothesis of equal revenue variabilities.

a. greater; accept b. less; reject c. greater; reject
d. less; accept e. only (c) and (d) of the above

16. Suppose that you are to verify if the Chicago shop has a higher revenue variability than the Rolling Meadows shop. Which of the following is the appropriate set of hypothesis?

a. $H_0: \sigma^2_{CH} \leq \sigma^2_{RM}$ vs $H_1: \sigma^2_{CH} > \sigma^2_{RM}$
b. $H_0: \sigma^2_{CH} < \sigma^2_{RM}$ vs $H_1: \sigma^2_{CH} \geq \sigma^2_{RM}$
c. $H_0: \sigma^2_{CH} \geq \sigma^2_{RM}$ vs $H_1: \sigma^2_{CH} < \sigma^2_{RM}$
d. $H_0: \sigma^2_{CH} > \sigma^2_{RM}$ vs $H_1: \sigma^2_{CH} \leq \sigma^2_{RM}$
e. $H_0: \sigma^2_{CH} = \sigma^2_{RM}$ vs $H_1: \sigma^2_{CH} \neq \sigma^2_{RM}$

17. At a 5% significance level, if you are to test the Chicago shop has a higher revenue variability than the Rolling Meadows shop, the critical (=table) value of the F-statistic is ____ with the numerator degree of freedom of _____ and the denominator degree of freedom of _____.

a. 1.96; 21; 31 b. 1.96; 20; 30 c. 2.20; 20; 30
d. 1.93; 20; 30 e. 2.04; 30; 20

18. If the calculated F-statistic is 5 and the critical (=table) value of F is 3 at the correct significance level, you would _____ the null hypothesis of the Chicago shop having a lower or equal revenue variability than the Rolling Meadows shop.

a. accept b. reject c. fail to reject
d. only (a) and (c) of the above e. either (b) or (c) of the above

19. Suppose that you are to verify if the Chicago shop has a lower revenue variability than the Rolling Meadows shop. Which of the following is the appropriate set of hypothesis?

a. $H_0: \sigma^2_{CH} \leq \sigma^2_{RM}$ vs $H_1: \sigma^2_{CH} > \sigma^2_{RM}$
b. $H_0: \sigma^2_{CH} < \sigma^2_{RM}$ vs $H_1: \sigma^2_{CH} \geq \sigma^2_{RM}$
c. $H_0: \sigma^2_{CH} \geq \sigma^2_{RM}$ vs $H_1: \sigma^2_{CH} < \sigma^2_{RM}$
d. $H_0: \sigma^2_{CH} > \sigma^2_{RM}$ vs $H_1: \sigma^2_{CH} \leq \sigma^2_{RM}$
e. $H_0: \sigma^2_{CH} = \sigma^2_{RM}$ vs $H_1: \sigma^2_{CH} \neq \sigma^2_{RM}$

20. If the calculated F-statistic is 2 and the critical (=table) value of F is 3 at the correct significance level, you would _____ the null hypothesis of the Chicago shop having an equal revenue variability as the Rolling Meadows shop.

a. accept
b. reject
c. fail to reject
d. only (a) and (c) of the above
e. either (b) or (c) of the above

Chapters 13. Simple Regression Analysis

Because we have now mastered the fundamental concepts and techniques of statistics such as the probability theory, probability distributions, confidence interval construction, and hypothesis testing, we are now switching gears and going to study regression analysis which is probably the most important topic in applied statistics because it enables one to establish a statistically important relationship between two or more variables of one's interest. It allows one to quantify a possible causal relationship that may exist between two or more variables and to examine its validity based on a probabilistic possibility.

Because this topic is so important, two chapters – this and the next – are devoted to it. While a more complex case of multiple regression is discussed in the next chapter, a simpler case of simple regression is covered in this chapter which is divided into 3 sections. Section 1 describes the theoretical background information about regression analysis by comparing it to the correlations analysis, describing the different regression models, and providing a theoretical framework of what regression analysis tries to accomplish. Section 2 delves deeper into the theoretical modeling of regression analysis in order to apply the concepts to more complex cases. Finally, in Section 3, examples of simple regression analysis are presented.

Section 1. Background Information

In statistical analysis, it is often very important to understand and establish a relationship between two variables. The correlation analysis and the simple regression analysis provide such methods. However, there is a fundamental difference between the two, which needs to be examined in detail. The correlation analysis will be explained first and the simple regression analysis will soon follow thereafter.

A. Correlation Analysis

Suppose that one wishes to know what factors or characteristics determine a person's wealth. To probe into this inquiry, people may wish to identify factors or characteristics – be they qualitative or quantitative – that make logical sense. Among many, let's assume that we have identified a salary of an individual as the most important factor[129].

Because salary and wealth are variable in the sense that they can take any dollar value from 0 to billions, we often refer to them as variables. Furthermore, because we are interested in the fact if salary determines wealth[130], we are implicitly assuming that wealth is **dependent** upon salary or wealth is determined by salary. Therefore, we call

---

[129] Whether one wishes to identify only one vs. many factors depends on the nature of the inquiry and the need of the analyst's. Furthermore, whether one prefers salary over income as the major determinant of wealth can be debated. However, let's just assume that salary is the only factor in this example for simplicity.

[130] It should be noted that the causality runs from salary to wealth, not from wealth to salary. That is, one's wealth normally does not determine one's salary but salary may determine wealth.

wealth a **dependent variable** and salary an **independent variable**. We can express this in an algebraic form as follows:

$$W = f(S)$$

where W = wealth and S = salary. We read this functional expression as "W is a function of S" and it means, in a layman's term, that "W is determined by S." That is, this expression implies that if we know the value of S, we can determine the value of W. That is why W is a dependent variable and S, an independent variable.

If we are to prove the existence of this relationship, we must collect data on W and S either as a primary data via surveys or a secondary data via public sources. Suppose the following is what we had collected from 15 individuals by an unbiased sampling method.

Individual (i)	**Wealth (W_i)**	**Salary (S_i)**
1	100	50
2	200	75
3	150	45
4	300	100
5	205	80
6	100	30
7	150	35
8	270	50
9	280	55
10	700	90
11	500	30
12	300	25
13	650	110
14	600	50
15	600	30

Wealth and Salary are measured in $1,000

One simple way to quantify the relationship between W and S is to measure the degree of their **linear co-movements** by calculating their correlation coefficient. Either by calculating manually or using the Excel command of "=correl(X,Y)", we can find the value of the correlation coefficient for the above data as:

Given the general formula for a correlation coefficient, r_{xy}, as follows:

$$r_{XY} = \frac{S_{XY}}{S_X \cdot S_Y} = \frac{\sum (X_i - \overline{X})(Y_i - \overline{Y})}{\sqrt{\sum (X_i - \overline{X})^2} \cdot \sqrt{\sum (Y_i - \overline{Y})^2}}$$

we can calculate the sample correlation coefficient between S and W as:

$$r_{SW} = \frac{S_{SW}}{S_S \cdot S_W} = \frac{1621}{27.5032 \cdot 211.0219} = 0.29925$$

Given that $-1 \le r_{SW} \le +1$, a possible interpretation of 0.29925 is that there is a **weak positive** relationship between W and S. Even though we label it weak, we are not sure how weak "weak" is. In fact, this conclusion is very subjective, because it is based on the analyst's experience and hunch. In order to eliminate this subjectivity, we must introduce objective decision criteria to judge how weak the correlation is, based on the probability theory – namely the confidence level or the significance level. That is, we can examine if the correlation coefficient between W and S is statistically equal to zero (= very, very weak that there is no relationship) or not at a 5% significance level[131] as follows:

The hypotheses to be tested are:

$$H_0 : \rho_{XY} = 0 \qquad \text{vs.} \qquad H_0 : \rho_{XY} \neq 0$$

The test procedure, based on the t-statistic[132], is as follows:

(1) Identify the table t-value = $t_{\alpha/2,n-2} = t_{0.025,13} = 2.1604$

(2) Calculate the t-value → the calculated t-value =

$$t_c = \frac{r_{XY} - \rho_{XY}}{S_{r_{XY}}} = \frac{r_{XY} - \rho_{XY}}{\sqrt{\frac{1-r_{XY}^2}{n-2}}} = \frac{r_{XY}\sqrt{n-2}}{\sqrt{1-r_{XY}^2}} = \frac{0.29925\sqrt{15-2}}{\sqrt{1-0.29925^2}} = \frac{1.07896}{0.95417} = 1.1308$$

(3) Because ($t_c = 1.1308$) < ($t_{0.025,13} = 2.1604$), we accept $H_0 : \rho_{XY} = 0$ → there is NO statistically significant linear relationship between W and S at a 5% significance level, given the data we collected[133] → X(=S) and Y(=W) are not (linearly) correlated.

As evidenced in the graph below, it is hard to see a definite upward- or downward-sloping linear line that passes through the observed values of W and S. That is, a visual inspection of the data can concur with the above statistical conclusion that there seems to be NO strong linear relationship between W and S.

---

[131] Please note that any significance level can be chosen. However, the 5% significance level is most often used. Furthermore, in expressing the correlation coefficient formula or the null and alternative hypotheses, no specific attempt to use S for X and W for Y are made herein. That is, we are using the generic form for simplicity.

[132] We must use a t-test because a population standard error of the correlation coefficient is not known.

[133] Note once again, that this acceptance of the null hypothesis only means that there is no statistically significant LINEAR relationship. However, they can be a significant NONLINEAR relationship.

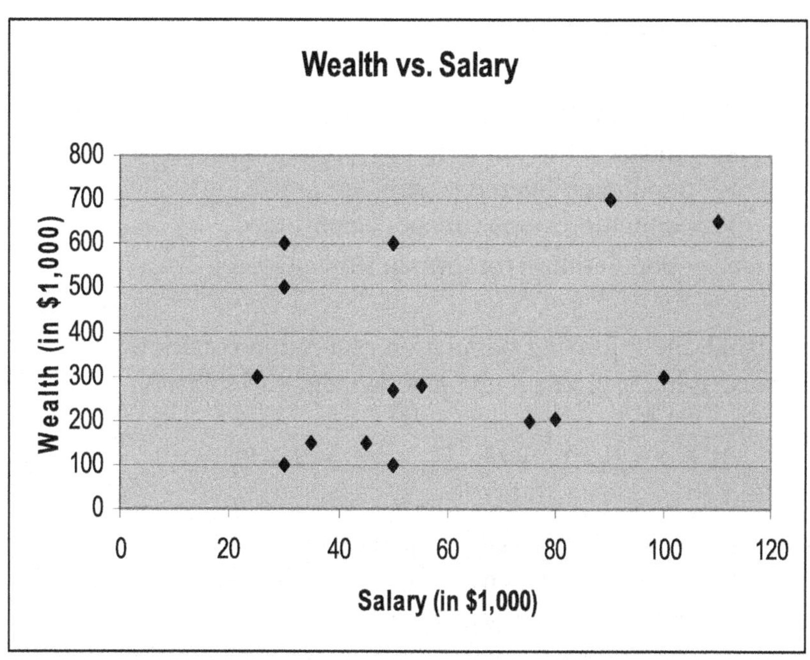

Now, let's boldly assume a new scenario. That is, let's suppose that we rejected the above $H_0 : \rho_{XY} = 0$ and thus, accepted the alternative hypothesis of $H_0 : \rho_{XY} \neq 0$. If this was the case, we know that W and S have a statistically significant linear relationship. Given this conclusion, can we estimate how much W will change if S changes by $1? Unfortunately, the correlation analysis can not answer this question. That is, it can not quantify the exact change in W caused by a change in S. The regression analysis can now outshine the correlation analysis by being able to provide answers to these types of questions. That is, the regression analysis can provide the amount of change in the dependent variable, Y(=W), caused by a change in the independent variable, X(=S). In fact, because the regression analysis supplements and complements the correlation analysis, it is more frequently used in data analysis.

B. The Types of Regression Analysis

The regression analysis is of two types – simple and multiple regression analyses. Their main difference lies in the number of independent variables being used in the regression equation. If there is only one independent variable[134], it is called a simple regression analysis. If there are more than one independent variables, it is called a multiple regression analysis.

For example, the following population regression equation is a simple regression equation because it has only one independent variable, X:

[134] The intercept term is not considered as an independent variable.

$$Y_t = \alpha + \beta \cdot X_t + \varepsilon_t$$

where Y_t = dependent variable[135] = actual, observed value of Y at time period t
X_t = independent variable[136] = actual, observed value of X at time period t
α = population intercept term
β = population slope (or coefficient) term
ε_t = population error term at time period t

On the other hand, the following population regression equation is a multiple regression equation because it has more than one independent variables, X_1, X_2, , , , and X_k:

$$Y_t = \alpha + \beta_1 \cdot X_{1t} + \beta_2 \cdot X_{2t} + ... + \beta_k \cdot X_{kt} + \varepsilon_t$$

where Y_t = dependent variable = actual, observed value of Y at time period t
X_{it} = the i-th independent variable actually observed at time period t
α = population intercept term
β_i = population slope (or coefficient) term associated with X_{it}
ε_t = population error term at time period t

Simple regression analysis will be discussed herein while multiple regression will be discussed in Chapter 14.

C. The Purpose of Simple Regression Analysis

The purpose of regression analysis – whether it is simple or multiple regression analysis – is to find (or estimate) a regression equation that will best capture the true (but hidden) relationship between two or more variables – one dependent variable and one or more independent variables. This true but unknown relationship is called the population regression equation and must be estimated on the basis of data collected. On the other hand, a relationship estimated on the basis of a sample data is called the sample regression equation or estimated regression equation. The simple regression equation will be used for an explanation purpose herein.

A simple regression model tries to establish a statistically significant relationship between two variables. One of them, called an independent variable, is believed to explain (or determine or influence) the other variable, called a dependent variable. This causal relationship must be established prior to data collection on the basis of economic theory, business intuition, life experience, etc.

---

135 It is also known as an endogenous, to-be-determined, or response variable because the value of this variable is determined by the value(s) of an independent variable(s).

136 It is also known as an exogenous, pre-determined, or explanatory variable because the value of this variable is determined independently of the regression model and used to explain (or understand) the nature of the dependent variable.

Given the population simple regression equation[137] of:

$$Y_t = \alpha + \beta \cdot X_t + \varepsilon_t$$

where Y_t = dependent variable = actual, observed value of Y at time period t
X_t = independent variable = actual, observed value of X at time period t
α = population intercept term
β = population slope or coefficient term
ε_t = population error term at time period t

we are implicitly saying that X causes or explains Y. Furthermore, we must estimate (or guess) this population regression equation by:

$$Y_t = a + b \cdot X_t + e_t$$

where a = intercept term calculated from a sample data → an estimate for α
b = slope (or coefficient) term calculated from a sample data
→ an estimate for β
e_t = error term at time period t based on a sample data
→ an estimate for ε_t

Because we will never know the true population regression equation, we must find the sample[138] regression equation that will give us a high confidence to call it a good estimate of the population regression equation.

When estimating a sample regression equation, there arise two major issues.

First, we must understand the nature of the task at hand by making the distinction between the actual value of Y, denoted as Y, and the estimated value of Y, denoted as $\hat{Y}$ or Y-hat[139]. If we recognize that there is a difference between Y and $\hat{Y}$, then it is easy to see that the purpose of regression analysis is to find $\hat{Y}$ that is closest to Y so that the estimation error, e, would be the smallest. That is, given $e = Y - \hat{Y}$, we wish to come up with $\hat{Y}$ that will be very close to Y. Consequently, the closer $\hat{Y}$ is to Y, the smaller e will be. Note that there are many Y's in the data set and thus, there are many corresponding $\hat{Y}$'s. Consequently, there are many e's as well. This means that we must somehow come up with a general and robust way to evaluate the size of all estimation errors

---

[137] Some textbooks and reference books also use the following form to represent the population regression equation: $Y_t = \beta_0 + \beta_1 \cdot X_t + \varepsilon_t$ where β_0 is the intercept and β_1 is the slope. In this case, b_0 and b_1, or $\hat{\beta}_0$ and $\hat{\beta}_1$, are often used as their respective estimated coefficients.

[138] "Sample" regression is an estimated regression equation based on a sampled data set whereas "simple" regression means a regression equation with one dependent variable.

[139] The correct name for $\hat{Y}$ is Y-caret but for convenience and familiarity, it is often called Y-hat.

embedded in a regression equation. The best criterion is to find $\hat{Y}$'s such that the sum of squared errors (SSE) would be the smallest. Mathematically, it can be expressed as:

Find $\hat{Y}$ that will

$$\text{minimize SSE} = \sum_{i=1}^{n} e_i^2 = \sum_{i=1}^{n} (Y_i - \hat{Y}_i)^2$$

The real task, however, is to find a way to come up with $\hat{Y}$'s. That is, we need an equation that will estimate $\hat{Y}$'s. Among many possible forms of regression models (or equations), we choose the simplest linear form of:

$$\hat{Y}_i = a + b \cdot X_i$$

Given this equation, then, we need to find the appropriate values of a and b. We may wish to resort to guessing them but obviously, that will not be the optimal solution. As shown in the graph below, many linear lines can fit the data by eye-balling but will not be reliable and accurate.

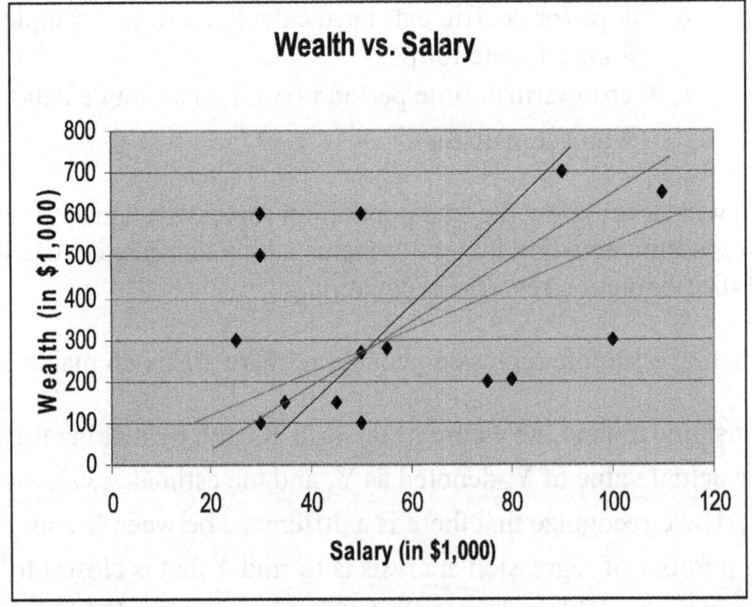

A preferred way is to develop a mathematical model that can yield the smallest sum of squared errors (SSE)[140]. Given the generic liner equation of $\hat{Y}_i = a + b \cdot X_i$, we can see that finding $\hat{Y}_i$ that minimizes SSE is the same as finding a and b that minimizes SSE, given X_i. In other words, the above minimization problem can be rewritten as:

---

140 Instead of 'smallest," the word, "least" is more often used. Thus, the "smallest sum of squared errors" is known as "LEAST sum of SQUARED errors," which is further shortened as "Least Squares." Therefore, a group of estimation methods that rely on the smallest or minimum sum of squared errors is generally known as a Least Squares (LS) method. This LS method is further divided into the Ordinary Least Squares (OLS) method and the Generalized Least Squares (GLS) method. For simplicity, we will only study herein the less complex estimation method of OLS.

Find a and b that will

$$\text{minimize SSE} = \sum_{i=1}^{n} e_i^2 = \sum_{i=1}^{n} (Y_i - \hat{Y}_i)^2 = \sum_{i=1}^{n} (Y_i - a - b \cdot X_i)^2$$

because $\hat{Y}_i = a + b \cdot X_i$

In fact, if the above minimization problem is solved, the values of a and b are found as:

$$a = \overline{Y} - b \cdot \overline{X}$$

$$b = \frac{S_{XY}}{S_X^2} = \frac{\sum (X_i - \overline{X})(Y_i - \overline{Y})}{\sum (X_i - \overline{X})^2}$$

Therefore, when the above equations are used for the Wealth and Salary data, we find:

$$b = \frac{1736.786}{(27.5032)^2} = 2.296 \qquad \text{and} \qquad a = 340.33 - 2.296 \cdot 57 = 209.45$$

Because these values of a and b minimize the sum of squared errors (SSE), this method of calculating a and b are known as the Ordinary Least Squares (OLS) method. Also, a and b are identified as the BLUEs (best linear unbiased estimators) for population values of α and β. The estimated (or sample) regression equation for the wealth and salary data is, therefore, identified as:

$$\hat{Y}_i = a + b \cdot X_i = 209.45 + 2.296X_i$$

This estimated regression equation can be equivalently rewritten as follows:

Because $\quad e = Y - \hat{Y} \quad \rightarrow \quad Y = \hat{Y} + e \quad$ and $\quad \hat{Y} = a + b \cdot X$

we find: $\quad Y = \hat{Y} + e = a + b \cdot X + e$

Therefore, the above estimated regression equation can be alternatively written as:

$$Y_i = \hat{Y} + e = a + b \cdot X_i + e = 209.45 + 2.296X_i + e$$

However, many textbooks and reference books do not make this distinction and sloppily use them interchangeably. Therefore, if we write

$$Y = 209.45 + 2.296X$$

instead of $\hat{Y} = 209.45 + 2.296X$ or $Y = 209.45 + 2.296X + e$, we will accept and understand it to mean the estimated regression equation. The following graph shows this estimated regression line for the Wealth and Salary data.

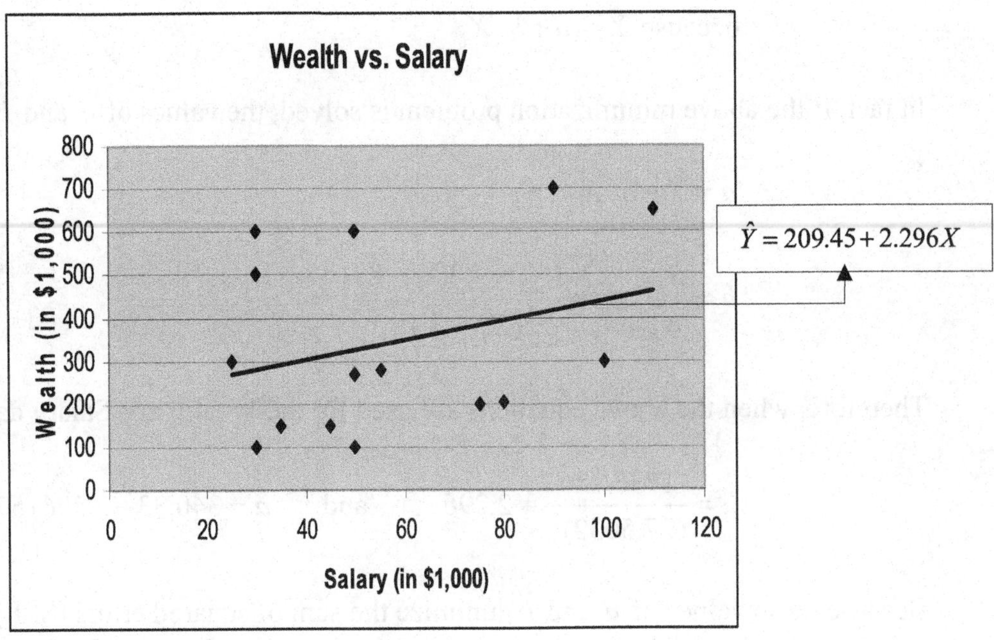

D. The Usefulness of a Regression Equation

Once an estimated regression equation is identified, such as for our case of Wealth (W) and Salary (S) of:

$$\hat{W}_i = 209.45 + 2.296S_i$$

we can use it to estimate or predict W, given S. That is, if we are given S_i as 50, for example, we can obtain $\hat{W}_i$ as:

$$\hat{W}_i = 209.45 + 2.296S_i = 209.45 + 2.296(50) = 324.25$$

Therefore, we have succeeded in estimating the value of wealth, W, by a regression equation that utilizes the value of salary, S. Generically, we can say this fact by numerous equivalent ways as follows:

(1) W is estimated by S → S estimates W.
(2) W is explained by S → S explains W.
(3) W is predicted or forecast by S → S predicts or forecasts W[141].

---

[141] Even though they are often used interchangeably without any distinction, one must be careful in using the terms of "estimate," "explain," and "predict" in regression analysis. Because most often X and Y data are collected contemporaneously – that is, they are collected for the same time period, it is wrong to say

Furthermore, we note that the slope coefficient of the S variable measures the change in W given a change in S. That is, because

$$\text{Slope} = \frac{\Delta Y}{\Delta X} \rightarrow \frac{\Delta W}{\Delta S} = \frac{dW}{dS} = 2.296 = \frac{2.296}{1}$$

if $\Delta S = +1$, then $\Delta W = +2.296$. Similarly, if $\Delta S = -1$, then $\Delta W = -2.296$. Therefore, the slope sign being positive indicates that S and W have a positive relationship → that is, S and W tend to move up or down together. In this case, if salary (S) increases (decreases), it will increase (decreases) wealth (W). Of course, if the slope sign is negative, it indicates that S and W have a negative or inverse relationship → that is, S and W tend to move up or down in an opposite or inverse direction.

Therefore, the usefulness of a regression equation can be summarized as:

(1) It can estimate the value of the dependent variable, W, by an independent variable, S.

(2) The sign and magnitude of its slope coefficient can describe a specific relationship between W and S → the positive sign means that W and S move together and the value of 2.296 means that if S increases by \$1, W will increase by \$2.296.

E. Statistical Validity of a Regression Equation

It is nice to have a regression equation with its intercept and slope coefficient all identified by the Ordinary Least Squares (OLS) method where the sum of squared errors was minimized. However, the real issue is if the estimated regression equation and its coefficients will hold in light of the probability theory. That is, the statistical validity of the estimated regression equation must be verified. **The tool to be used for this purpose is the t-test of individual slope coefficients and an F-test for the all slope coefficients being jointly equal to zero.** The regression theory that deals with these and other issues will be described in Section 2 and its applications via examples in Section 3 of this Chapter. Specific topics on multiple regression will be covered in Chapter 14.

Section 2. Regression Theory

---

that the estimated value of Y, Y-hat, for time period t is predicted or forecast by X observed in the same time period of t. Therefore, it is often better to use the terms of "estimate" or "explain." However, if X for time period t is known (possibly by a forecast), then Y for time period t can be forecast based on the forecast value of X. This is the reason why some people say that regression is better suited for explaining a relationship between X and Y and should not be used for forecasting Y based on X. We will not indulge ourselves with this philosophical issue too much. We simply understand that "estimate" or "explain" can be interchangeably used for "forecast" or "predict" in regression analysis to satisfy our simple current need.

This Section 2 of simple regression analysis is further devoted to the theoretical exposition of how regression analysis is formed and used to examine a statistically important and meaningful relationship between two or more variables. Even though this topic is quite dry, we should not lose focus on the logic and the methodology of determining what constitutes a significant relationship between variables and thus, a good regression model or equation to be used for one's prediction purpose. After studying the theoretical aspects of regression analysis here in Section 2, we will look at some examples and applications in Section 3.

A. Assumptions of Regression Analysis

Once you estimate a and b, the next issue is to know how reliable, trustworthy, and confident these estimates are in relation to the population parameters of α and β.

Thus, we must conduct tests to determine its statistical significance.

However, before we do the statistical tests, we must understand what assumptions will allow us to conduct statistical tests in regression analysis. The major assumptions are as follows:

1. the values of the independent variable, X_t, are measured without error; determined exogenously outside the model; and known.

2. the value of dependent variable, Y_t, are normally distributed given X_t → Equivalently, e_t are normally distributed around a mean of zero → $E(e_t) = 0$.

3. e_t are identically distributed → have the same constant variance, σ_e^2, for every e_t → that is, **homoscedasticity** must exist.

4. e_t are independently distributed in relation to other error terms such as e_{t-1}, e_{t-2}, …., and e_{t-k} → no serial- or auto-correlation exists.

These assumptions are concisely and simply expressed as follows:

$$e_t \sqcup \text{n.i.i.d.}\ (0,\ \sigma_e^2)$$

That is, it means that "errors (or residuals) are **n**ormally, **i**dentically, and **i**ndependently **d**istributed with the mean of 0 and the constant variance of σ_e^2. If any of these assumptions are violated, then the estimated regression equation may be unreliable and invalid for the following reasons:

1. Without the normal distribution assumption of errors, a significance test of individual slope coefficients by a t-test is not possible.

2. The case of identical (or even) error distribution is known as homoscedasticity whereas that of unidentical (or uneven) error distribution is known as **heteroscedasticity**. Homoscedasticity allows the regression estimation by the OLS method trouble-free whereas heteroscedasticity creates a major estimation problem by OLS. In the presence of heteroscedasticity, the Generalized Least Squares (GLS) method needs to be used instead of OLS.

3. When errors are not independent among themselves, an autocorrelation or serial correlation problem is said to exist. If an **autocorrelation** problem exists, then errors are correlated with each other. This additional information has to be incorporated into the estimated regression equation by the GLS method. Otherwise, the estimation equation is not complete and needs to be improved.

Whenever heteroscedasticity or autocorrelation problems are encountered, the OLS method will not yield the optimal solution and thus, advanced methods – such as the generalized least squares (GLS) method – should be used[142].

B. Significance Testing of the Slope Coefficient

The most important and most frequently asked question in regression analysis is if there is a (statistically) significant relationship between the dependent and the independent variables. That is, given the estimated regression equation of:

$$\hat{Y}_t = a + b \cdot X_t$$

Or equivalently[143],

$$Y_t = a + b \cdot X_t + e_t$$

if b is judged to be statistically equal to zero, then X does not explain (or determine or influence) Y → there is NO relationship between X and Y → X is NOT significant (or insignificant) → b (or β) is NOT significant (or insignificant).

On the other hand, if b is judged to be statistically NOT equal to zero, then X does explain (or determine or influence) Y → there is a SIGNIFICANT relationship between X and Y → X is significant → b (or β) is significant → the value of b is different from zero.

Following the tradition of expressing this type of questions into a null and an alternative hypotheses, we can formally state it in a layman's term and then, formally express it as a null and alternative hypotheses as follows:

---

[142] However, due to its complexity, the GLS method will not be covered herein.

[143] because $Y_t = \hat{Y}_t + e_t = a + b \cdot X_t + e_t$

The Layman's Question:

Is Y determined by X?
= Is there a (statistically significant) relationship between X and Y?
= Is the slope coefficient, b, statistically equal to zero (or significant)?

Equivalently, the corresponding formal null and alternative hypotheses are stated as:

Test the significance of H_0: $\beta = 0$ vs. H_1: $\beta \neq 0$ at a α significance level.

In order to conduct a significance test of these null and alternative hypotheses at a significance level, we must obtain the estimated value of β, b, and its standard error, S_b.

It is important to know that the intercept and the slope coefficient themselves are not a permanently fixed number. Their values can always change given different sets of sample data. **Because the intercept and the slope coefficient values can change with different data, it is only logical to have a measure of their variability known as a standard error of the coefficient.** The standard error of the intercept, S_a, and the standard error of the slope coefficient, S_b, are calculated as follows[144]:

$$S_a = \text{standard error of } a = \sqrt{S_e^2 \cdot [\frac{1}{n} + \frac{\overline{X}^2}{\sum (X - \overline{X})^2}]}$$

$$S_b = \text{standard error of } b = \sqrt{\frac{S_e^2}{\sum (X - \overline{X})^2}}$$

where $S_e^2 = \dfrac{\sum e_t^2}{n-k-1}$ and Error Sum of Squares (ESS) =
$\sum e_t^2 = \sum (Y_t - \hat{Y}_t)^2$

Note: the Standard Error of the Regression = Standard Error of the Estimate (SEE):

$$\text{SEE} = \sqrt{\frac{\sum e_t^2}{n-k-1}} = \sqrt{\frac{ESS}{n-k-1}} = \sqrt{MS_e} = \sqrt{S_e^2} = S_e$$

[144] No need to memorize this formula for the exam. The standard error formula for the intercept is not shown here because we will not use it to manually calculate its value. That is, in this course, standard error of the coefficient formulas will not be invoked for a calculation purpose.

Given the population regression equation of $Y_t = \alpha + \beta \cdot X_t + \varepsilon_t$, we obtain its estimated equation as $Y_t = a + b \cdot X_t + e_t$, based on the sample data[145]. Once the values of a and b are obtained along with their respective standard errors, the formal statistical significance testing is conducted as follows:

Step 1. Establish the hypotheses to be tested as:

$$H_0: \beta = 0 \qquad \text{vs.} \qquad H_1: \beta \neq 0$$

Step 2. Select a significance level[146], α, and identify the corresponding critical t-value → always a t-value is used in regression analysis because population standard error of the estimated regression coefficient is UNKNOWN.

The appropriate table value of t, $t_{table} = t_{\alpha/2,(n-k-1)}$, can be found from the t-table at an α/2% significance level because this hypothesis testing is a two-tailed test. That is, the test statistic to use is t with (n-k-1) degrees of freedom where k = the number of independent variables used in the regression equation → In the case of simple regression, k=1[147].

Step 3. Calculate the t-value by:

$$t_{calculated} = t_c = \frac{b - \beta}{S_b}$$

where b = the slope value from the estimated regression equation

β = the hypothesized value from the population regression equation → most often 0 because a statistical significance test tests if it is zero or not.

S_b = the standard error of b ← calculated by the formula above or generated by a computer software such as Excel Spreadsheet.

Step 4. Apply the decision criteria as follows:

a) If $|t_c| \leq t_{\alpha/2,(n-k-1)}$, then accept H_0: β=0.

Note: This conclusion means any of the following expressions:

---

145 When the population and the sample regression equations are compared, α is the population parameter for a and β is the population parameter for b. Therefore, when a hypothesized value of a is asked, α is identified and likewise, when a hypothesized value of b is asked, β is identified.

146 Note that this alpha represents the significance level and thus, is different from the alpha that represents the intercept term in the population regression equation.

147 The reason why the degrees of freedom is (n-k-1) for a t-value is because, given n observations, there are k slope coefficients and 1 intercept term to estimate. Thus, we lose (k+1) degrees of freedom in regression analysis.

Accepting H_0 means that X does NOT explain (or determine or influence) Y → there is NO relationship between X and Y → X is NOT significant (or insignificant) → b (or β) is NOT significant (or insignificant).

b) Otherwise, reject H_0: $\beta=0$ → accept H_1: $\beta\neq0$

Note: This conclusion means any of the following expressions:

Rejecting H_0 means that X does explain (or determine or influence) Y → there is a SIGNIFICANT relationship between X and Y → X is significant → b (or β) is significant → the value of b is different from zero.

Note: The above decision criteria and testing method for the significance of a regression slope coefficient is exactly the same as previously discussed in Section 2 of Chapter 11 where t-tests for $\bar{x}$ were described. Just replace $\bar{x}$ with b ; μ with β; and $S_{\bar{X}}$ with S_b.

C. Significance Testing of the Intercept Term

The answer to a question on the significance of an intercept term involves whether the relationship between X and Y starts from the origin or not. That is, if the intercept term, α, is judged to be statistically equal to zero, then the line between X and Y must pass through the origin. **The formal statement of this question and test thereof is identical to that of testing the slope coefficient** as soon be seen here below:

Step 1. For the question of if the intercept coefficient is statistically equal to zero, we establish the hypotheses to be tested as:

$$H_0: \alpha = 0 \quad \text{vs.} \quad H_1: \alpha \neq 0$$

Step 2. Select a significance level[148], α, and identify the corresponding critical t-value → always a t-value is used in regression analysis because population standard error of the coefficient is UNKNOWN.

The appropriate table value of t, $t_{table} = t_{\alpha/2,(n-k-1)}$, can be found from the t-table at an α/2% significance level because this hypothesis testing is a two-tailed test. That is, the test statistic to use is t with (n-k-1) degrees of freedom where k = the number of independent variables used in the regression equation → In the case of simple regression, k=1.

Step 3. Calculate the t-value by:

---

148 Note that this alpha represents the significance level and thus, is different from the alpha that represents the intercept term in the population regression equation.

$$t_{calculated} = t_c = \frac{a-\alpha}{S_a}$$

where a = estimated value from the regression equation
α = the hypothesized value → 0 in this case
S_a = standard error of a ← calculated by the formula above or generated by a computer software such as Excel Spreadsheet.

Step 4. Apply the decision criteria as follows:

If $|t_c| \leq t_{\alpha/2,(n-k-1)}$, then accept H_0: $\alpha=0$ → α is statistically equal to zero → the intercept (or α) is NOT significant → X and Y start from the origin.

Otherwise, reject H_0: $\alpha=0$ → accept H_1: $\alpha \neq 0$ → α is statistically different from zero → the intercept (or α) is significant → X and Y do NOT start from the origin.

D. The Upper-Tail Test of Regression Coefficients

We can ask whether the intercept term or the coefficient term is greater (or smaller) than a specific value such as zero → This implies that we are hypothesizing either $\alpha > 0$, $\alpha < 0$, $\beta > 0$ or $\beta < 0$.

Here, let's choose the following case of a upper-tail test at a α% significance level[149]:

H_0: $\beta \leq 0$ vs. H_1: $\beta > 0$

If so, we must set up the following:

Step 1. Test at a α% significance level H_0: $\beta \leq 0$ vs. H_1: $\beta > 0$

Step 2. Find the table value of t, $t_{table} = t_{\alpha,(n-k-1)}$, from the t-table at a α% significance level. (Note that we are using α, not $\alpha/2$, because this is a one-tail test.)

Step 3. Calculate the t-value by:

$$t_{calculated} = t_c = \frac{b-\beta}{S_b}$$

[149] Note that this alpha represents the significance level and thus, is different from the alpha that represents the intercept term in the population regression equation.

where b = estimated value from the regression equation
β = the hypothesized value → 0 in this case.
S_b = standard error of b

Step 4. Apply the decision criteria as follows[150]:

If $t_c \leq t_{\alpha,(n-k-1)}$, then accept H_0: $\beta \leq 0$

If $t_c > t_{\alpha,(n-k-1)}$, then reject H_0: $\beta \leq 0$ → accept H_1: $\beta > 0$

Note:

The above decision criteria and testing method for the upper-tail significance of a regression slope coefficient is exactly the same as previously discussed in Section 2 of Chapter 11 where t-tests for $\overline{x}$ were described.

E. The Lower-Tail Test of Regression Coefficients

We can ask whether the intercept term or the coefficient term is greater (or smaller) than a specific value such as zero → This implies that we are hypothesizing either $\alpha > 0$, $\alpha < 0$, $\beta > 0$ or $\beta < 0$.

Here, let's choose the following case of a lower-tail test at a α% significance level:

H_0: $\beta \geq 0$ vs. H_1: $\beta < 0$

If so, we must set up the following:

Step 1. Test at a α% significance level H_0: $\beta \geq 0$ vs. H_1: $\beta < 0$

Step 2. Find the table value of t, $t_{table} = t_{\alpha,(n-k-1)}$, from the t-table at a α% significance level. (Note that we are using α, not $\alpha/2$, because this is a one-tail test.)

Step 3. Calculate the t-value by:

$$t_{calculated} = t_c = \frac{b-\beta}{S_b}$$

where b = estimated value from the regression equation
β = the hypothesized value → 0 in this case.

---

[150] Because the critical t-value approach is most often used for one-tail tests, only this approach is introduced and described herein.

S_b = standard error of b

Step 4. Apply the decision criteria as follows:

If $t_c \geq -t_{\alpha,(n\text{-}k\text{-}1)}$, then accept H_0: $\beta \geq 0$

If $t_c < -t_{\alpha,(n\text{-}k\text{-}1)}$, then reject H_0: $\beta \geq 0 \rightarrow$ accept H_1: $\beta < 0$

Note:

The above decision criteria and testing method for the lower-tail significance of a regression slope coefficient is exactly the same as previously discussed in Section 2 of Chapter 11 where t-tests for $\overline{X}$ were described.

F. The Confidence Intervals for Y_t, and Slope Coefficient, β.

1. For a 95% C.I. for Y_t

$$Y_t = \hat{Y}_t \pm t_{\alpha/2,(n-k-1)} \cdot S_{\hat{Y}}$$

where $\hat{Y}_t$ is obtained from $\hat{Y}_t = a + bX_t$ and

$$S_{\hat{Y}} = SEE \cdot \sqrt{1 + \frac{1}{n} + \frac{(X_i - \overline{X})^2}{(n-1)S_X^2}}$$

2. For a 95% C.I. for β

$$\beta = b \pm t_{\alpha/2,(n-k-1)} \cdot S_b$$

Note:

In multiple regression, the confidence intervals for Y_t and β can not be manually calculated unless matrix algebra is used. Thus, they are rarely constructed unless a computer software is used.

G. The Decomposition and the ANOVA

The regression analysis provides a very convenient way to understand how much the independent variable explains the variability (=movement) of the dependent variable.

This can be accomplished by decomposing the relationship and then organizing it into an **AN**alysis **O**f **VA**riance (ANOVA) table as follows:

Given $Y_t = \hat{Y}_t + e_t$, we note that the following relationships hold:

$$(Y_t - \overline{Y}) = (\hat{Y}_t - \overline{Y}) + e_t$$

$$\sum(Y_t - \overline{Y})^2 = \sum(\hat{Y}_t - \overline{Y})^2 + \sum e_t^2$$

Note that these sums of squares can be named as follows:

TSS = RSS + ESS

where

TSS = Total Sum of Squares $= \sum(Y_t - \overline{Y})^2$

RSS = Regression Sum of Squares $= \sum(\hat{Y}_t - \overline{Y})^2$

ESS = Error (or Residual) Sum of Squares[151] $= \sum e_t^2$

That is, we can construct the following ANOVA table for regression analysis:

Source of Variation	Sum of Squares	Degrees of Freedom	Mean Square **(=Variance)**	Calculated F Value = F_c
Due to Regression	RSS	k	$MS_R = \dfrac{RSS}{k}$	$F_c = \dfrac{MS_r}{MS_e}$
Due to Error (or Residual)	ESS	n–k–1	$MS_E = \dfrac{ESS}{n-k-1}$	
Total	TSS	n–1	$Var(Y) = \dfrac{TSS}{n-1}$	

H. The relationship between Multiple R, R^2, and Adjusted R^2

1. Multiple R = the correlation coefficient between the actual values of Y and their estimated values of $\hat{Y}$ based on a regression equation[152] → a low multiple R means a poor predictability of the regression equation for Y. It is always a positive number!

$$0 \le \text{Multiple R} \le 1$$

2. R^2 = R-squared = R-square = (Multiple R)^2 = the coefficient of determination = a measure of explanatory power = a measure of goodness of fit → measures a proportion of a total variation in Y explained by the regression equation → the higher the R^2, the better is the fit (or explanation) of the Y data by the regression equation.

---

[151] Note that we called this the SSE (sum of squared errors) in Section 1 of this Chapter 13.

[152] For simple regression, the Multiple R is also an absolute value of the correlation coefficient between X and Y = $|r_{XY}|$. However, this is not the case for multiple regression because of more than one independent variables in the regression equation. That is, for multiple regression, the Multiple R is the correlation coefficient between Y and Y-hat, not a correlation coefficient between Y and X's.

$$R^2 = \frac{RSS}{TSS} = 1 - \frac{ESS}{TSS}$$

$$0 \le R^2 \le 1$$

Usually expressed as %. Thus, if $R^2 = 0.6$, it is referred as 60% and interpreted as **"60% of total variation (or movement or fluctuation) in Y is explained by the estimated regression equation."**

Note: r_{XY} = the correlation coefficient between X and Y = a measure of a linear relationship between two variables, X and Y.

For the simple regression, $R^2 = (r_{XY})^2$ because $R^2 = (\text{Multiple R})^2$ and Multiple $R = |r_{XY}|$. However, this relationship does not hold for multiple regression due to there being more than one independent variables.

3. Adjusted $R^2 = \overline{R}^2$ = a measure of model adequacy = a measure of model efficiency → It determines if adding an independent variable improves the regression model or not. → A small to no increase in an adjusted R-squared is not desirable because this means that a newly added independent variable is not improving the model. → A decrease in $\overline{R}^2$ clearly indicates that a recently added independent variable does not add value in explaining the variation in Y and thus, should be dropped from the regression model.

$$\text{Adjusted } R^2 = \overline{R}^2 = 1 - \left[(1 - R^2) \cdot \frac{n-1}{n-k-1}\right]$$

Note: $\overline{R}^2$ can be negative because $\overline{R}^2 \le 1$.

Example 1>

Assume that initially a multiple regression equation with 2 independent variables was estimated and yielded the following adjusted R-squared value:

That is, given $n = 10$, $k = 2$, and $R^2 = 0.3$,

$$\text{Adjusted } R^2 = \overline{R}^2 = 1 - \left[(1 - R^2) \cdot \frac{n-1}{n-k-1}\right] = 1 - \left[(1 - 0.3) \cdot \frac{10-1}{10-2-1}\right] = 0.1$$

Example 2>

Now, assume that a new multiple regression equation with one more independent variables – a total of 3 – is estimated and yielded the following result:

That is, given n = 10, k = 3, and $R^2 = 0.32$,

$$\text{Adjusted } R^2 = \overline{R}^2 = 1 - \left[(1-R^2)\cdot\frac{n-1}{n-k-1}\right] = 1 - \left[(1-0.32)\cdot\frac{10-1}{10-3-1}\right] = -0.02$$

Note 1: In comparing the two examples above, one can see that the R-squared had increased but the adjusted R-squared decreased and in fact, it turned out to be negative. This means that the 3rd variable should not be included in the model.

Note 2: The problem of a negative $\overline{R}^2$ or an insignificant increase in $\overline{R}^2$ often occurs because there are too many slope coefficients (for too many independent variables) to be estimated based on too few observations. Therefore, one major way to alleviate this problem is to increase the number of observations. Of course, not adding new independent variables into the regression equation is the other way to minimizing the chance of this problem from occurring.

Note 3: If the change in R-squared values does not increase significantly[153], the change in adjusted R-squared values will not increase significantly either. Thus, often by judging the changed amount in the R-squared value, one can tell if adding an additional independent variable to the regression model is a wise thing to do or not.

I. The F-test

Often, one wishes to know the significance of all slope coefficients being simultaneously or jointly equal to zero. That is, we can conduct a joint or composite significance test of H_0: All Slope Coefficients = 0, using an F-test. As was discussed in Section 2 of Chapter 12, the sensitivity of an F-statistic depends on 2 different degrees of freedom: The numerator degrees of freedom is determined by the number of independent variables (k) and the denominator degrees of freedom, by (n – k – 1). The decision criteria is similar to any significance test such that:

If $F_c = \dfrac{MS_R}{MS_E} \le F^k_{n-k-1,\alpha}$, then fail to reject H_0 → Accept H_0.

Otherwise, reject H_0 → accept H_1 → not all slope coefficients are zero.

If this inquiry (or null hypothesis) is judged to be significant, it means that not all slope coefficients are equal to zero. On the other hand, **if the null hypothesis is judged to be**

[153] There is an advanced statistical method such as the Chow test that can determine if an additional independent variable is justified or not. However, one may, for simplicity and convenience, use personal judgment in determining how much of a change is significant or not.

insignificant – that is, if the null hypothesis is accepted – then, it means that the estimated regression equation is not good at all because their slope coefficients are statistically all equal to zero. However, this inquiry is not that meaningful in simple regression because there is only one slope coefficient[154]. It becomes a lot more important and interesting under the multiple regression analysis which will be discussed in detail in the next chapter.

The intuitive meaning of a calculated F-value, F_c, can be obtained by examining the regression sum of squares (RSS) and the error sum of squares (ESS) in the ANOVA table above. Because F_c is a ratio of MS_R to MS_E, it is, in essence, a ratio of RSS to ESS adjusted by their respective degrees of freedom. Therefore, the larger the RSS is in relation to ESS, the larger F_c is. This observation is also very closely linked to the fact that the larger the RSS is in relation to the total sum of squares (TSS), the larger the R^2 is. Therefore, a large RSS means a large F_c, which in turn means a large R^2, and thus, a very good model which has a high degree of explanatory power.

J. The Durbin-Watson (DW) Statistic for Auto-correlation Detection

Autocorrelation is a type of serial correlation, arising from the fact that one error term and its lagged term are serially correlated. Thus, it is often generically called serial correlation for simplicity.

1. The DW Formula:

$$DW = \frac{\sum_{t=2}^{n}(e_t - e_{t-1})^2}{\sum_{t=1}^{n} e_t^2} \approx 2\cdot(1 - r_{e_t, e_{e-1}})$$

   Note that $r_{e_t, e_{t-1}}$ is called the (first-order) autocorrelation coefficient between two successive error terms of e_t and e_{t-1}. Its values range between -1 and $+1$.

2. Testing the First-degree Autocorrelation:

   a. Positive Autocorrelation

      → is detected when $\text{DW} \approx 0$ due to $r_{e_t, e_{t-1}} \approx +1$.

      → error terms tend to show **protracted upward movements followed by protracted downward movements** or vice versa.

   b. Negative Autocorrelation

[154] Therefore, in simple regression analysis, the conclusion about the significance of the β slope coefficient is the same if a t-test or an F-test is used. In fact, a calculated F = a calculated t-value squared.

→ is detected when DW ≈ 4 due to $r_{e_t, e_{t-1}} \approx 1$.

→ error terms tend to show **zigzag or saw-tooth** movements.

c. No Autocorrelation

→ is detected when DW ≈ 2 due to $r_{e_t, e_{t-1}} \approx 0$.

→ error terms are **scattered randomly** and thus, there is no identifiable pattern.

However, a more precise determination about the presence of (the first-order) autocorrelation must be based on a statistical test known as the Durbin-Watson test whose decision criteria are described below.

3. The Critical Durbin-Watson Statistics

The Durbin-Watson (DW) statistic is originally reported by J. Durbin and G. S. Watson in an academic journal called *Biometrika*, Vol. 41 (1951), Pages 173-175, to examine the presence of the first-order autocorrelation – i.e., the degree of co-movements between two successive error terms. Thus, it is not capable of examining the presence of a higher-order autocorrelation. It is comprised of a Durbin-Watson lower bound (expressed as DW_L, D_L, or d_L) and a Durbin-Watson upper bound (expressed as DW_U, D_U, or d_U) and arranged by the significance level, α; the number of slope coefficients estimated, k; and the number of observations, n.

Many statistics books report their exact values but we shall report approximation of that information here below for an illustration purpose only. That is, for more accurate DW statistic, refer to other sources.

Durbin-Watson Statistic for the Significance Level of 5%

Number of Obs.	Number of Independent Variables, k									
	k=1		k=2		k=3		k=4		k=5	
n	DW_L	DW_U	DW_L	DW_U	DW_L	DW_U	DW_L	DW_U	DW_L	DW_U
15 - 19	1.08	1.36	0.95	1.54	0.82	1.75	0.69	1.97	0.56	2.21
20 - 24	1.27	1.45	1.19	1.55	1.00	1.68	1.01	1.78	0.93	1.90
25 - 29	1.34	1.48	1.27	1.56	1.20	1.65	1.12	1.74	1.05	1.84
30 - 35	1.39	1.51	1.33	1.58	1.21	1.65	1.21	1.73	1.15	1.81
36 - 40	1.43	1.54	1.38	1.60	1.33	1.66	1.27	1.72	1.22	1.79
41 - 45	1.48	1.57	1.43	1.62	1.38	1.67	1.34	1.72	1.29	1.78
46 - 50	1.50	1.59	1.46	1.63	1.42	1.67	1.38	1.72	1.34	1.77
51 - 55	1.53	1.60	1.49	1.64	1.45	1.68	1.41	1.72	1.38	1.77
56 - 60	1.55	1.62	1.51	1.65	1.48	1.69	1.44	1.73	1.41	1.77

Durbin-Watson Statistic for the Significance Level of 1%

Number of Obs.	Number of Independent Variables, k									
	k=1		k=2		k=3		k=4		k=5	
n	DW_L	DW_U	DW_L	DW_U	DW_L	DW_U	DW_L	DW_U	DW_L	DW_U
15 - 19	0.93	1.13	0.83	1.26	0.74	1.41	0.65	1.58	0.56	1.77
20 - 24	1.04	1.20	0.96	1.30	0.88	1.41	0.80	1.53	0.72	1.66
25 - 29	1.12	1.25	1.05	1.33	0.99	1.42	0.92	1.51	0.85	1.61
30 - 35	1.18	1.30	1.13	1.36	1.07	1.43	1.01	1.51	0.95	1.59
36 - 40	1.24	1.34	1.19	1.39	1.14	1.45	1.09	1.52	1.03	1.58
41 - 45	1.29	1.38	1.24	1.42	1.20	1.48	1.16	1.53	1.11	1.58
46 - 50	1.32	1.40	1.28	1.45	1.24	1.49	1.20	1.54	1.16	1.59
51 - 55	1.36	1.43	1.32	1.47	1.28	1.51	1.25	1.55	1.21	1.59
56 - 60	1.38	1.45	1.35	1.48	1.32	1.52	1.28	1.56	1.25	1.60

4. The Decision Criteria for Testing the First-Order Autocorrelation via DW.

   The decision criteria are better depicted in the following diagram:

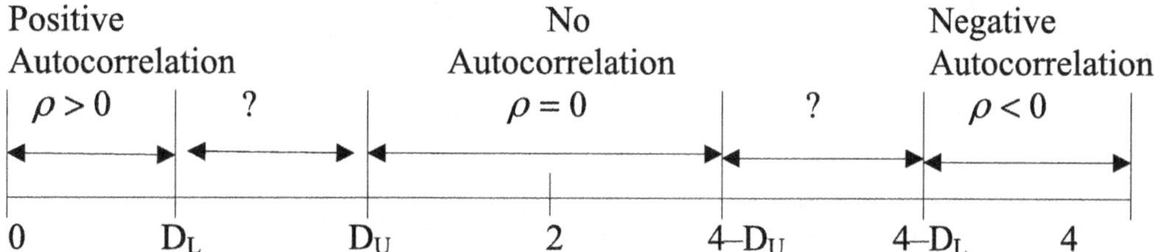

   Note that DW-Values that fall in the "?" range indicates an "inconclusive" or "uncertain" presence of autocorrelation.

Section 3. Uses and Applications

In this Section 3, uses and applications of simple regression analysis are shown on the basis of a typical research situation where an analyst is given a task to identify, explain, and predict the level of a certain variable such as an individual's wealth[155]. The examples below show the step-by-step research progression that one may follow in accomplishing such a task. Because a large part of regression analysis is asking the right questions and following specific steps to answer those, it is essential that you must understand what questions to ask and what specific procedures to follow to answer them. The following discussion shows those questions and procedures applied to a hypothetical case on analyzing the determinants of one's wealth level. Of course, one can always be imaginative and inventive by changing the situation and the variables to suit one's need. For example, one can replace the "wealth" variable by "sales" or "profits" to find an application that fits one's need.

[155] The subject of wealth building has been well explained and documented by Thomas J. Stanley and William D. Danko in their book, The Millionaire Next Door, Pocket Books, 1996, ISBN:0-671-01520-6.

A. Simple Regression Analysis For Wealth Determined by Salary

Suppose that your boss asked you to identify a relationship that can explain, estimate, and predict the level of an individual's wealth. Let's assume that you are to use simple regression analysis to accomplish this task for now.

1. What would you do to accomplish this task?

Answer:

We need to establish a regression model by identifying the dependent and the independent variables on the basis of economic theory, business intuition, common sense, etc. Then, we must collect data for the variables of our interest.
Suppose you had collected the following data:

Determination of Factors Affecting Wealth

Wealth	Salary	Age
100	50	20
200	75	25
150	45	29
300	100	40
205	80	35
100	30	24
150	35	25
270	50	35
280	55	32
700	90	60
500	30	70
300	25	80
650	110	55
600	50	65
600	30	76

Note: Wealth and Salary are measured in $1,000.

2. Given the nature of the estimation problem, which variable would be the dependent variable and which would be the independent variable?

Answer:

Because we wish to determine the wealth, it has to be the dependent variable. However, we can identify any number of independent variables. For now, we identify Salary and Age as the two independent variables. However, we arbitrarily choose Salary as the only independent variable because we wish to construct a simple regression equation.

3. Determine a regression equation (or model) that relates to Salary to Wealth.

Answer:

A regression equation can be estimated by using an Excel Spreadsheet as follows:

Step 1. Input data on Excel Spreadsheet with the first row labeled with variable names such as Wealth, Salary, Age, or Profits, Prices, Sales, etc.

Step 2. In Excel Spreadsheet, click "Tools" → "Data Analysis"[156] → Scroll down to "Regression" → Click "OK" → Define the values of the dependent variable, Y, in the "Input Y Range" by highlighting the values of Y in the Excel Spreadsheet → Move your cursor to "Input X Range" → Define the values of the independent variables, X_1, X_2, X_3, etc., in the "Input X Range" by highlighting the values of Xs in the Excel Spreadsheet → If labels for Y and X were highlighted when these values were input, check-mark "Labels" → Check "Confidence Level" and select any level such as 99% → (95% is a default level) → Check-mark "Residuals" to obtain the values of error (=residual) terms → Click "OK" → You should see the regression estimates in the Excel Spreadsheet (in Sheet4) → Now, manually calculate Durbin-Watson Statistic, if needed.

If you successfully followed these Excel Spreadsheet regression estimation procedures, you would obtain the following results as shown in Table 1 herein:

Table 1. Simple Regression of Wealth on Salary

Regression Statistics		
Multiple R	0.29925	**<-- Correlation coefficient between Y and Y-hat**
R Square	0.089551	**<-- Explanatory power or goodness of fit**
Adjusted R Square	0.019516	**<-- Model adequacy and efficiency**
Standard Error	208.9526	**<-- Standard Error of the Estimate = SQRT{ESS/(n-k-1)}**
Observations	15	**<-- n, the number of observations**

ANOVA

	df	*SS*	*MS*	*F* (**Calculated F**)	*Significance F*	**←p-value for an F-test**
Regression	1	55828.07	55828.07	1.278666	0.278578	
Residual	13	567595.3	43661.17			
Total	14	623423.3				

	Coefficients	*Standard Error*	*t Stat*	*P-value*	*Lower 95%*	*Upper 95%*
Intercept	209.4594	127.6947	1.640314	0.1249	-66.4083	485.3271
Salary	2.296034	2.030485	1.130781	0.278578	-2.09056	6.68263
			Calculated t assuming H_0: β=0			

RESIDUAL OUTPUT

Observation	*Predicted Wealth*	**Error=e=** *Residuals*	*e(-1)*	*[e-e(-1)]^2*	*e^2*

[156] If you do not see "Data Analysis," you need to install the statistical package by going to "Tools" → "Add-Ins" → Check mark on "Analysis TookPak" and "Analysis ToolPak-VBA" → "OK"

1	324.2611	-224.261			50293.04
2	381.6619	-181.662	-224.261	1814.688	33001.06
3	312.7809	-162.781	-181.662	356.4929	26497.63
4	439.0628	-139.063	-162.781	562.5497	19338.46
5	393.1421	-188.142	-139.063	2408.78	35397.46
6	278.3404	-178.34	-188.142	96.07332	31805.3
7	289.8206	-139.821	-178.34	1483.777	19549.8
8	324.2611	-54.2611	-139.821	7320.426	2944.266
9	335.7413	-55.7413	-54.2611	2.190903	3107.089
10	416.1025	283.8975	-55.7413	115354.5	80597.82
11	278.3404	221.6596	283.8975	3873.564	49132.97
12	266.8602	33.13975	221.6596	35539.73	1098.243
13	462.0231	187.9769	33.13975	23974.53	35335.3
14	324.2611	275.7389	187.9769	7702.176	76031.94
15	278.3404	321.6596	275.7389	2108.709	103464.9
			321.6596		
			Sum =	202598.2	567595.3 **<-- Error SS**
			DW =	0.356941	

The first column of the 3rd paragraph from the top of this regression output shows the names of the variables estimated such as "Intercept" and "Salary." The values of an intercept and a slope coefficient are given in the next column under "Coefficients." While an intercept is labeled as "Intercept," and the slope coefficient carries its own label such as "Salary" in this case, the Excel Spreadsheet does NOT identify and give the dependent variable's label in its print-out and thus, you must be aware of which variable is used as the dependent variable. Based on Table 1, the estimated regression equation is identified as:

$$\text{Wealth(W)} = a + b \cdot Salary = 209.46 + 2.296 \text{ Salary(S)}$$

4. Given an economic interpretation of the intercept (a) term.

Answer:

When Salary is equal to zero, Wealth = 209.46 (x \$1,000) = \$209,460. This means that there are other factors than just Salary that determines or influences the level of Wealth. Thus, the intercept term captures the influence of various factors other than Salary that still determine the wealth level.

5. At a 5% significance level, test the significance of the intercept term[157] by using the critical value approach.

Answer:

The frequent use of "test the significance of an intercept term or a slope coefficient" in regression analysis simply means that you should conduct a

[157] This is a statistics jargon in that it is same as saying, "test the null hypothesis that the intercept is statistically equal to zero."

hypothesis test, assuming that the hypothesized value of an intercept or a slope is zero. That is, you are to conduct a hypothesis test of the following form:

Test $H_0: \alpha = 0$ vs. $H_1: \alpha \neq 0$ at a chosen significance level such as 5%.

Because you must use a t-test in regression analysis[158], you can test the significance of an intercept term by following exactly the same hypothesis testing procedures that we studied in earlier chapters. That is,

(1) identify the table value of t as: $t_{0.05/2,15-1-1} = t_{0.025,13} = 2.1604$

(2) calculate the calculated t-value[159] as:

$$t_c = \frac{a-\alpha}{S_a} = \frac{209.459-0}{127.69} = \frac{a}{S_a} = \frac{209.459}{127.69} = 1.6403$$

(3) draw the conclusion as:

Because $t_c = \frac{a-\alpha}{S_a} = \frac{209.459-0}{127.69} = 1.6403 < t_{0.05/2,15-1-1} = t_{0.025,13} = 2.1604$

we accept $H_0: \alpha = 0$ → the intercept term is statistically equal to zero at a 5% significance level → the intercept term is NOT significant at a 5% significance level.

6. At a 5% significance level, test the significance of the intercept term by using the p-value approach.

Answer:

Because the p-value can be best obtained by the Excel command (or any computer output) as shown in the 5th column under "P-value," we use the information as is:

That is, given that $H_0: \alpha = 0$ and $H_1: \alpha \neq 0$,
because (p-value = 0.1249) > (the chosen significance level, α, = 0.05),
we accept $H_0: \alpha = 0$ → the intercept term is statistically equal to zero at a 5% significance level → the intercept term is NOT significant.

7. At a 5% significance level, test the significance of the intercept term by using the confidence interval approach.

Answer:

Given that $H_0: \alpha = 0$ and $H_1: \alpha \neq 0$

the 95% Confidence Interval (CI) for α = $a \pm t_{0.025,13} \cdot S_a$

= 209.459 ± 2.1604 x 127.69 = [–66.41, 485.32]

---

158 Because the population standard errors of the coefficients are not known, you must use a t-test.

159 Excel Spreadsheet shows the (sample) standard errors in the third column and the calculated t-values in the fourth column under "t Stat" of the 3rd paragraph of its output.

Because this confidence interval captures α being equal to zero, we accept $H_0: \alpha = 0 \rightarrow$ the intercept term is statistically equal to zero at a 5% significance level → the intercept term is NOT significant.

Note that the Excel Spreadsheet shows the 95% confidence interval as default information in the 6th and 7th columns. If you wish to select a different confidence level such as 99%, you can specify it by clicking the "Confidence Level" box in the regression command.

8. If the intercept is judged to be statistically significant, what does it mean in a layman's term?

Answer:

The intercept term is not equal to zero. → Thus, it is significant and important in defining the relationship between Wealth and Salary. → The estimated regression line does NOT start at the origin.

9. If the intercept is judged to be statistically insignificant, what does it mean in a layman's term?

Answer:

The intercept term is NOT significant and NOT important in defining the relationship between Wealth and Salary. It is statistically equal to zero and thus, insignificant. The estimated regression line starts at the origin.

10. Given an economic interpretation of the slope (b) coefficient.

Answer:

Because a slope measures the change in Y given a change in X, it can be calculated by the differential calculus as follows:

$$Slope = \frac{\Delta Y}{\Delta X} = \frac{\Delta Wealth}{\Delta Salary} = 2.296$$

→ Given a $1 increase (decrease) in Salary, Wealth will increase (decrease) by $2.296.

11. At a 5% significance level, test the significance of the slope coefficient[160] by using the critical value approach.

Answer:

The frequent use of "test the significance of a slope coefficient" or "test the significance of β" in regression analysis simply means that you should conduct a hypothesis test, assuming that the hypothesized value of a slope is zero. That is, you are to conduct a hypothesis test of the following form:

Test $H_0: \beta = 0$ vs. $H_1: \beta \neq 0$ at a chosen significance level such as 5%.

[160] This is a statistics jargon in that it is same as saying, "test the null hypothesis that the independent variable has no relationship with the dependent variable."

Because you must use a t-test in regression analysis[161], you can test the significance of a slope coefficient by following exactly the same hypothesis testing procedures that we studied in earlier chapters. That is,

(1) identify the table value of t as: $t_{0.05/2,15-1-1} = t_{0.025,13} = 2.1604$

(2) calculate the calculated t-value[162] as:

$$t_c = \frac{b-\beta}{S_b} = \frac{2.296-0}{2.03} = \frac{b}{S_b} = \frac{2.296}{2.03} = 1.137$$

(3) draw the conclusion as:

Because $t_c = \frac{b-\beta}{S_b} = \frac{2.296-0}{2.03} = 1.1307 < t_{0.05/2,15-1-1} = t_{0.025,13} = 2.1604$

we accept $H_0 : \beta = 0$ → the slope coefficient is statistically equal to zero at a 5% significance level → the slope coefficient is NOT significant → the slope coefficient is insignificant → β is NOT significant → β is insignificant → Salary does NOT determine Wealth → Wealth is NOT determined by Salary .

12. At a 5% significance level, test the significance of the slope coefficient by using the p-value approach.

Answer:

Because the p-value can be best obtained by the Excel command (or any computer output) as shown in the 5th column under "P-value," we use the information as is:

That is, given that $H_0 : \beta = 0$ vs. $H_1 : \beta \neq 0$,
because (p-value = 0.2785) > (the chosen significance level, α, = 0.05),
we accept $H_0 : \beta = 0$ → the slope coefficient is statistically equal to zero at a 5% significance level → the slope coefficient is NOT significant → the slope coefficient is insignificant → β is NOT significant → β is insignificant → Salary does NOT determine Wealth → Wealth is NOT determined by Salary .

13. At a 5% significance level, test the significance of the slope coefficient by using the confidence interval approach.

Answer:

Given that $H_0 : \beta = 0$ and $H_1 : \beta \neq 0$,

the 95% Confidence Interval (CI) for β = $b \pm t_{0.025,13} \cdot S_b$

= 2.296 ± 2.1604 x 2.0304 = [–2.09, 6.68]

---

[161] Because the population standard errors of the coefficients are not known, you must use a t-test.

[162] Excel Spreadsheet shows the (sample) standard errors in the third column and the calculated t-values in the fourth column under "t Stat" of the 3rd paragraph of its output.

Because this confidence interval captures β being equal to zero, we accept $H_0 : \beta = 0$ → the slope coefficient is statistically equal to zero at a 5% significance level → the slope coefficient is NOT significant → the slope coefficient is insignificant → β is NOT significant → β is insignificant → Salary does NOT determine Wealth → Wealth is NOT determined by Salary .

Note that the Excel Spreadsheet shows the 95% confidence interval as default information in the 6th and 7th columns. If you wish to select a different confidence level such as 99%, you can specify it by clicking the "Confidence Level" box in the regression command.

14. If the independent variable is judged to be statistically significant, what does it mean in a layman's term?

Answer:

We reject $H_0 : \beta = 0$ → we accept $H_1 : \beta \neq 0$ → the slope coefficient is statistically NOT equal to zero at a 5% significance level → the slope coefficient is significant → β is significant → Salary does determine Wealth → Wealth is determined by Salary.

15. If the independent variable is judged to be statistically insignificant, what does it mean in a layman's term?

Answer:

We accept $H_0 : \beta = 0$ → the slope coefficient is statistically equal to zero at a 5% significance level → the slope coefficient is NOT significant → the slope coefficient is insignificant → β is NOT significant → β is insignificant → Salary does NOT determine Wealth → Wealth is NOT determined by Salary .

16. Calculate the coefficient of determination.

Answer:

$$R^2 = \frac{RSS}{TSS} = \frac{55828.07}{623423.3} = 0.08955 \rightarrow 8.9\%$$

Salary explains 8.9% of variation in Wealth → Not a very good model because this R-squared may be too low.

17. At a 5% significance level, conduct an F-test of the overall significance of the regression equation (or model).

Answer:

In simple regression, an F-test is the same as a t-test because there is only one independent variable. The null and the alternative hypotheses to be tested are the same as:

$$H_0 : \beta = 0 \text{ and } H_1 : \beta \neq 0$$

However, the F-test procedure is different from a t-test as shown below:

Because $F_c = \frac{MS_R}{MS_E} = \frac{55828.07}{43661.17} = 1.2786$ is less than $F^k_{n-k-1,\alpha} = F^1_{13,0.05} = 4.67$,

we accept $H_0 : \beta = 0$ → the slope coefficient is statistically equal to zero at a 5% significance level → the slope coefficient is NOT significant → the slope coefficient is insignificant → β is NOT significant → β is insignificant → Salary does NOT determine Wealth → Wealth is NOT determined by Salary .

More importantly, the acceptance of $H_0 : \beta = 0$ means that the regression model is NOT good enough or adequate enough to be used as is → Find another model.

Also, note that $F_c = (t_c)^2$ which is a unique property for the case of simple regression. That is, $[F_c = 1.2786] = [(t_c)^2 = (1.1307)^2]$. Thus, an F-test and a t-test must yield the same conclusion about the significance of the slope coefficient in the simple regression equation.

18. Is there an autocorrelation problem in the estimated regression equation? Test the presence of the autocorrelation at a 5% significance level.

Answer:

The calculated Durbin-Watson (DW) statistic is 0.3569 and the table values of DW at a 5% significance level with n = 15 and k = 1 are:

$DW_L = D_L = 1.08$ and $DW_U = D_U = 1.36$

Because $DW_{calculated} = DW_c = 0.3569 < DW_L = D_L = 1.08$,
we conclude that there is "positive" autocorrelation in the error (or residual) terms.

That is,

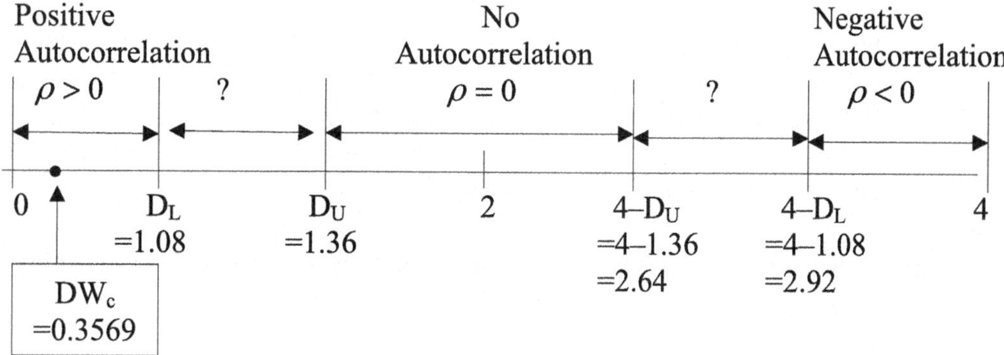

Note that DW-Values that fall in the "?" range indicates an "inconclusive" or "uncertain" presence of autocorrelation.

19. Suppose your 35 year-old friend has an annual salary of $43,000. What is the best estimate, based on the regression model, of her wealth?

Answer:

Given that Salary is measured in $1,000, the Salary of $43,000 = 43, and Wealth(W) = 209.46 + 2.296 Salary(S)

Wealth = 209.46 + 2.296 (43) = 308.188 → $308,188

Thus, the best estimate is $308,188.

20. Construct a 95 percent confidence interval for her wealth.

Answer:

The 95% Confidence Interval for her Wealth, W = $\hat{W} \pm t_{\alpha/2,n-k-1} \cdot S_{\hat{W}}$

where

$$S_{\hat{W}} = \sqrt{\frac{\sum e^2}{n-k-1} \cdot \left(1 + \frac{1}{n} + \frac{(S-\bar{S})^2}{\sum (S-\bar{S})^2}\right)} = SEE \cdot \sqrt{1 + \frac{1}{n} + \frac{(S_i - \bar{S})^2}{(n-1)S_S^2}}$$

Therefore, given $\hat{W} = 308.188$; $t_{0.05/2,15-1-1} = t_{0.025,13} = 2.1604$; SEE = 208.95; n = 15; S_S^2 = variance of Salary = 756.4286; S_i = 43; $\bar{S}$=57; and

$$S_{\hat{W}} = 208.95 \cdot \sqrt{1 + \frac{1}{15} + \frac{(43-57)^2}{(15-1)(756.4286)}} = 208.95 \text{ x } 1.0417 = 217.663$$

the 95% Confidence Interval for her Wealth,

W = $\hat{W} \pm t_{\alpha/2,n-k-1} \cdot S_{\hat{W}}$ =308.188 ± 2.1604 x 217.663 = [−162.051, 778.427]

Note: SEE is obtained from the regression estimation whereas $\bar{S}$ = the mean of Salary and S_S^2 = variance of Salary are calculated from the Salary data.

21. Suppose that your 50-year-old boss wants to know her wealth based on the estimated regression equation. Assuming that she is paid $1 million a year, what would be her estimated wealth?

Answer:

Given that Salary is measured in $1,000, the Salary of $1 million = 1,000, and Wealth(W) = 209.46 + 2.296 Salary(S)

Wealth = 209.46 + 2.296 (1000) = 2505.46 → $2,505,460

Thus, the best estimate is $2,505,460.

Note that her age of 50 has no role to play in the wealth estimation because the regression model does not recognize the Age variable in it.

22. Would the estimate of your boss's wealth above be reliable? Why or why not?

Answer:

Because the data used to estimate the above regression equation does not contain a salary level approaching $1 million, the use of the regression equation to estimate wealth based on the salary-outlier can not be reliable. The estimated regression equation is most valid when the forecast data are within the range of observed values.

23. Would you use this model as is?

Answer:

No. Because of the following reasons:

(1) The slope coefficient is NOT significant as judged by the t-test.
(2) The DW test shows a positive autocorrelation.
(3) R-squared of 0.0895 seems too low in this situation.

Some might still argue for the strengths of this regression model as:

(1) The sign of the slope coefficient makes an economic and intuitive sense. That is, the positive sign associated with the Salary variable indicates that it has a positive influence on the Wealth level, which makes an economic and intuitive sense → However, it is statistically equal to zero → Thus, this is a mute point.
(2) The simple regression model is concise and parsimonious enough to be useful → However, the model is NOT useful for the reasons given above.

B. Simple Regression Analysis For Wealth Determined by Age

Suppose that your boss asked you to identify the factors that may determine one's wealth and if possible, estimate a relationship between or among them.

1. What would you do to accomplish this task?

Answer:

The same as before. That is, we need to establish a regression model by identifying the dependent and the independent variables on the basis of economic theory, business intuition, common sense, etc. Then, we must collect data for the variables of our interest.

Suppose you had collected the following data:

Determination of Factors Affecting Wealth

Wealth	Salary	Age
100	50	20
200	75	25
150	45	29
300	100	40
205	80	35
100	30	24
150	35	25
270	50	35
280	55	32
700	90	60
500	30	70
300	25	80
650	110	55
600	50	65
600	30	76

Note: Wealth and Salary are measured in $1,000.

2 Given the nature of the estimation problem, which variable would be the dependent variable and which would be the independent variable?

Answer:

Because we wish to determine the wealth, it has to be the dependent variable. However, we can identify any number of independent variables. For now, we identify Salary and Age as the two independent variables. However, we arbitrarily choose Age as the only independent variable because we wish to construct a simple regression equation.

Table 2. Simple Regression of Wealth on Age

Regression Statistics		
Multiple R	0.782679	**<-- Correlation coefficient between Y and Y-hat**
R Square	0.612586	**<-- Explanatory power or goodness of fit**
Adjusted R Square	0.582785	**<-- Model adequacy and efficiency**
Standard Error	136.3037	**<-- Standard Error of the Estimate = SQRT{Error SS/(n-k-1)}**
Observations	15	**<-- n, the number of observations**

ANOVA				**Calculated F**		
	df	*SS*	*MS*	*F*	*Significance F*	**<-- p-value for an F-test**
Regression	1	381900.3	381900.3	20.55582	0.000561	
Residual	13	241523.1	18578.7			
Total	14	623423.3				

	Coefficients	*Standard Error*	*t Stat*	*P-value*	*Lower 95%*	*Upper 95%*
Intercept	-15.0484	85.92229	-0.17514	0.863669	-200.672	170.5754
Age	7.944449	1.752252	4.533852	0.000561	4.15894	11.72996
			Calculated t			

assuming H_0: β=0

RESIDUAL OUTPUT

		Error = e			
Observation	*Predicted Wealth*	*Residuals*	*e(-1)*	*[e-e(-1)]^2*	*e^2*
1	143.8406	-43.8406			1922
2	183.5629	16.43713	-43.8406	3633.408	270.1794
3	215.3407	-65.3407	16.43713	6687.608	4269.402
4	302.7296	-2.72961	-65.3407	3920.144	7.450751
5	263.0074	-58.0074	-2.72961	3055.63	3364.854
6	175.6184	-75.6184	-58.0074	310.1493	5718.145
7	183.5629	-33.5629	-75.6184	1768.669	1126.466
8	263.0074	6.99264	-33.5629	1644.749	48.89702
9	239.174	40.82599	6.99264	1144.695	1666.761
10	461.6186	238.3814	40.82599	39028.14	56825.7
11	541.0631	-41.0631	238.3814	78089.22	1686.177
12	620.5076	-320.508	-41.0631	78089.22	102725.1
13	421.8963	228.1037	-320.508	300974.3	52031.28
14	501.3408	98.65916	228.1037	16755.88	9733.63
15	588.7298	11.27022	98.65916	7636.827	127.0178
			11.27022		
			Sum =	542738.6	241523.1 **<-- Error SS**
			DW =	2.24715	

3. Determine a regression equation (or model) that relates to Age to Wealth.

Answer:

Per regression estimation shown in the above Table 2,

$$\text{Wealth(W)} = a + b \cdot Age = -15.04 + 7.94\ \text{Age(A)}$$

4. Given an economic interpretation of the intercept (a) term.

Answer:

When Age is equal to zero, Wealth = –15.04 (x $1,000) = –$15,040. This means that when a person is born, he (or she) has a negative wealth of $15,040. Thus, it does not make much common sense. One should not, for this reason, read into the meaning of the intercept term too much.

5. At a 5% significance level, test the significance of the intercept term[163] by using the critical value approach.

Answer:

Given that $H_0 : \alpha = 0$ and $H_1 : \alpha \neq 0$

[163] This is a statistics jargon in that it is same as saying, "test the null hypothesis that the intercept is statistically equal to zero."

because

$$\text{Absolute value of } t_c = \frac{a-\alpha}{S_a} = \frac{-15.04-0}{85.922} = -0.1751 \rightarrow 0.1751$$

$$0.1751 < t_{0.05/2,15-1-1} = t_{0.025,13} = 2.1604$$

we accept $H_0 : \alpha = 0 \rightarrow$ the intercept term is statistically equal to zero at a 5% significance level → the intercept term is NOT significant.

6. At a 5% significance level, test the significance of the intercept term by using the p-value approach.

Answer:

Given that $H_0 : \alpha = 0$ and $H_1 : \alpha \neq 0$,
the p-value can be best obtained by the Excel command or the computer output, we use the information as is:

Because (p-value = 0.8636) > (the chosen significance level, α, = 0.05), we accept $H_0 : \alpha = 0 \rightarrow$ the intercept term is statistically equal to zero at a 5% significance level → the intercept term is NOT significant.

7. At a 5% significance level, test the significance of the intercept term by using the confidence interval approach.

Answer:

Given that $H_0 : \alpha = 0$ and $H_1 : \alpha \neq 0$

the 95% Confidence Interval (CI) for α = $a \pm t_{0.025,13} \cdot S_a$
$= -15.04 \pm 2.1604 \times 85.922 = [-200.67, 170.57]$

Because this confidence interval captures α being equal to zero, we accept $H_0 : \alpha = 0 \rightarrow$ the intercept term is statistically equal to zero at a 5% significance level → the intercept term is NOT significant.

8. If the intercept is judged to be statistically significant, what does it mean in a layman's term?

Answer:

The intercept term is significant and important in defining the relationship between Wealth and Age. It is not equal to zero and thus, significant. The estimated regression line does NOT start at the origin.

9. If the intercept is judged to be statistically insignificant, what does it mean in a layman's term?

Answer:

The intercept term is NOT significant and NOT important in defining the relationship between Wealth and Age. It is statistically equal to zero and thus, insignificant. The estimated regression line starts at the origin.

10. Given an economic interpretation of the slope (b) coefficient.

Answer:

Because a slope measures the change in Y given a change in X, it can be calculated by the differential calculus as follows:

$$Slope = \frac{\Delta Y}{\Delta X} = \frac{\Delta Wealth}{\Delta Age} = 7.9444$$

→ Given a 1-year increase (decrease) in Age, Wealth will increase (decrease) by 7.9444 (x \$1000) = \$7,944.40. Note how the measurement unit will change the interpretation.

11. At a 5% significance level, test the significance of the slope coefficient[164] by using the critical value approach.

Answer:

Given that $H_0 : \beta = 0$ and $H_1 : \beta \neq 0$

because

$$t_c = \frac{b-\beta}{S_b} = \frac{7.9444-0}{1.7522} = 4.5338 > t_{0.05/2,15-1-1} = t_{0.025,13} = 2.1604$$

We reject $H_0 : \beta = 0$ → we accept $H_1 : \beta \neq 0$ → the slope coefficient is statistically NOT equal to zero at a 5% significance level → the slope coefficient is significant → β is significant → Age does determine Wealth → Wealth is determined by Age.

12. At a 5% significance level, test the significance of the slope coefficient by using the p-value approach.

Answer:

Given that $H_0 : \beta = 0$ and $H_1 : \beta \neq 0$,

the p-value can be best obtained by the Excel command or the computer output, we use the information as is:

Because (p-value = 0.0005) < (the chosen significance level, α, = 0.05), we reject $H_0 : \beta = 0$ → we accept $H_1 : \beta \neq 0$ → the slope coefficient is statistically NOT equal to zero at a 5% significance level → the slope coefficient

[164] This is a statistics jargon in that it is same as saying, "test the null hypothesis that the independent variable has no relationship with the dependent variable."

is significant → β is significant → Age does determine Wealth → Wealth is determined by Age.

13. At a 5% significance level, test the significance of the slope coefficient by using the confidence interval approach.

Answer:

Given that $H_0: \beta = 0$ and $H_1: \beta \neq 0$,

the 95% Confidence Interval (CI) for β = $b \pm t_{0.025,13} \cdot S_b$

= 7.9444 ± 2.1604 x 1.7522 = [4.1589, 11.7299]

Because this confidence interval does NOT capture β of zero, we reject $H_0: \beta = 0$ → we accept $H_1: \beta \neq 0$ → the slope coefficient is statistically NOT equal to zero at a 5% significance level → the slope coefficient is significant → β is significant → Age does determine Wealth → Wealth is determined by Age.

14. If the independent variable is judged to be statistically significant, what does it mean in a layman's term?

Answer:

It means that we reject $H_0: \beta = 0$ → we accept $H_1: \beta \neq 0$ → the slope coefficient is statistically NOT equal to zero at a 5% significance level → the slope coefficient is significant → β is significant → Age does determine Wealth → Wealth is determined by Age.

15. If the independent variable is judged to be statistically insignificant, what does it mean in a layman's term?

Answer:

It means that we accept $H_0: \beta = 0$→ the slope coefficient is statistically equal to zero at a 5% significance level → the slope coefficient is NOT significant → the slope coefficient is insignificant → β is NOT significant → β is insignificant → Age does NOT determine Wealth → Wealth is NOT determined by Age.

16. At a 5% significance level, test **the significance of the slope coefficient being equal to 8** by using the critical value approach.

Answer:

Because the hypothesized value of β is now equal to 8, instead of 0, the corresponding null and alternative hypotheses are:

$$H_0: \beta = 8 \text{ and } H_1: \beta \neq 8$$

Because

$$\text{Absolute value of } (t_c = \frac{b - \beta}{S_b} = \frac{7.9444 - 8}{1.7522} = -0.0317) \rightarrow 0.0317$$

therefore, $(t_c = 0.0317) < (t_{0.05/2,15-1-1} = t_{0.025,13} = 2.1604)$

we accept $H_0 : \beta = 8 \rightarrow$ the slope coefficient is statistically equal to 8 at a 5% significance level[165].

17. Calculate the coefficient of determination.

Answer:

$$R^2 = \frac{RSS}{TSS} = \frac{381900.3}{623423.3} = 0.6125 \rightarrow 61.25\%$$

Age explains 61.25% of variation in Wealth → A good model because this R-squared seems to be high → if a single independent variable – just like in this case – explains a significant[166] portion of variation in the dependent variable, then it should be considered a good model.

18. At a 5% significance level, conduct an F-test of the overall significance of the regression equation (or model)[167].

Answer:

In simple regression, an F-test is the same as a t-test because there is only one independent variable. The null and the alternative hypotheses to be tested are the same as:

$$H_0 : \beta = 0 \text{ and } H_1 : \beta \neq 0$$

However, the F-test procedure is different from a t-test as shown below:

Because $F_c = \frac{MS_R}{MS_E} = \frac{381900.3}{18578.7} = 20.5558$ is greater than $F^k_{n-k-1,\alpha} = F^1_{13,0.05} = 4.67$,

we reject $H_0 : \beta = 0 \rightarrow$ we accept $H_1 : \beta \neq 0 \rightarrow$ the slope coefficient is statistically NOT equal to zero at a 5% significance level → the slope coefficient is significant → β is significant → Age does determine Wealth → Wealth is determined by Age.

More importantly, the rejection of $H_0 : \beta = 0$ means that the regression model is GOOD enough or adequate enough to be used as is.

Also, note that $F_c = (t_c)^2$ which is a unique property for the case of simple regression. That is, $[F_c = 20.5558] = [(t_c)^2 = (4.5338)^2]$. Thus, an F-test and a t-test must yield the same conclusion about the significance of the slope coefficient in the simple regression equation.

---

[165] This method allows one to test any hypothesized value of the slope coefficient, even though the case of the hypothesized value of zero is the most often tested.

[166] What percentage constitutes a "significant proportion" is still subjective and thus, must be evaluated on a case-by-case basis.

[167] This is a statistics jargon in that it is same as saying, "test the null hypothesis that all slope coefficients are simultaneously equal to zero."

19. Is there an autocorrelation problem in the estimated regression equation? Test the presence of the autocorrelation at a 5% significance level.

Answer:

The calculated Durbin-Watson (DW) statistic is 2.2471 and the table values of DW at a 5% significance level with n = 15 and k = 1 are:

$DW_L = D_L = 1.08$ and $DW_U = D_U = 1.36$

Because $(DW_U = 1.36) < (DW_{calculated} = DW_c = 2.2471) < (4 - DW_U = 2.64)$, we conclude that there is "NO" autocorrelation in the error (or residual) terms.

That is,

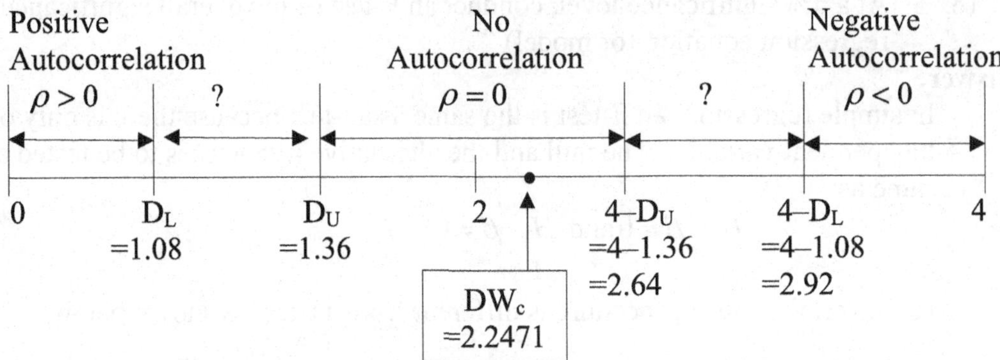

Note that DW-Values that fall in the "?" range indicates an "inconclusive" or "uncertain" presence of autocorrelation.

20. Suppose your 35 year-old friend has an annual salary of $43,000. What is the best estimate, based on the regression model, of her wealth?

Answer:

Given that Age = 35, and Wealth(W) = –15.0484 + 7.9444 Age(A)

Wealth = –15.0484 + 7.9444 (35) = 263.0056 → $263,005.60

Thus, the best estimate is $263,005.60.

21. Construct a 95 percent confidence interval for her wealth.

Answer:

The 95% Confidence Interval for her Wealth, W = $\hat{W} \pm t_{\alpha/2, n-k-1} \cdot S_{\hat{W}}$

$$\text{where } S_{\hat{W}} = \sqrt{\frac{\sum e^2}{n-k-1} \cdot \left(1 + \frac{1}{n} + \frac{(A-\overline{A})^2}{\sum (A-\overline{A})^2}\right)} = SEE \cdot \sqrt{1 + \frac{1}{n} + \frac{(A_i - \overline{A})^2}{(n-1)S_A^2}}$$

Therefore, given $\hat{W} = 263.0056$; $t_{0.05/2,15-1-1} = t_{0.025,13} = 2.1604$; SEE = 136.3037; n = 15; S_A^2 = variance of Age = 432.2095; A_i = 35; $\overline{S}$ = 44.7333; and

$$S_{\hat{W}} = 136.3037 \cdot \sqrt{1 + \frac{1}{15} + \frac{(35-44.7333)^2}{(15-1)(432.2095)}} = 136.3037 \text{ x } 1.0403 = 141.80$$

the 95% Confidence Interval for her Wealth,

$$\text{W} = \hat{W} \pm t_{\alpha/2,n-k-1} \cdot S_{\hat{W}} = 263.0056 \pm 2.1604 \text{ x } 141.80 = [-43.449,\ 569.350]$$

Note: SEE is obtained from the regression estimation whereas $\overline{A}$ (= the mean of Age) and S_A^2 (= variance of Age) are calculated from the Age data.

22. Suppose that your 50-year-old boss wants to know her wealth based on the estimated regression equation. Assuming that she is paid $1 million a year, what would be her estimated wealth?

Answer:

Given that Age = 50, and Wealth(W) = –15.0484 + 7.9444 Age(A)

Wealth = –15.0484 + 7.9444 (50) = 382.1716 → $382,171.60

Thus, the best estimate is $382,171.60.
Note that her salary of $1 million has no role to play in the wealth estimation because the regression model does not recognize the Salary variable in it.

23. Would the estimate of your boss's wealth above be reliable? Why or why not?

Answer:

Because the data used to estimate the above regression equation does contain an age level of 50, the use of this regression equation to estimate wealth can be reliable. The use of the estimated regression equation seems to be valid because (1) the forecast data are within the range of observed values, (2) R-squared is reasonably high, (3) the slope coefficient is significant based on the t- and the F-tests, (4) there is no autocorrelation, and (5) the relationship makes an economic common sense.

24. Would you use this model as is?

Answer:

Yes. Because of the following reasons:

(1) The slope coefficient is significant as judged by the t-test.
(2) The DW test shows no autocorrelation.
(4) R-squared of 0.5827 is reasonably high enough.
(5) The sign of the slope coefficient makes an economic and intuitive sense. That is, the positive sign associated with the Age variable indicates that it

has a positive influence on the Wealth level, which makes an economic and intuitive sense.

(6) The simple regression model is concise and parsimonious enough to be useful.

Note:

This does not mean that you found a perfect model and thus, no other model can be better than this one. It only means that it is adequate to be used. However, other models such as multiple regression models should be considered and examined before finalizing your pick. This is our next topic.

Exercise Problems on
Chapter 13. Simple Regression Analysis

This set of exercise problems has 23 problems, worth 25 points.

1. In simple regression analysis, the dependent variable is also known as a/an _____ variable and the independent variable is also known as a/an _____ variable.

   a. exogenous; endogenous
   b. response; explanatory
   c. to-be-determined; predetermined
   d. all of the above
   e. only (b) and (c) of the above

2. In simple regression analysis, you will find _____ independent variable(s) and ____ dependent variable(s).

   a. one; one
   b. one; two
   c. two; one
   d. two; two
   e. more than one; one

3. Given the relationship of: $Y = \beta_0 + \beta_1 X + \varepsilon$,
   β_0 and β_1 are both called regression coefficients. More specifically, however, β_0 is called a/an _____ and β_1 is called a/an _____.

   a. base; intercept
   b. intercept; slope coefficient
   c. slope coefficient; intercept
   d. slope coefficient; slope coefficient
   e. Y-intercept; X-intercept

4. The least-squares method for estimating regression coefficients is named as such because it tries to find the coefficient values by minimizing ______.

   a. the product of coefficients squared
   b. the sum of coefficients squared
   c. the sum of errors
   d. the sum of errors squared
   e. the sum of the dependent variable squared

5. Given the following data of:
   Covariance between X and Y = $S_{XY} = 40$
   Variance of X = $S_X^2 = 50$
   Number of observations = n = 10
   $\overline{X} = 5$ and $\overline{Y} = 8$
   the intercept value is _____ and the slope coefficient is _____.

   a. 4; 0.8
   b. 0.8; 4
   c. 1.25; 1.75
   d. 1.75; 1.25
   e. 0.4; 0.08

6. A positive slope coefficient means that _____.

a. X and Y are positively and directly related.
b. as X decreases, Y decreases as well.
c. as X increases, Y increases as well.
d. all of the above is true.
e. only (a) and (c) are true.

7. If a slope coefficient is statistically equal to zero, it means that _____.

a. X and Y are positively and directly related.
b. as X decreases, Y increases as well.
c. as X increases, Y decreases as well.
d. X can determine the value of Y.
e. none of the above.

8. The analysis-of-variance (ANOVA) table utilizes the partition of the various sums of squares. Given the total sum of squares (TSS), the regression sum of squares (RSS) and the error or residual sum of squares (ESS), which of the following relationship is true?

a. TSS = RSS + ESS
b. ESS = TSS – RSS
c. RSS = TSS – ESS
d. all of the above
e. only (a) and (c) of the above

9. The regression sum of squares (RSS) is also called _____ variation and the error or residual sum of squares (ESS) is also called _____ variation.

a. explained; unexplained
b. unexplained; explained
c. total; uncertain
d. good; bad
e. only (c) and (d) of the above

10. Given the following data:
The regression sum of squares (RSS) = 104
The error or residual sum of squares (ESS) = 56
The coefficient of determination is _____ and the correlation coefficient between X and Y is _____.

a. 0.5384; 0.7338
b. 0.7338; 0.5384
c. 0.65; 0.8062
d. 0.8062; 0.65
e. 0.65; 0.4225

11. A statistic that measures how much actual values of Y vary around the predicted values of Y on the basis of a regression equation is called the _____.

a. standard deviation of Y
b. standard error of the estimate
c. variance of the predicted Y
d. variance of Y
e. any of the above

12. Which of the following is an assumption of regression analysis?

a. errors (or residuals) are independent of one another.
b. errors are normally distributed.
c. errors are identically distributed and thus, have an equal variance.
d. all of the above.
e. only (a) and (b) of the above.

13. Autocorrelation means that lagged values of _____ with one another.

a. errors (or residuals) are correlated
b. an independent variable are correlated
c. errors (or residuals) are not correlated
d. an independent variable are not correlated
e. errors (or residuals) are identical.

14. Autocorrelation problem can be detected by the use of the _____.

a. Z-statistic
b. t-statistic
c. F-statistic
d. Durbin-Watson statistic
e. either (b) or (d) of the above

15. The significance test for a slope coefficient in an estimated regression equation can be done by the use of the _____.

a. Z-statistic
b. t-statistic
c. F-statistic
d. Durbin-Watson statistic
e. either (a) or (b) of the above

Given that the following demand relationship between the quantity demanded (Q) and its per-unit price (P) is estimated by a simple regression equation on the basis of 20 observations, answer Questions 16 through 21:

	Coefficient	Standard Error	p-value
Intercept	100	45	0.03
P	-5	2	0.005

The corresponding ANOVA table shows only the sums of squares as follows:

	Sum of Squares
Regression	200
Error (or Residual)	150

16. Which of the following represents the estimated regression equation on the basis of the above information?

a. Q = 100 – 5P b. Q = 45 + 2P c. P = 100 – 5Q
d. P = 45 + 2Q e. Q = 2.22 – 2.5P

17. If you are to conduct a significance test for the above regression coefficients at a 5% significance level, you would find the calculated t-value for the intercept to be _____ and that for the slope coefficient to be _____.

a. 0; 0 b. 100; -5 c. 45; 2
d. 2.22; 2.5 e. 2.22; -2.5

18. If you are to conduct a significance test for each of the estimated regression coefficients at a 5% significance level, you would conclude that the intercept term is _____ and the slope coefficient is _____.

a. insignificant; insignificant b. insignificant; significant
c. significant; significant d. significant; insignificant
e. normal; normal

19. If the calculated t- value for the slope coefficient is -1.9, you would find the table value of t to be _____ and thus, conclude to _____ the null hypothesis of the slope coefficient being equal to zero at the 5% significance level.

a. 1.7247; reject b. 1.7341; accept c. 1.7341; reject
d. 2.1009; accept e. 2.1009; reject

20. Based on the sums of squares given, the coefficient of determination is _____ and the corresponding calculated F-statistic is _____.

a. 0.5714; 24 b. 0.4286; 24 c. 0.5714; 1.333
d. 0.4286; 1.333 e. 0.75; 1.333

21. Given that an F-test is to be done for the joint significance of coefficients, you would find that the degrees of freedom for the numerator and the denominator are respectively _____ and _____.

a. 1; 19 b. 1; 18 c. 2; 20
d. 2; 19 e. 2; 18

The following are worth 2 points each.

22. On the basis of 15 observations, you found a correlation coefficient between the inflation rate change (via the CPI) and your salary change to be 0.65. If you are to test the significance of this correlation coefficient, you would use a/an _____ and conclude that there is a/an _____ relationship between the inflation rate change and your salary change at a 5% significance level.

    a. Z-statistic; significant
    b. Z-statistic; insignificant
    c. t-statistic; significant
    d. t-statistic; insignificant
    e. F-statistic; significant

23. If the slope coefficient is found to be 5 in a simple regression equation and its standard error to be 3 on the basis of 10 observations, the 95% confidence interval for this slope coefficient lies between _____ and _____.

    a. -1; 11
    b. -1.918; 11.918
    c. -0.5785; 10.5785
    d. -0.4375; 10.4375
    e. 0; 12

Chapter 14. Multiple Regression Analysis

Multiple regression analysis is largely an extension of simple regression analysis in that many concepts discussed in simple regression is still valid in multiple regression. For example, the method of using a t-test for the significance of an individual regression coefficient, the interpretation of R-squared and adjusted R-squared, that of the Durbin-Watson Statistics and the ANOVA table – all remain the same.

The major differences are, however, that multiple regression analysis allows (1) more than one independent variables in the regression equation, (2) various functional equations to be estimated, (3) violations of basic assumptions on regression analysis are often encountered.

How various functional forms can be estimated by multiple regression is first discussed below:

A. Multiple Regression

1. Various functional forms

   a. The basic Linear equation: $Y = a + b X + c P + d Q$
   where X, P, and Q are independent variables.

   b. A Quadratic equation: Given $Y = a + b X + c X^2$
   if we treat $Z = X^2$, we can transform this quadratic equation into a multiple regression of the following form: $Y = a + b X + c Z$.

   c. A Polynomial equation: Given $Y = a + b X + c X^2 + d X^3$
   if we treat $Z = X^2$, and $W = X^3$, we can transform this quadratic equation into a multiple regression of the following form: $Y = a + b X + c Z + d W$.

   d. A Logarithmic or an Exponential equation: Given $Y = AX^bZ^c$
   we can convert this exponential function into a logarithmic function of:
   $$\log Y = \log A + b \log X + c \log Z$$

   Now, by treating $\log Y = Y^*$; $\log A = a$; $\log X = X^*$; and $\log Z = Z^*$, we can transform this logarithmic equation into a multiple regression of the following form: $Y^* = a + b X^* + c Z^*$.

2. Dummy variable regression

   a. Event analysis
   → On-off variables or dummy variables

   b. Seasonality or Quarterly analysis

→ Seasonal or quarterly dummy variables → There can only be a maximum of 3 (= S – 1) dummy variables in the 4-season (or quarterly) model.

3. Stepwise regression

   a. Forward stepwise → Adding independent variables one at a time

   b. Backward stepwise → Deleting independent variables one at a time

   c. Mixed stepwise → A combination of forward and backward stepwise regression → Data-mining

B. Statistical Significance Tests

1. The t-tests of individual coefficients

All significance tests of individual regression coefficients using the t-test remain exactly the same as the simple regression case.

2. The Durbin-Watson Statistic

The use and interpretation of the Durbin-Watson statistic remain the same as the simple regression case.

3. The Interpretation of an ANOVA table

All interpretation of the ANOVA table, as shown below, remains the same and the joint (or composite or global) significance test for all coefficients using the F-test remains the same as in the simple regression case.

Source of Variation	Sum of Squares	Degrees of Freedom	Mean Square **(Variance)**	Calculated F Value = F_c
Due to Regression	RSS	k	$MS_r = \frac{RSS}{k}$	$F_c = \frac{MS_r}{MS_e}$
Due to Error	ESS	n-k-1	$MS_e = \frac{ESS}{n-k-1}$	
Total	TSS	n-1	$Var(Y) = \frac{TSS}{n-1}$	

where

$$R^2 = \frac{RSS}{TSS} = 1 - \frac{ESS}{TSS}$$

$$R^2 = \frac{RSS}{TSS} = 1 - \frac{ESS}{TSS}$$

and

$$\text{Adjusted } R^2 = \overline{R}^2 = 1 - \left[(1-R^2)\cdot\frac{n-1}{n-k-1}\right]$$

As shown, the calculation and interpretation of R^2 and adjusted R^2 are the same as the simple regression case.

4. The F-test for Joint Hypothesis Testing

While t-tests allow the examination of statistical significance of individual coefficients one at a time, an F-test in regression analysis allows the examination of statistical significance of all coefficients in the equation at the same time. That is why an F-test is called a joint hypothesis test.

a. Formal Hypotheses to Be Tested

The hypotheses to be tested by an F-test are formally written as:

H_0: All Slope Coefficients = 0 vs. H_1: Not All Slope Coefficients = 0

For the case of a simple regression equation, the null and the alternative hypotheses mean:

H_0: All Slope Coefficients = 0 → H_0: $\beta=0$
H_1: Not All Slope Coefficients = 0 → H_1: $\beta \neq 0$

That is, the t-test on β should yield the same conclusion as an F-test because there is only one slope coefficient. In fact, we note that $t_c^2 = F_c$ for a simple regression.

However, for a multiple regression, because there are more than one independent variables – let's say, 2 – as shown below, the same null hypothesis stated for a simple regression model has a different meaning.

That is, given a population regression equation of

$$Y = \alpha + \beta_1 \cdot X + \beta_2 Z + \varepsilon$$

suppose we obtained the following estimated regression equation:

$$\hat{Y} = a + b_1 \cdot X + b_2 Z$$

The appropriate joint hypothesis for this multiple regression equation is:

H_0: All Slope Coefficients = 0 → H_0: $\beta_1=\beta_2=0$
H_1: Not All Slope Coefficients = 0 → H_1: Not H_0

Note that if one accepts H_1, it means of the following:

Either $\beta_1 \neq 0$ or $\beta_2 \neq 0$ or both $\beta_1 \neq 0$ and $\beta_2 \neq 0$

b. F-test Procedures

An F-test of a joint hypothesis can be conducted in two different ways of (1) the critical F-value approach and (2) the p-value approach. The confidence interval approach is not possible because the F-test is always an upper-tail test[168].

While the interpretation of the p-value approach for an F-test remains exactly the same as that for a t-test, the critical F-value approach is different because an F-distribution requires two types of degrees of freedom – the numerator degrees of freedom and the denominator degrees of freedom.

The following describes the Critical F-value and the p-value approaches:

i. Calculate the F value, F_c, from the ANOVA table, as a ratio of MS_R (Mean Squares due to Regression) to MS_E (Mean Squares due to Error):

$$F_c = \frac{MS_R}{MS_E} = \frac{\frac{RSS}{k}}{\frac{ESS}{n-k-1}}$$

→ Note: the numerator degrees of freedom (d.f.) = k and the denominator degrees of freedom (d.f.) = n – k – 1

ii. Decision Criteria[169].

(a) The Critical F-value Approach

Find the table F value, F_t, from the F-table → Note: $F_t = F^{k}_{n-k-1,\alpha}$

[168] Note that a calculated F value is a ratio of MS_R (Mean Squares due to Regression) to MS_E (Mean Squares due to Error) where both of them are positive values. Thus, the F probability distribution is bounded by the value of zero at the lower end and thus, its shape is not normal but in fact, skewed to the right.
[169] Note that these decision criteria are exactly the same as those for the Z- and t-tests.

Note how the numerator d.f. = k and the denominator d.f. = n – k – 1 are written into the F_t along with the significance level of α.

If $F_c > F_t$, reject H_0: All Slope Coefficients = 0 → That is, accept H_1: Not All Slope Coefficients = 0.

Otherwise, accept H_0: All Slope Coefficients = 0.

(b) The p-value Approach ← Assume that the p-value is given for an F test via an Excel or other regression estimation tool.

If p-value ≥ a chosen α, accept H_0: All Slope Coefficients = 0.

If p-value < a chosen α, reject H_0: All Slope Coefficients = 0 → That is, accept H_1: Not All Slope Coefficients = 0.

C. The major problems in regression analysis

The following problems can be present in any regression analysis – whether simple or multiple regression. However, they are more often found and more problematic in multiple regression analysis than simple regression analysis.

1. Multicollinearity

a. Nature of the problem → high correlation among independent variables and their linear combinations.

For example, given a multiple regression of $\hat{Y} = 10 + 5X_1 + 3X_2$, if X_1 and X_2 are highly correlated such that $X_1 = X_2 = X$, then we would not be able to distinguish the difference among the following equations because all of them are the same:

$$\hat{Y} = 10 + 5X_1 + 3X_2 = 10 + 8X$$
$$\hat{Y} = 10 + 2X_1 + 6X_2 = 10 + 8X$$
$$\hat{Y} = 10 + 4X_1 + 4X_2 = 10 + 8X$$

In situations like this, a computer may not know which equation has the right coefficient values for X_1 and X_2. Therefore, it will not be able to compute them or may spit out any combination of these numbers. Thus, multicollinearity can result in an inaccurate estimation problem.

b. How to detect the problem → Examine the correlation coefficients among independent variables that are larger than 0.7 → However, some will say that a correlation coefficient of 0.5 or higher can be a possible source of a multicollinearity problem.

c. Possible solutions → drop one of the correlated independent variables or take a ratio/difference of the two correlated variables.

2. Heteroscedasticity

a. Nature of the problem → error terms are NOT distributed identically (at a given value of an independent variable, X.) → the magnitude of error terms are NOT the same → errors are either increasing in size as the value of X increases or vise versa →For example, errors can be dispersed like a shot-gun shot as X increases → error terms are from different populations and thus, have different variances → The different variances of errors is called a heteroscedasticity[170].

b. How to detect the problem → (1) visual inspection of the error plot (by plotting errors against an independent variable) and (2) conduct the Breusch-Pagan (Chi-Square) test

c. Possible solutions → need to use the Hansen Method or the Generalized Least Squares (GLS) method.

3. Autocorrelation or Serial Correlation

a. Nature of the problem → error terms are NOT distributed independently from each other → high correlation among successive error terms.

i. Negative autocorrelation
→ error terms tend to show zigzag or saw-tooth movements.

ii. Positive autocorrelation
→ error terms tend to show protracted upward movements followed by protracted downward movements or vice versa.

iii. No autocorrelation
→ no identifiable pattern exists among errors because they are randomly scattered around zero.

---

[170] The word "hetero" means "different" whereas "homo" means "same." Because homoscedasticity means the same variances for errors, this does not cause any estimation problem in regression analysis and thus, strongly desired. However, heteroscedasticity creates problem in that its results do not incorporate all available information that the data presents and thus, requires a new estimation of the regression equation via advanced methods such as the generalized least squares (GLS) method.

b. How to detect the problem → the Durbin-Watson test

c. Possible solutions → need to use the Hansen Method or the Generalized Least Squares (GLS) method.

D. An Example of Multiple Regression Analysis: Estimation of Wealth Determined by Salary and Age

Suppose that your boss asked you to identify the factors that may determine one's wealth and if possible, estimate a multiple regression relationship among them.

1. What would you do to accomplish this task?

Answer:

Do the same as in the case of simple regression. That is, we need to establish a regression model by identifying the dependent and the independent variables on the basis of economic theory, business intuition, common sense, etc. Then, we must collect data for the variables of our interest.

Suppose you had collected the following data:

Determination of Factors Affecting Wealth

Wealth	Salary	Age
100	50	20
200	75	25
150	45	29
300	100	40
205	80	35
100	30	24
150	35	25
270	50	35
280	55	32
700	90	60
500	30	70
300	25	80
650	110	55
600	50	65
600	30	76

Note: Wealth and Salary are measured in $1,000.

2. Given the nature of the estimation problem, which variable would be the dependent variable and which would be the independent variables if you are to construct a multiple regression model?

Answer:

Because we wish to determine the wealth, it has to be the dependent variable. Intuitively, we can identify Salary and Age as the two independent variables.

3. Determine a regression equation (or model) that relates to Salary and Age to Wealth.

Answer:

Assume that you had successfully obtained the following regression output, using the data above.

Table 3. Multiple Regression of Salary and Age on Wealth

Regression Statistics		
Multiple R	0.883005	**<-- Correlation coefficient between Y and Y-hat**
R Square	0.779698	**<-- Explanatory power or goodness of fit**
Adjusted R Square	0.742981	**<-- Model adequacy and efficiency**
Standard Error	106.9819	**<-- Standard Error of the Estimate = SQRT{Error SS/(n-k-1)}**
Observations	15	**<-- n, the number of observations**

ANOVA

	df	*SS*	*MS*	**Calculated F** *F*	*Significance F*	**<-- p-value for an F test**
Regression	2	486081.9	243041	21.23534	0.000114	
Residual	12	137341.4	11445.12			
Total	14	623423.3				

	Coefficients	*Standard Error*	*t Stat*	*P-value*	*Lower 95%*	*Upper 95%*
Intercept	-220.807	95.91117	-2.3022	0.040037	-429.78	-11.8347
Salary	3.165568	1.049219	3.017071	0.010721	0.879516	5.451619
Age	8.510502	1.388043	6.131296	5.09E-05	5.486216	11.53479

Calculated t assuming H_0: $\beta=0$

RESIDUAL OUTPUT

Observation	*Predicted Wealth*	**Error = e** *Residuals*	*e(-1)*	*[e-e(-1)]^2*	*e^2*
1	107.6813	-7.68129			59.00216
2	229.373	-29.373	-7.68129	470.5299	862.7725
3	168.448	-18.448	-29.373	119.3562	340.3273
4	436.1697	-136.17	-18.448	13858.41	18542.19
5	330.3058	-125.306	-136.17	118.0235	15701.55
6	78.41194	21.58806	-125.306	21577.82	466.0444
7	102.7503	47.24972	21.58806	658.5208	2232.536
8	235.3388	34.66119	47.24972	158.4711	1201.398
9	225.6351	54.36486	34.66119	388.2344	2955.538

10	574.7241	125.2759	54.36486	5028.382	15694.06
11	469.895	30.10499	125.2759	9057.51	906.3104
12	539.1722	-239.172	30.10499	72510.2	57203.33
13	595.4829	54.51709	-239.172	86253.39	2972.113
14	490.6539	109.3461	54.51709	3006.225	11956.58
15	520.958	79.04198	109.3461	918.3422	6247.635
			79.04198		
			Sum =	214123.4	137341.4 **<-- Error SS**
			DW =	1.55906	

Per regression result shown above,

$$\text{Wealth} = a + b_1 \cdot Salary + b_2 \cdot Age$$

Therefore, Wealth = –220.807 + 3.1655 Salary + 8.5105 Age

4. Given an economic interpretation of the intercept (a) term.

Answer:

When Salary and Age are both equal to zero, Wealth = –220.87 (x $1,000) = –$220,807. This means that when a person is born and has no salary, he (or she) has a negative wealth of $220,807. Thus, it does not make much common sense, except that it is used to define a mathematical relationship. One should not, for this reason, read into the meaning of the intercept term too much.

5. At a 5% significance level, test the significance of the intercept term[171] by using the critical value approach.

Answer:

Given that $H_0 : \alpha = 0$ and $H_1 : \alpha \neq 0$

because

$$\text{Absolute value of } t_c = \frac{a - \alpha}{S_a} = \frac{-220.807 - 0}{95.9111} = -2.3022 \rightarrow 2.3022$$

$$2.3022 > t_{0.05/2,15-2-1} = t_{0.025,12} = 2.1788$$

we reject $H_0 : \alpha = 0$ → accept $H_1 : \alpha \neq 0$ → the intercept term is statistically NOT equal to zero at a 5% significance level → the intercept term is significant → α is significant →the intercept term is needed in the regression equation → the estimated regression line does NOT start at the origin.

[171] This is a statistics jargon in that it is same as saying, "test the null hypothesis that the intercept is statistically equal to zero."

6. At a 5% significance level, test the significance of the intercept term by using the p-value approach.

Answer:

Given that $H_0 : \alpha = 0$ and $H_1 : \alpha \neq 0$,

the p-value can be best obtained by the Excel command or the computer output, we use the information as is:

Because (p-value = 0.040037) < (the chosen significance level, α, = 0.05), we reject $H_0 : \alpha = 0$ → accept $H_1 : \alpha \neq 0$ → the intercept term is statistically NOT equal to zero at a 5% significance level → the intercept term is significant → α is significant →the intercept term is needed in the regression equation → the estimated regression line does NOT start at the origin.

7. At a 5% significance level, test the significance of the intercept term by using the confidence interval approach.

Answer:

Given that $H_0 : \alpha = 0$ and $H_1 : \alpha \neq 0$

the 95% Confidence Interval (CI) for α = $a \pm t_{0.025,12} \cdot S_a$

$$= -220.807 \pm 2.1788 \text{ x } 95.9111 = [-429.78, -11.8347]$$

Because this confidence interval does NOT capture the hypothesized value of α being equal to zero, we reject $H_0 : \alpha = 0$ → accept $H_1 : \alpha \neq 0$ → the intercept term is statistically NOT equal to zero at a 5% significance level → the intercept term is significant → α is significant →the intercept term is needed in the regression equation → the estimated regression line does NOT start at the origin.

8. If the intercept is judged to be statistically significant, what does it mean in a layman's term?

Answer:

The intercept term is not equal to zero and thus, significant and important in defining the relationship between Wealth, and Salary and Age.

9. If the intercept is judged to be statistically insignificant, what does it mean in a layman's term?

Answer:

The intercept term is statistically equal to zero and thus, insignificant and unimportant in defining the relationship between Wealth, and Salary and Age. If possible, it can be dropped from the model specification.

10. Given an economic interpretation of the slope coefficients.

Answer:

Because a slope measures the change in Y given a change in X, it can be calculated by the partial differential calculus as follows:

$$Slope = \frac{\Delta Y}{\Delta X} = \frac{\Delta Wealth}{\Delta Salary} = \frac{\partial Wealth}{\partial Salary} = 3.1655$$

→ Given a $1 increase (decrease) in Salary, Wealth will increase (decrease) by $3.1655.

Likewise the partial differential calculus shows:

$$Slope = \frac{\Delta Y}{\Delta X} = \frac{\Delta Wealth}{\Delta Age} = \frac{\partial Wealth}{\partial Age} = 8.5105$$

→ Given a 1-year increase (decrease) in Age, Wealth will increase (decrease) by 8.5105 (x $1000) = $8,510.50. Note how the measurement unit will change the interpretation.

11. At a 5% significance level, test the significance of each of the slope coefficients[172] by using the critical value approach.

Answer:

Given that the population regression equation is:

$$Wealth = \alpha + \beta_1 \cdot Salary + \beta_2 \cdot Age + \varepsilon$$

and its estimated regression equation, as was found in (3) above, is:

$$\text{Wealth} = a + b_1 \cdot Salary + b_2 \cdot Age$$

where $a = -220.807$; $b_1 = 3.1655$; and $b_2 = 8.5105$. Therefore,

$$\text{Wealth} = -220.807 + 3.1655 \text{ Salary} + 8.5105 \text{ Age}$$

Given this information, the null and the alternative hypotheses for the significance test of the Salary coefficient, β_1, can be stated as:

$$H_0 : \beta_1 = 0 \text{ vs. } H_1 : \beta_1 \neq 0$$

Because

$$t_c = \frac{b_1 - \beta_1}{S_{b_1}} = \frac{3.1655 - 0}{1.0492} = 3.0170 > t_{0.05/2,15-2-1} = t_{0.025,12} = 2.1788$$

---

[172] This is a statistics jargon in that it is same as saying, "test the null hypothesis that the independent variable has no relationship with the dependent variable."

we reject $H_0 : \beta_1 = 0$ → we accept $H_1 : \beta_1 \neq 0$ → the Salary (slope) coefficient is statistically NOT equal to zero at a 5% significance level → the Salary (slope) coefficient is significant → β_1 is significant → Salary is an important variable in determining the level of Wealth → Salary does determine Wealth → Wealth is determined by Salary.

As for the null and the alternative hypotheses for the significance test of the Age coefficient, β_2, can be stated as:

$$H_0 : \beta_2 = 0 \text{ vs. } H_1 : \beta_2 \neq 0$$

Because

$$t_c = \frac{b_2 - \beta_2}{S_{b_2}} = \frac{8.5105 - 0}{1.3880} = 6.1312 > t_{0.05/2,15-2-1} = t_{0.025,12} = 2.1788$$

we reject $H_0 : \beta_2 = 0$ → we accept $H_1 : \beta_2 \neq 0$ → the Age (slope) coefficient is statistically NOT equal to zero at a 5% significance level → the Age (slope) coefficient is significant → β_2 is significant → Age is an important variable in determining the level of Wealth → Age does determine Wealth → Wealth is determined by Age.

Note: these two t-tests only verify the significance of each individual coefficient one at a time, independently of each other.

12. At a 5% significance level, test the significance of each of the slope coefficients by using the p-value approach.

Answer:

For the Salary coefficient, β_1, the null and the alternative hypotheses are:

$$H_0 : \beta_1 = 0 \text{ vs. } H_1 : \beta_1 \neq 0$$

The p-value can be best obtained by the Excel command or the computer output.

Because (p-value = 0.0107 for Salary Coefficient) < (the chosen significance level, α, = 0.05), we reject $H_0 : \beta_1 = 0$ → we accept $H_1 : \beta_1 \neq 0$ → the Salary (slope) coefficient is statistically NOT equal to zero at a 5% significance level → the Salary (slope) coefficient is significant → β is significant → Salary does determine Wealth → Wealth is determined by Salary.

For the Age coefficient, β_2, the null and the alternative hypotheses are:

$$H_0 : \beta_2 = 0 \text{ vs. } H_1 : \beta_2 \neq 0$$

The p-value can be best obtained by the Excel command or the computer output.

Because (p-value = 5.09E-05 = 5.09 x 10^{-5} = 0.0000509 for Age Coefficient) is less than (the chosen significance level, α, = 0.05), we reject $H_0 : \beta_2 = 0$ → we accept $H_1 : \beta_2 \neq 0$ → the Age (slope) coefficient is statistically NOT equal to zero at a 5% significance level → the Age (slope) coefficient is significant → β is significant → Age does determine Wealth → Wealth is determined by Age.

13. At a 5% significance level, test the significance of each of the slope coefficients by using the confidence interval approach.

Answer:

For the Salary coefficient, β_1, the null and the alternative hypotheses are:

$$H_0 : \beta_1 = 0 \text{ vs. } H_1 : \beta_1 \neq 0$$

the 95% Confidence Interval (CI) for $\beta_1 = b_1 \pm t_{0.025,12} \cdot S_{b_1}$

$= 3.1655 \pm 2.1788 \text{ x } 1.0492 = [0.8795, 5.4516]$

Because this confidence interval does NOT capture β_1 of zero, we reject $H_0 : \beta_1 = 0$ → we accept $H_1 : \beta_1 \neq 0$ → the Salary (slope) coefficient is statistically NOT equal to zero at a 5% significance level → the Salary (slope) coefficient is significant → β_1 is significant → Salary does determine Wealth → Wealth is determined by Salary.

For the Age coefficient, β_2, the null and the alternative hypotheses are:

$$H_0 : \beta_2 = 0 \text{ vs. } H_1 : \beta_2 \neq 0$$

the 95% Confidence Interval (CI) for $\beta_2 = b_2 \pm t_{0.025,12} \cdot S_{b_2}$

$= 8.5105 \pm 2.1788 \text{ x } 1.3880 = [5.4862, 11.5347]$

Because this confidence interval does NOT capture β_2 of zero, we reject $H_0 : \beta_2 = 0$ → we accept $H_1 : \beta_2 \neq 0$ → the Age (slope) coefficient is statistically NOT equal to zero at a 5% significance level → the Age (slope) coefficient is significant → β is significant → Age does determine Wealth → Wealth is determined by Age.

14. Calculate the coefficient of determination.

Answer:

$$R^2 = \frac{RSS}{TSS} = \frac{486081.9}{623423.3} = 0.7796 \rightarrow 77.96\%$$

Salary and Age jointly explain 77.96% of variation in Wealth → A good model because this R-squared seems reasonably high.

15. Calculate the adjusted R-squared.

Answer:

Adjusted $R^2 = \overline{R}^2 = 1-\left[(1-R^2)\cdot\frac{n-1}{n-k-1}\right] = 1-\left[(1-0.7796)\cdot\frac{15-1}{15-2-1}\right] = 0.7429$

16. At a 5% significance level, conduct an F-test of the overall significance of the regression equation (or model)[173].

Answer:

The null and the alternative hypotheses to be tested are:

$$H_0 : \beta_1 = \beta_2 = 0 \qquad \text{vs.} \qquad H_1 : not\ H_0$$

Note that "$H_1 : not\ H_0$" means any of the following:

Either	(1)	$\beta_1 \neq 0$	or	
	(2)	$\beta_2 \neq 0$	or	
	(3)	Both $\beta_1 \neq 0$	and	$\beta_2 \neq 0$

The F-test procedure uses the Mean Squares as follows:

Because $F_c = \frac{MS_R}{MS_E} = \frac{\left(\frac{RSS}{k}\right)}{\left(\frac{ESS}{n-k-1}\right)} = \frac{\left(\frac{486081.9}{2}\right)}{\left(\frac{137341.4}{15-2-1}\right)} = \frac{243041}{11445.12} = 21.2353$

is greater than $F^{k}_{n-k-1,\alpha} = F^{2}_{12,0.05} = 3.89$,

we reject $H_0 : \beta_1 = \beta_2 = 0$ → we accept $H_1 : not\ H_0$ → NOT all slope coefficients are jointly or simultaneously equal to zero at a 5% significance level → some slope coefficients are significant → some β's are significant

Note: the F-test result does not say which slope coefficient(s) is significant. It simply states that NOT ALL coefficients are simultaneously equal to zero. In this case, as judged by the t-tests on individual coefficients, both Salary and Age coefficients are significant, which is reinforced by this F-test[174] → Both Salary and Age jointly determine Wealth → Wealth is determined by Salary and Age.

---

[173] This is a statistics jargon in that it is same as saying, "test the null hypothesis that all slope coefficients are simultaneously equal to zero."

[174] The conclusions drawn from the F-test and the t-test will reinforce each other if all underlying assumptions of the regression analysis hold. However, when some of them are violated, the F-test and t-test results can conflict each other. This is when advanced estimation techniques such as the Generalized Least Squares method can be useful.

17. Is there an autocorrelation problem in the estimated regression equation? Test the presence of the autocorrelation at a 5% significance level.

Answer:

The calculated Durbin-Watson (DW) statistic is 1.5590 and the table values of DW at a 5% significance level with n = 15 and k = 2 are:

$DW_L = D_L = 0.95$ and $DW_U = D_U = 1.54$

Because $(DW_U = 1.54) < (DW_{calculated} = DW_c = 1.5590) < (4 - DW_U = 2.46)$, we conclude that there is "NO" autocorrelation in the error (or residual) terms.

That is,

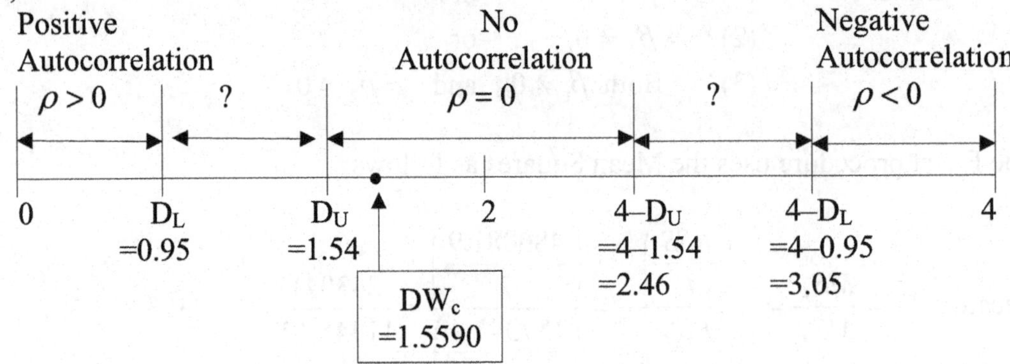

Note that DW-Values that fall in the "?" range indicates an "inconclusive" or "uncertain" presence of autocorrelation.

18. Suppose your 35 year-old friend has an annual salary of $43,000. What is the best estimate, based on the regression model, of her wealth?

Answer:

Given Salary = $43,000 → 43
Age = 35, and
Wealth = –220.807 + 3.1655 Salary + 8.5105 Age
= –220.807 + 3.1655 x 43 + 8.5105 x 35 = 213.177 → $213,177

19. Suppose that your 50-year-old boss wants to know her wealth based on the estimated regression equation. Assuming that she is paid $1 million a year, what would be her estimated wealth?

Answer:

Given Salary = $1,000,000 → 1,000
Age = 50, and
Wealth = –220.807 + 3.1655 Salary + 8.5105 Age
= –220.807 + 3.1655 x 1000 + 8.5105 x 50 = 3370.218
→ $3,370,218

20. Would the estimate of your boss's wealth above be reliable? Why or why not?

Answer:

The age of 50 is within the range of the observed values and thus, it creates no alarm. However, the salary of $1 million is NOT within the range of the observed values, it makes the estimation (or forecast) unreliable. Thus, the analyst must exercise an extreme care in this type of situations.

21. Do you think this multiple regression model is good to use? Do you think it is superior to the previous two simple regression models? That is, justify the reason(s) for the model you would prefer to use.

Answer:

The multiple regression model is superior to the two simple regression models for the following reasons:

(1) All slope coefficients are significant as judged by the t-tests.
(2) This fact has been reinforced by the significant F-test result.
(3) The DW test shows no autocorrelation.
(4) R-squared of 0.7796 is reasonably high enough.
(5) Adjusted R-squared of 0.7429 is reasonably larger than 0.0195 for the Salary Model and 0.5827 of the Age Model → Adding one more independent variable to the regression model is justified.
(6) The signs of all slope coefficients make an economic and intuitive sense. That is, the positive signs associated with the Salary and Age variables indicate that they have a positive influence on the Wealth level, which makes an economic and intuitive sense.
(7) The multiple regression model with 2 independent variables is still concise and parsimonious enough to be useful.

E. Review of Basic Concepts in Multiple Regression Analysis

Assume that a regression equation is estimated by a computer software. However, due to printing problem some information are left blank.

1. Fill in the missing information (blanks) in the following multiple regression computer output.

Dependent Variable: SALES

	Coefficient Estimates	Standard Error	t-Stat
INTERCEPT	A?	3.00	2.205
PRICE	–0.35	B?	–0.05
INCOME	C?	0.5	2.88
ADVERTISING	–6.25	5.00	D?

ANOVA Table:

Source of Variation	df	Sum of Squares	Mean Squares
REGRESSION	E?	248.625	F?
RESIDUAL	G?	H?	25.5
TOTAL	25		

R-Squared = I? Adjusted R-Squared = J?

F-ratio = $F_{calculated}$ = K?

Answer:

A = 6.615

Because t-Stat = $\dfrac{Coefficient-\alpha}{Std.\,Error} = \dfrac{A-0}{3} = 2.205 \rightarrow A = 2.205 \times 3 = 6.615$

B = 7

Because t-Stat = $\dfrac{Coefficient-\beta}{Std.\,Error} = \dfrac{-0.35-0}{B} = -0.05 \rightarrow B = -0.35/(-0.05) = 7$

C = 1.44

Because t-Stat = $\dfrac{Coefficient-\beta}{Std.\,Error} = \dfrac{C-0}{0.5} = 2.88 \rightarrow C = 2.88 \times 0.5 = 1.44$

D = –1. 25

Because D = t-Stat = $\dfrac{Coefficient-\beta}{Std.\,Error} = \dfrac{-6.25-0}{5} = -1.25$

E = 3

Because k = 3. That is, there are 3 independent variables of Price, Income, and Advertising.

F = 82.875

Because Mean Squares Due to Regression = $MS_R = \dfrac{RSS}{k} = \dfrac{248.625}{3} = 82.875$

G = 22

Because (n – k – 1) = (26 – 3 – 1) = 22. Note that "Total" = n – 1.

Therefore, n = Total + 1 = 25 + 1 = 26.

H = 561

Because Mean Squares Due to Error = $MS_E = \dfrac{ESS}{n-k-1} = \dfrac{H}{22} = 25.5$.

Therefore, H = 25.5 x 22 = 561

I = 0.3071

Because $R^2 = \dfrac{RSS}{TSS} = \dfrac{RSS}{RSS + ESS} = \dfrac{248.625}{248.625 + 561} = 0.30708$

J = 0.2126

Because J = Adjusted $R^2 = \overline{R}^2$

$$= 1 - \left[(1 - R^2) \cdot \frac{n-1}{n-k-1}\right] = 1 - \left[(1 - 0.3071) \cdot \frac{26-1}{26-3-1}\right] = 0.2126$$

K = 3.25

Because $F_c = \dfrac{MS_R}{MS_E} = \dfrac{82.875}{25.5} = 3.25$

2. Determine which of the independent variables (if any) are statistically significant at a 0.05 significance level.

Answer:

Given that the table value of t, $t_{\alpha/2,n-k-1}$, is equal to: $t_{\alpha/2,n-k-1} = t_{0.025,22} = 2.0739$,

we note that only the INCOME coefficient is significant because of its calculated t-value of 2.88 is greater than the table t-value of 2.0739. However, the PRICE and ADVERTISING variables are NOT significant because their respective calculated t-values are less than the table t-value of 2.0739.

3. Determine whether the independent variables explain a significant proportion of the variation in the dependent variable at a 5% significance level.

Answer:

Note that this question asks for the significance test of how well independent variables explain the variation in the dependent variable. Therefore, an F-test for all coefficients being jointly equal to zero must be examined. If this is the case – that is, if we accept the null hypothesis of all coefficients being simultaneously equal to zero, then we can say that the independent variables are NOT adequately explaining the variation in the dependent variable. Otherwise, we can state that the independent variables are explaining some portion of the variation in the dependent variable.

The null and the alternative hypotheses can be stated as:

$H_0: \beta_1 = \beta_2 = \beta_3 = 0$ vs. H_1: not H_0

Note that "H_1: not H_0" means of the following:

Either (1) $\beta_1 \neq 0$, or (2) $\beta_2 \neq 0$, or (3) $\beta_3 \neq 0$, or (4) $\beta_1 \neq \beta_2 \neq \beta_3 \neq 0$, or any combinations of these.

Given the calculated F-value = F_c = 3.25 as shown in (K) above and the table value of F = $F^k_{n-k-1,\alpha} = F^3_{26-3-1,0.05} = F^3_{22,0.05} = 3.05$, we reject $H_0: \beta_1 = \beta_2 = \beta_3 = 0$ because (F_c = 3.25) > (F_{table} = 3.05) → accept H_1: not H_0 → This conclusion is drawn due to the significant INCOME variable → That is, $\beta_2 \neq 0$.

4. What is SALES if PRICE = 10; INCOME = 100; and ADVERTISING = 5? (Disregard the units.)

Answer:

SALES = 6.615 – 0.35 PRICE + 1.44 INCOME – 6.25 ADVERTISING
= 6.615 – 0.35 (10) + 1.44 (100) – 6.25 (5) = 115.865

5. Do you think this is a good model to use? Justify your answer.

Answer:

This model is NOT a good model because

(1) Only the INCOME (slope) coefficient is significant as judged by the t-test.
(2) R-squared of 0.3071 may be too low given that there are 3 independent variables in the equation[175].
(3) Adjusted R-squared of 0.2126 is considered too low[176].
(4) The sign of the ADVERTISING variable being negative does not make an economic and intuitive sense.

6. What other statistics do you wish to have to make a better decision about the appropriateness of this estimated regression equation?

Answer:

A calculated Durbin-Watson statistic would be valuable to judge the presence of an autocorrelation.

[175] This is a subjective judgment based on one's hunch. In some cases, this level of R-squared may be acceptable.
[176] Just like the case of the R-squared, this claim is also subjective based on one's hunch. In some cases, this level of an adjusted R-squared may be acceptable.

F. An Application of Regression to Demand Equation Estimation

The following demand equation is estimated by regression analysis on the basis of 30 observations:

$$\begin{array}{ccccc} Q = & 10 & -5P & +2I & +3P_c \\ & (5) & (2) & (0.5) & (6) \end{array}$$

where

Q = quantity demanded of your product
P = price of your product
I = average income of the consumers of your product
P_c = price of a very similar product that your competitor sells

The standard error of each of the above coefficients is given in parenthesis and the calculated Durbin-Watson statistic is 2.5.

1. Determine which of the independent variables (if any) are statistically significant at the 0.05 significance level.

Answer:

In order to test $H_0 : \beta_i = 0$ vs. $H_1 : \beta_i \neq 0$, we can use either the critical value approach, the p-value approach, or the confidence interval approach. However, because there are no p-values given, it is simpler to use the critical t-value approach as follows:

The table t-value = $t_{\alpha/2,n-k-1} = t_{0.05/2,30-3-1} = t_{0.025,26} = 2.0555$

The calculated t-value for P = $t_c = \dfrac{b_i - \beta_i}{S_{b_i}} = \dfrac{-5-0}{2} = -2.5 \rightarrow 2.5$

The calculated t-value for I = $t_c = \dfrac{b_i - \beta_i}{S_{b_i}} = \dfrac{2-0}{0.5} = 4$

The calculated t-value for P_c = $t_c = \dfrac{b_i - \beta_i}{S_{b_i}} = \dfrac{3-0}{6} = 0.5$

Because the calculated t-values for P and I are greater than the table t-value of 2.0555, we reject the null hypothesis for P and I → P and I are significant. However, P_c is NOT significant because the calculated t-value for P_c is less than the table t-value of 2.0555.

2. Given P = 1; I = 5; and P_c = 2,

   a. Calculate the own price elasticity of demand at P = 1.

Answer:

Because $Q = 10 - 5P + 2I + 3P_c = 10 - 5(1) + 2(5) + 3(2) = 21$

the own price elasticity of demand $= \varepsilon_p = \frac{\partial Q}{\partial P} \cdot \frac{P}{Q} = -5 \cdot \frac{1}{21} = -0.2381 \rightarrow$ inelastic

b. Calculate the income elasticity of demand at I = 5.

Answer:

Because $Q = 10 - 5P + 2I + 3P_c = 10 - 5(1) + 2(5) + 3(2) = 21$

The income elasticity of demand $= \varepsilon_I = \frac{\partial Q}{\partial I} \cdot \frac{I}{Q} = 2 \cdot \frac{5}{21} = 0.4762 \rightarrow$ inelastic

c. Calculate the cross price elasticity of demand at $P_c = 2$.

Answer:

Because $Q = 10 - 5P + 2I + 3P_c = 10 - 5(1) + 2(5) + 3(2) = 21$

The cross price elasticity of demand $= \varepsilon_{XY} = \frac{\partial Q}{\partial P_c} \cdot \frac{P_c}{Q} = 3 \cdot \frac{2}{21} = 0.2857 \rightarrow$ inelastic

3. Draw a conclusion about the meaning of each of these elasticities for your business. Should you worry about your competitor for now?

Answer:

As shown above, $\varepsilon_{XY} = \frac{\partial Q}{\partial P_c} \cdot \frac{P_c}{Q} = \frac{\%\Delta Q}{\%\Delta P_c} = 0.2857 = \frac{0.2857}{1}$

That is, given $\%\Delta P_c = +1 \rightarrow$ that is, as the competitor increases his price by 1%, the resulting percentage change in your quantity sold, $\%\Delta Q = 0.2857$. That is, there is an increase in sales by 0.2857% as the competitor increases his price by 1%. This means that your product and your competitor's product are SUBSTITUTES for each other because as your competitor increases his price - and consequently, his sales will go down - his customers will come and buy more of your product.

Note: if the sign of the cross price elasticity is negative, it means that your product and your competitor's product are COMPLEMENTS for each other. Thus, he is not a competitor, but a friend in business because as your competitor decreases his price - and consequently his sales will go up - his customers will come and buy more of your product as well. This means that customers are consuming your competitor's product and your product together.

4. What conclusions can be drawn from the data about the possible presence of autocorrelation at the 5% significance level?

Answer:

Because the calculated DW of 2.5 lies between $(4 - D_L) = (4 - 1.21) = 2.79$ and $(4 - D_U) = (4 - 1.65) = 2.35$, the presence of autocorrelation is INCONCLUSIVE.

Note that given n = 30, and k = 3, D_L = 1.21 and D_U = 1.65 at the 5% significance level.

G. An Advanced Topic: Dummy-Variable Regression

When qualitative variables are to be introduced into a regression model, they can be handled by the use of dummy variables. Dummy variables take the values of only 0 or 1 for a given characteristic. For that reason, they are often called qualitative variables, on-and-off variables, 0 – 1 variables, seasonal variables, etc.

The key point in using dummy variables in regression analysis is to avoid a possible multicollinearity problem which is created when a linear combination of independent variables are highly correlated among themselves. The following example will demonstrate the construction of a dummy-variable model and the issue of multicollinearity.

1. Simple Regression Model Estimation

Suppose that you had collected a following data set on quarterly sales (in $ million) and the number of salespersons (SP) for your company:

Year	Quarter	Sales	Salespersons (SP)
2X01	1	10	4
	2	5	4
	3	7	5
	4	13	5
2X02	1	15	5
	2	8	5
	3	17	6
	4	18	6
2X03	1	18	6
	2	16	7
	3	20	6
	4	24	7
2X04	1	25	8
	2	20	8
	3	27	9
	4	28	9
2X05	1	28	9
	2	25	8
	3	30	9
	4	31	10

Given the following regression estimation, you can now examine the validity of the relationship between the quarterly sales and the number of salespersons (SP).

SUMMARY OUTPUT

Regression Statistics	
Multiple R	0.93755
R Square	0.879
Adjusted R Square	0.872278
Standard Error	2.837525
Observations	20

ANOVA

	df	*SS*	*MS*	*F*	*Significance F*
Regression	1	1052.822	1052.822	130.7602	1.09E-09
Residual	18	144.9279	8.051551		
Total	19	1197.75			

	Coefficients	*Standard Error*	*t Stat*	*P-value*	*Lower 95%*	*Upper 95%*
Intercept	-8.07515	2.472399	-3.26612	0.00429	-13.2695	-2.88084
SP	4.018405	0.351412	11.43504	1.09E-09	3.280117	4.756693

(1) Which coefficient(s) is/are significant?
(2) What is the model (= estimated equation)?
(3) What are the pros and cons of using this model?
(4) Is the model a good one to use?

2. Dummy-Variable Regression Model Estimation

Suppose that you are now interested in knowing if any seasonal factor[177] has influenced the sales in addition to the number of salespersons (SP) hired. In order to do this, you are to construct a seasonal dummy-variable model as follows:

$$\text{Sales} = a + b_1 \cdot SP + b_2 \cdot Q1 + b_3 \cdot Q2 + b_4 \cdot Q3 + e$$

where a is the intercept and b_i's are slope coefficients. Up to now, nothing unusual is found in this model construction. Now, the important distinction comes from the construction and interpretation of the dummy variables, Q1, Q2, and Q3.

[177] Because the data is collected quarterly, the seasonal factor being examined is a "quarterly" effect. If the data was collected monthly, the seasonal factor being examined is a "monthly" effect.

Note:

Q1 is the First Quarter Dummy variable which takes the value of 1, otherwise, 0.
Q2 is the Second Quarter Dummy variable which takes the value of 1, otherwise, 0.
Q3 is the Third Quarter Dummy variable which takes the value of 1, otherwise, 0.

Q4 for the Fourth Quarter Dummy variable **MUST be eliminated from the model** because if it is included, then the intercept term will be perfectly correlated with the sum of 4 dummy variables of Q1, Q2, Q3, and Q4. That is, under this construction, adding Q4 in the model will create a perfect multicollinearity. Therefore, always a maximum of (k – 1) seasonal dummy variables are allowed in a regression model where an intercept term is present[178]. The specific data structure of a quarterly dummy variable model looks as follows:

Year	Quarter	Sales	SP	Q1	Q2	Q3
2X01	1	10	4	1	0	0
2X02	2	5	4	0	1	0
2X03	3	7	5	0	0	1
2X04	4	13	5	0	0	0
2X05	1	15	5	1	0	0
2X06	2	8	5	0	1	0
2X07	3	17	6	0	0	1
2X08	4	18	6	0	0	0
2X09	1	18	6	1	0	0
2X10	2	16	7	0	1	0
2X11	3	20	6	0	0	1
2X12	4	24	7	0	0	0
2X13	1	25	8	1	0	0
2X14	2	20	8	0	1	0
2X15	3	27	9	0	0	1
2X16	4	28	9	0	0	0
2X17	1	28	9	1	0	0
2X18	2	25	8	0	1	0
2X19	3	30	9	0	0	1
2X20	4	31	10	0	0	0

[178] In order to avoid a multicollinearity problem in a dummy-variable regression model where an intercept term is included, the maximum number of dummy variables allowed is (k – 1) where k is the number of seasons or months. Therefore, a maximum number of monthly dummy variables allowed in a model is 11 and that of quarterly dummy variables is 3. Furthermore, the missing season (=quarter or month) is represented by the intercept that serves as the reference point of comparison.

Based on this data set, a dummy-variable regression model is estimated as follows:

SUMMARY OUTPUT

Regression Statistics	
Multiple R	0.96378
R Square	0.928872
Adjusted R Square	0.909904
Standard Error	2.383193
Observations	20

ANOVA

	df	*SS*	*MS*	*F*	*Significance F*
Regression	4	1112.556	278.139	48.97149	1.96E-08
Residual	15	85.19416	5.67961		
Total	19	1197.75			

	Coefficients	*Standard Error*	*t Stat*	*P-value*	*Lower 95%*	*Upper 95%*
Intercept	-6.19935	2.486941	-2.49276	0.024858	-11.5001	-0.89856
SP	3.918831	0.303647	12.90588	1.59E-09	3.271623	4.566039
Q1	0.318831	1.537545	0.207364	0.838515	-2.95837	3.596031
Q2	-4.08117	1.537545	-2.65434	0.018034	-7.35837	-0.80397
Q3	-1.03247	1.51215	-0.68278	0.505154	-4.25554	2.190603

(1) Which coefficient(s) is/are significant?
(2) What is the model (= estimated equation)?
(3) Which quarter tends to have the lowest sales given this estimated model?
(4) What are the pros and cons of using this model?
(5) Is the model a good one to use in comparison to the non-seasonal dummy-variable model estimated earlier?

In the absence of Q4 like in this case, however, the intercept term represents Q4, the Fourth Quarter, and serves the model as the base reference point of comparison. Therefore, because the Q1 coefficient is positive, it means that Q1 is higher than Q4. Also because the Q2 and Q3 coefficients are negative, it means that Q2 and Q3 are lower than Q4. Overall, Q1 has the highest sales and Q2 has the lowest sales in consideration of the salespersons' role in the company.

In addition to these points, a complete analysis, however, must include the discussion of a Durbin-Watson test along with a common-sense validity of model specification.

Exercise Problems on
Chapter 14. Multiple Regression Analysis

This set of exercise problems has 22 problems, worth a total of 25 points.

1. Multiple regression analysis involves ____ independent variable(s) and _____ dependent variable(s).

   a. one; one
   b. one; more than one
   c. more than one; one
   d. more than one; more than one
   e. no; no

2. If the error (or residual) terms are correlated within themselves, we would find a _____ problem in regression estimation.

   a. autocorrelation or serial correlation
   b. homoscedasticity
   c. heteroscedasticity
   d. multicollinearity
   e. any of the above

3. If the assumption of equal variance of error terms is violated, we conclude that there is a _____ problem in regression estimation.

   a. homoscedasticity
   b. heteroscedasticity
   c. multicollinearity
   d. autocorrelation
   e. only (c) and (d) of the above

4. The R-squared in multiple regression is ______ the corresponding adjusted R-squared.

   a. greater than
   b. less than
   c. greater than or equal to
   d. less than or equal to
   e. the same as

5. If independent variables are correlated among themselves in multiple regression, there is a ______ problem in estimation.

   a. homoscedasticity
   b. heteroscedasticity
   c. multicollinearity
   d. autocorrelation
   e. only (c) and (d) of the above

6. If there is a multicollinearity problem, you can alleviate this problem by _____.

   a. deleting one of the correlated independent variables.
   b. taking a ratio of the correlated independent variables.

c. averaging the correlated variables.
d. doing any of the above
e. doing none of the above.

7. If the intercept term is judged to be statistically equal to zero, it means that the intercept term is _____ in defining the regression equation.

a. important and significant
b. unimportant but significant
c. important but insignificant
d. unimportant and insignificant
e. useful

8. If a slope coefficient in a regression equation is estimated to be 5 and its standard error, 3, you would _____ the null hypothesis of this coefficient being equal to zero if the appropriate critical (table)value is 2.

a. reject
b. accept

9. Given 25 data points, if you found a calculated F value of 2.5 for a regression equation that has an intercept term and 4 independent variables, you would _____ the null hypothesis of all coefficients being zero at a 5% significance level because the appropriate critical (table) value is _____.

a. accept; 2.87 b. reject; 2.87 c. accept; 5.80
d. reject; 5.80 e. accept; 3.51

Given the following description, answer Questions 10 – 22.

Suppose that your boss asked you to analyze the impact of various factors on your company's quarterly profits. In order to write an analytical report to your boss, you collected the following quarterly data (in million dollars):

Time period	Profits	Advertisement Expenses	Sales	Salaries Paid
Y1 Q1	100	10	200	50
Y1 Q2	90	5	190	45
Y1 Q3	95	10	200	50
Y1 Q4	120	12	240	60
Y2 Q1	110	6	210	55
Y2 Q2	100	7	200	53
Y2 Q3	120	8	250	60

Y2 Q4	130	10	260	65
Y3 Q1	120	10	250	65
Y3 Q2	100	5	200	60
Y3 Q3	110	8	240	60
Y3 Q4	130	4	300	63
Y4 Q1	120	10	280	62
Y4 Q2	110	11	270	60
Y4 Q3	110	15	280	61
Y4 Q4	150	13	340	65
Y5 Q1	130	13	330	65
Y5 Q2	120	12	320	67
Y5 Q3	140	14	350	70
Y5 Q4	160	17	370	70

10. If you are to conduct a regression analysis, the endogenous variable(s) in this case should be _____.

a. Profits
b. Advertisement Expenses
c. Sales
d. Salaries Paid
e. all of the above

11. If you are to conduct a regression analysis, the exogenous variable(s) in this case could be _____.

a. Profits
b. Advertisement Expenses
c. Sales
d. Salaries Paid
e. any or all of the above except (a) Profits.

12. If you estimate a regression analysis, using the above data for Profits (P) as a dependent variable and Advertisement Expenses (AD), Sales (S), and Salaries Paid (SP) as independent variables, you would find the following as the estimated regression equation.

a. P = 15.988 + 20.236 AD + 0.7900 S + 0.4410 SP
b. P = 15.988 – 0.1676 AD + 0.2184 S + 0.7672 SP
c. P = 15.988 + 0.1676 AD - 0.2184 S - 0.7672 SP
d. P = 15.988 – 0.1676 AD - 0.2184 S - 0.7672 SP
e. P = 10.734 – 0.1676 AD + 0.2184 S + 0.7672 SP

13. The independent variable(s) that is/are significant at a 5% significance level is/are:

a. Advertisement Expenses
b. Sales
c. Salaries Paid
d. all of the above
e. only (b) and (c) of the above

14. If you are to use the critical-value approach to a hypothesis testing of a regression coefficient, the appropriate test statistic to be used is the ____ statistic.

a. Z　　b. t　　c. F
d. Chi-square　　e. either t or F

15. If you are to find the appropriate degrees of freedom for significance testing of each of the 3 independent variables and an intercept with 20 data points (=observations), it would be _____.

a. 20　　b. 19　　c. 17
d. 16　　e. none of the above

16. If you are to conduct a significance testing of a regression coefficient via the critical-value approach, the (critical) table value at a 5% significance level will be _____.

a. 2.0930　　b. 1.7291　　c. 2.1199
d. 1.7459　　e. 1.96

17. The coefficient of determination of the regression model based on the above data set is _____.

a. 0.9110　　b. 0.8300　　c. 0.7982
d. 0.8155　　e. 8.1557

18. If you are test the joint significance of all regression coefficients, you would use a/an _____ test.

a. Z　　b. t　　c. F
d. Chi-square　　e. either t or F

19. If you are to test the joint significance of all coefficients in the regression equation, you would ______ the null hypothesis of all coefficients being zero at a 5% significance level.

a. reject　　b. accept　　c. fail to reject
d. only (b) and (c) of the above　　e. none of the above

The following questions are worth 2 points each.

20. The Durbin-Watson (DW) statistic is _____.

a. 1.1235　　b. 1.5973　　c. 2.5489
d. 3.2717　　e. 3.9921

21. If you are to conduct a hypothesis testing of no autocorrelation for this estimated regression equation, you would find the critical (table) lower DW value at a 5% significance level to be _____ and the corresponding upper DW value to be _____.

a.	1.10; 1.54	b.	1.00; 1.68	c.	0.90; 1.83
d.	0.77; 1.41	e.	0.68; 1.57		

22. If the calculated DW statistic is 2.88, and the critical lower and upper DW values are 0.5 and 1.54, respectively, then you would _____ the null hypothesis of no autocorrelation.

a.	reject	b.	accept	c.	fail to reject
d.	be inconclusive about	e.	only (b) and (c) of the above		

Chapter 15. The Chi-Square Test

In all of the previous chapters, Z-, t-, and F-tests were studied for examining the statistical validity of the mean and the variance that are calculated on the basis of sample data. Since these tests are based on parameter values of the mean and the variance, they are called "parametric" tests.

However, when a statistical test does not specifically require information on parameter values of the mean and the variance, they are called "nonparametric" tests. Among them, the most prominent is the Chi-square test[179]. It is used to examine the validity of one's expectation against the obtained information. If the observed data is the same as one's expectation, then it is said that there is a good fit. Thus, the Chi-square test is known as a "goodness-of-fit" test. Also, if the observed data has multiple characteristics, then the Chi-square can allow us to test the independence (or dependence) among these different characteristics. For this reason, a Chi-square test is also known as the test of "independence."

More specifically, the Chi-square test will allow us to examine the validity of various questions of a qualitative and categorical nature. Following are a few examples:

- When a coin is flipped 100 times, do heads appear 50% of time?
- Does the start position matter in an auto race or a horse race?
- Do cars made on Mondays tend to be worse than those made on Wednesdays?
- Are there more accidents in the morning than in the afternoon?
- Are male shoppers more prone to impulse buying than female shoppers?

Because nonparametric tests, like the Chi-square, do not rely on the mean and the variance, it often carries less respect than parametric tests in the academic world. However, this weakness of not relying on the mean and the variance can be a major strength in that it is more "flexible" and "forgiving" about the assumed underlying probability distributions. Therefore, Chi-square tests should not be looked down. In fact, they are frequently used in the analysis of categorical data. The Chi-square statistic, χ_c^2, is calculated as[180]:

$$\chi_c^2 = \sum_{i=1}^{c}\left(\frac{(E_i - O_i)^2}{E_i}\right)$$

where E_i = the expected frequency, O_i = the observed frequency, c = the number of characteristics. Note that the numerator of $(E_i - O_i)$ measures a deviation or an error between what is expected and what is observed.

[179] This does not mean that the Chi-square test is used only for nonparametric tests. It is also used for parametric tests such as the test of a variance value, which has not been described in this book.

[180] Note that it does not matter if the numerator is calculated as $(E_i - O_i)^2$ or $(O_i - E_i)^2$.

A nonparametric Chi-square test is in essence an upper tail test as is the case with an F-test because the calculated Chi-square can never be a negative number[181]. In fact, if the observed values are very similar to the expected values, then the calculated Chi-square will be very close to zero. Thus, a small Chi-square value will most likely support the acceptance of the expectation based on the data. Once again, because a chi-square test examines how well the collected data fits one's expectation, it is often known as a "goodness-of-fit" test.

The official decision criteria for a Chi-square test are, however, as follows:

If $\chi_c^2 \le \chi_{table}^2$, then accept the null hypothesis.

If $\chi_c^2 > \chi_{table}^2$, then reject the null hypothesis.

where χ_c^2 is the calculated Chi-square and χ_{table}^2 is the critical or table value of Chi-square. χ_{table}^2 is defined by the significance level and the degrees of freedom.

A. The Table of Chi-Square Values

When conducting a hypothesis testing via a Chi-square distribution, the critical value approach is most often used. That is, as shown above, the decision criteria is based on comparing the calculated Chi-square against a critical (or table) value of a Chi-square distribution.

The Excel command of "=Chidist(X, DF)" where X = the calculated Chi-square value and DF = the number of degrees f freedom will yield the critical upper-tail probability.

The Excel command of "=Chiinv(Probability, DF)" where Probability = the upper-tail probability or the significance level; and DF = the number of degrees f freedom will yield the critical Chi-square value, χ_{table}^2.

The Excel command of "=Chiinv(Probability, DF)" was used to generate the following table that shows these critical values of a Chi-square distribution at various significance levels and degrees of freedom.

[181] This is because the expected occurrences are positive and the differences between the expected and the observed are squared and summed. Furthermore, when a Chi-square test is used as a parametric test for a hypothesized variance value, for example, it can handle a two-tail test.

Table of the Critical Chi-Square Values for a Given Upper-Tail Significance Level

	Upper-Tail Significance Level							
DF	0.99	0.975	0.95	0.9	0.1	0.05	0.025	0.01
1	0.0002	0.0010	0.0039	0.0158	2.7055	3.8415	5.0239	6.6349
2	0.0201	0.0506	0.1026	0.2107	4.6052	5.9915	7.3778	9.2103
3	0.1148	0.2158	0.3518	0.5844	6.2514	7.8147	9.3484	11.3449
4	0.2971	0.4844	0.7107	1.0636	7.7794	9.4877	11.1433	13.2767
5	0.5543	0.8312	1.1455	1.6103	9.2364	11.0705	12.8325	15.0863
6	0.8721	1.2373	1.6354	2.2041	10.6446	12.5916	14.4494	16.8119
7	1.2390	1.6899	2.1673	2.8331	12.0170	14.0671	16.0128	18.4753
8	1.6465	2.1797	2.7326	3.4895	13.3616	15.5073	17.5345	20.0902
9	2.0879	2.7004	3.3251	4.1682	14.6837	16.9190	19.0228	21.6660
10	2.5582	3.2470	3.9403	4.8652	15.9872	18.3070	20.4832	23.2093
11	3.0535	3.8157	4.5748	5.5778	17.2750	19.6751	21.9200	24.7250
12	3.5706	4.4038	5.2260	6.3038	18.5493	21.0261	23.3367	26.2170
13	4.1069	5.0088	5.8919	7.0415	19.8119	22.3620	24.7356	27.6882
14	4.6604	5.6287	6.5706	7.7895	21.0641	23.6848	26.1189	29.1412
15	5.2293	6.2621	7.2609	8.5468	22.3071	24.9958	27.4884	30.5779
16	5.8122	6.9077	7.9616	9.3122	23.5418	26.2962	28.8454	31.9999
17	6.4078	7.5642	8.6718	10.0852	24.7690	27.5871	30.1910	33.4087
18	7.0149	8.2307	9.3905	10.8649	25.9894	28.8693	31.5264	34.8053
19	7.6327	8.9065	10.1170	11.6509	27.2036	30.1435	32.8523	36.1909
20	8.2604	9.5908	10.8508	12.4426	28.4120	31.4104	34.1696	37.5662
21	8.8972	10.2829	11.5913	13.2396	29.6151	32.6706	35.4789	38.9322
22	9.5425	10.9823	12.3380	14.0415	30.8133	33.9244	36.7807	40.2894
23	10.1957	11.6886	13.0905	14.8480	32.0069	35.1725	38.0756	41.6384
24	10.8564	12.4012	13.8484	15.6587	33.1962	36.4150	39.3641	42.9798
25	11.5240	13.1197	14.6114	16.4734	34.3816	37.6525	40.6465	44.3141
26	12.1981	13.8439	15.3792	17.2919	35.5632	38.8851	41.9232	45.6417
27	12.8785	14.5734	16.1514	18.1139	36.7412	40.1133	43.1945	46.9629
28	13.5647	15.3079	16.9279	18.9392	37.9159	41.3371	44.4608	48.2782
29	14.2565	16.0471	17.7084	19.7677	39.0875	42.5570	45.7223	49.5879
30	14.9535	16.7908	18.4927	20.5992	40.2560	43.7730	46.9792	50.8922
40	22.1643	24.4330	26.5093	29.0505	51.8051	55.7585	59.3417	63.6907
50	29.7067	32.3574	34.7643	37.6886	63.1671	67.5048	71.4202	76.1539
60	37.4849	40.4817	43.1880	46.4589	74.3970	79.0819	83.2977	88.3794
70	45.4417	48.7576	51.7393	55.3289	85.5270	90.5312	95.0232	100.4252
80	53.5401	57.1532	60.3915	64.2778	96.5782	101.8795	106.6286	112.3288
90	61.7541	65.6466	69.1260	73.2911	107.5650	113.1453	118.1359	124.1163
100	70.0649	74.2219	77.9295	82.3581	118.4980	124.3421	129.5612	135.8067

B. Chi-Square Tests for Single-Characteristic

The Chi-Square test for a single characteristic is often called the "goodness-of-fit test" because in essence, we are trying to examine how well the observed data fit our *a priori* expectation.

Example 1.

In order to test if a coin is unbiased and fair, a coin was tossed 200 times. The result was that a head was observed 90 times and a tail, 110 times. Is the coin a fair[182] one? Draw a conclusion at a 95% confidence level.

Answer:

The null hypothesis, H_0, is "The coin is a fair one."
The alternative hypothesis, H_a, is "The coin is NOT a fair one."

Step 1. Establish the expected probability[183] → That is, the probability of a head and a tail appearing is expected to be 50%-50% → This implies that we are trying to fit the observed data into an expected probability distribution that follows a uniform probability distribution of 50%-50% → Thus, we are conducting a goodness-of-fit test.

Step 2. Calculate the corresponding expected occurrences → That is, 50% of 200 tosses should be heads and the remaining 50% should be tails → 100 heads and 100 tails.

Step 3. Organize the information into a table[184] as follows:

Outcome (i)	Expected Occurrences (E_i)	Actual Observed Occurrences (O_i)
Head	200 x 0.5 = 100	90
Tail	200 x 0.5 = 100	110
Total	200	200

Step 4. Calculate the Chi-square statistic as follows:

$$\chi_c^2 = \sum_{i=1}^{c}\left(\frac{(E_i - O_i)^2}{E_i}\right) = \frac{(100-90)^2}{100} + \frac{(100-110)^2}{100} = 2$$

Step 5. Identify the degree of freedom as (c - 1) where c is the number of outcomes (or characteristics) being analyzed → Because there are only two outcomes of a head vs. a tail, c = 2 → Thus, the degree of freedom is c – 1 = 2 – 1 = 1.

---

[182] The word, "fair," in probability theory means "unbiased" and "true" to a theoretical outcome. Therefore, it describes the case where a theoretical probability (or outcome) is statistically supported by an empirical probability (or outcome).

[183] Note that the expected probability can be based on a theory or one's prior belief or hunch.

[184] Note that this table is not a true contingency table of c columns and r rows of different characteristics. In fact, it is of only 2 outcomes (or characteristics) – a head or no head.

Step 6. Identify the critical value of the Chi-square, $\chi^2_{c-1,\alpha}$, at a 5% significance level from the above table → $\chi^2_{table} = \chi^2_{c-1,\alpha} = \chi^2_{1,0.05} = 3.8415$.

Step 7. The decision is:
Because $(\chi_c^2 = 2) \le (\chi^2_{table} = 3.8415)$, **accept** the null hypothesis that the coin is unbiased and fair.

Example 2.

In order to test if a coin is unbiased and fair, a coin was tossed 1000 times. The result was that a head was observed 450 times. Is the coin a fair one? Draw a conclusion at a 99% confidence level.

Answer:

The null hypothesis, H_0, is "The coin is a fair one."
The alternative hypothesis, H_a, is "The coin is NOT a fair one."

Step 1. Establish the expected probability → That is, the probability of a head and a tail appearing is expected to be 50%-50% → A goodness-of-fit test.

Step 2. Calculate the corresponding expected occurrences → That is, 50% of 1000 tosses should be heads and the remaining 50% should be tails → 500 heads and 500 tails.

Step 3. Organize the information into a contingency table as:

Outcome (i)	Expected Occurrences (E_i)	Actual Observed Occurrences (O_i)
Head	1000 x 0.5 = 500	450
Tail	1000 x 0.5 = 500	550
Total	1000	1000

Step 4. Calculate the Chi-square statistic as follows:

$$\chi_c^2 = \sum_{i=1}^{c}\left(\frac{(E_i - O_i)^2}{E_i}\right) = \frac{(500-450)^2}{500} + \frac{(500-550)^2}{500} = 10$$

Step 5. Identify the degree of freedom as (c - 1) where c is the number of outcomes being analyzed → Because there are only two outcomes of a head vs. a tail, c = 2 → Thus, the degree of freedom is c – 1 = 2 – 1 = 1.

Step 6. Identify the critical value of the chi-square, $\chi^2_{c-1,\alpha}$, at a 1% significance level from the table → $\chi^2_{table} = \chi^2_{c-1,\alpha} = \chi^2_{1,0.01} = 6.6349$.

Step 7. The decision is:

Because $(\chi_c^2 = 10) > (\chi_{table}^2 = 6.6349)$, **reject** the null hypothesis that the coin is unbiased and fair → That is, the coin is NOT fair and it is biased.

Example 3.

In a car race, some people believe that the start position/order/gate of a car is very important in winning a race. Suppose that you obtained the data on four-car races where the numbers of winning finishes were found as follows:

Gate/Order	Number of Winning Finishes
1	17
2	9
3	14
4	10
Total	50

Given the above information, draw a conclusion at a 5% significance level that the start position (or gate/order) does not matter in which car has the winning finish.

Answer:

Before we conduct any statistical test, it is very important to know what we are to test. That is, we must understand what the null and the alternative hypotheses are saying and thus, know what accepting or rejecting the null hypothesis means. In this regard, the question asked herein can be viewed from many different angles. First, we are interested in knowing how well the observed data fit our expectation where our expectation is that a starting position does not determine the winning outcome and thus, a starting position independent of the winning outcome. Equivalently, we reason that if a starting position does not determine the winning outcome, the probability of winning from any position is the same. That is, the expected probability for each starting position is equal and uniform. Thus, we are once again fitting observed data into a uniform probability distribution in this case.

In summary, the null hypothesis, H_0, can be stated in many different ways as follows:

(1) The starting position does not determine the winning finish.
(2) The starting position is independent of the Winning Outcome.
(3) Any starting position has an equal (thus, uniform) probability of winning.

The alternative hypothesis, H_a, is "The null hypothesis, H_0, is NOT true."

Given this set of hypotheses, we can now follow the statistical testing procedures shown below:

Step 1. Establish the expected probability → That is, the expectation is that every starting position has the same probability of winning the race → 25% each since there are 4 gates (or orders) → A goodness-of-fit test via a uniform probability distribution of 25% for each of 4 outcomes.

Step 2. Calculate the corresponding expected occurrences → That is, 25% of 50 races is 12.5.

Step 3. Organize the information into a contingency table as:

Gate/Order (i)	Expected (E_i)	Observed(O_i)	$(E_i - O_i)^2$	$(E_i - O_i)^2/E_i$
1	12.5	17	20.25	1.62
2	12.5	9	12.25	0.98
3	12.5	14	2.25	0.18
4	12.5	10	6.25	0.5
Total/Sum	50	50		3.28

Step 4. Calculate the Chi-square statistic as follows:

$$\chi_c^2 = \sum_{i=1}^{c}\left(\frac{(E_i - O_i)^2}{E_i}\right) = 3.28$$

Step 5. Identify the degree of freedom as (c - 1) where c is the number of outcomes (or characteristics) being analyzed → **Because there are 4 outcomes or 4 choices of starting position, c = 4** → Thus, the degree of freedom is c – 1 = 4 – 1 = 3.

Step 6. Identify the critical value of the chi-square, $\chi^2_{c-1,\alpha}$, at a 5% significance level from the above Chi-square table → $\chi^2_{table} = \chi^2_{c-1,\alpha} = \chi^2_{3,0.05} = 7.8147$.

Step 7. The decision is:

Because $(\chi_c^2 = 3.28) \leq (\chi^2_{table} = 7.8147)$, then **accept** the null hypothesis that the starting position is independent of winning/losing a race → the starting position does not matter in determining the winner → the starting position does not determine or influence the race outcome → the observed data supports the expected outcome of all positions having the equal probability of winning (or losing, for that matter) → the observed data fits well a uniform probability distribution of 25% per each outcome.

Note: As these different expressions for the same conclusion are possible in a Chi-square test, an attempt to classify if a Chi-square test is for the goodness of fit or independence/dependence is not an important one in practice.

C. Chi-Square Tests for Multiple-Characteristic Tables → (r x c) Contingency Tables

In multiple-characteristic cases, the data is tabulated in the (r x c) contingency-table format. This can be best illustrated by using examples as follows.

1. An Example of a 2x2 Contingency Table

The following information about people watching the Chicago Bears football game on TV was obtained:

	Yes, Watch!	No, Not Watch!	Total
Male	80	40	120
Female	60	20	80
Total	140	60	200

Draw a conclusion if the gender affects football viewing on TV at a 5% significance level.

NOTE:

In this case, the null hypothesis can be stated as:

(1) The gender does NOT affect the football viewing on TV, or equivalently

(2) The football watching is independent of the gender, or equivalently

(3) There is NO difference between the proportion of male and that of female watching football on TV, or equivalently

(4) The proportion of male watching football on TV is the same as that of female watching it at 70%.

Answer:

Step 1. Establish the expected probability of male and female watching football on TV → That is, we can expect that an **equal** percentage of male and female would watch football on TV → If we do not assume an "equal" probability for male and female, we may already have accepted that there is a difference between male and female in TV viewing → i.e., we no longer have the null hypothesis of no gender difference to test.

Therefore, we can now calculate the average expected viewer-proportion as follows:

$$\bar{p} = \frac{All\ watching\ TV}{Total\ surveyed} = \frac{80+60}{200} = \frac{140}{200} = 0.7$$

Step 2. Calculate the corresponding expected occurrences → That is, 70% of males and 70% of females are expected to watch football on TV.

Step 3. Organize the information into a contingency table as:

	Yes, Watch!	No, Not Watch!	Total
Male: Observed	80	40	120
Expected	120 x 0.7 = 84	120 x 0.3 = 36	
Female: Observed	60	20	80
Expected	80 x 0.7 = 56	80 x 0.3 = 24	
Total	140	60	200

Step 4. Calculate the Chi-square statistic as follows:

$$\chi_c^2 = \sum_{i=1}^{c}\left(\frac{(E_i - O_i)^2}{E_i}\right) = \frac{(84-80)^2}{84} + \frac{(56-60)^2}{56} + \frac{(36-40)^2}{36} + \frac{(24-20)^2}{24} = 1.5873$$

Step 5. Identify the degree of freedom as [(c - 1) x (r - 1)] where c and r are the number of characteristics being analyzed → **For simplicity, c is the number of columns and r is the number of rows** → Thus, the degree of freedom is [(c – 1) x (r – 1)] = (2 – 1) x (2 – 1) = 1.

Step 6. Identify the critical value of the chi-square, $\chi^2_{(c-1)(r-1),\alpha}$, at a 5% significance level from the Chi-square table above → $\chi^2_{table} = \chi^2_{(c-1)(r-1),\alpha} = \chi^2_{1,0.05} = 3.8415$.

Step 7. The decision is:
Because $(\chi_c^2 = 1.5873) \le (\chi^2_{table} = 3.8415)$, we **accept** the null hypothesis that the gender does not affect/determine/explain the Chicago Bears football viewing on TV → Equivalently, there is no difference in Bears football viewing between male and female → Also, one can conclude that a statistically equal proportion of 70% of males and 70% of females watch Bears football on TV.

2. An Example of a 3x2 Contingency Table

GSB 420 is a required course for MBA students at DePaul University. The following data was obtained from 400 MBA students at DePaul who had declared one of the 3 concentration areas:

Concentration Field	Love GSB420	Hate GSB420	Total
Marketing	110	40	150
Accounting	150	50	200
Finance	40	10	50
Total	300	100	400

Draw a conclusion if the concentration fields affect the love/hate for GSB 420 at a 5% significance level.

Answer:

The null hypothesis, H_o, can be stated as follows:

(1) The concentration fields do not affect/determine/explain the love/hate of GSB420

(2) There is no difference in loving/hating GSB420 among concentration fields

(3) An equal proportion of MBA students with different concentration fields love/hate GSB420

Of course, the alternative hypothesis is "H_o is NOT true."

Step 1. Establish the expected probability of MBA students loving GSB 420 → That is, the average expected proportion of students loving GSB 420 can be calculated as follows:

$$\bar{p} = \frac{\textit{All students loving GSB420}}{\textit{Total students surveyed}} = \frac{110+150+40}{400} = \frac{300}{400} = 0.75$$

Step 2. Calculate the corresponding expected occurrences → That is, 75% of all MBA students love GSB420 and 25% hate it.

Step 3. Organize the information into a contingency table as:

Concentration Field	Love GSB420	Hate GSB420	Total
Marketing: Observed Expected	 110 150 x 0.75 = 112.5	 40 150 x 0.25 = 37.5	 150
Accounting: Observed Expected	 150 200 x 0.75 = 150	 50 200 x 0.25 = 50	 200
Finance: Observed Expected	 40 50 x 0.75 = 37.5	 10 50 x 0.25 = 12.5	 50
Total	300	100	400

Step 4. Calculate the Chi-square statistic as follows:

$$\chi_c^2 = \sum_{i=1}^{c}\left(\frac{(E_i - O_i)^2}{E_i}\right)$$

$$= \frac{(112.5-110)^2}{112.5} + \frac{(150-150)^2}{150} + \frac{(37.5-40)^2}{37.5}$$

$$+ \frac{(37.5-40)^2}{37.5} + \frac{(50-50)^2}{50} + \frac{(12.5-10)^2}{12.5} = 0.889$$

Step 5. Identify the degree of freedom as [(c - 1) x (r - 1)] where c and r are the number of characteristics being analyzed → **For simplicity, c is the number of columns and r is the number of rows** → Thus, the degree of freedom is [(c – 1) x (r – 1)] = (3 – 1) x (2 – 1) = 2.

Step 6. Identify the critical value of the chi-square, $\chi^2_{(c-1)(r-1),\alpha}$, at a 5% significance level from the table → $\chi^2_{table} = \chi^2_{(c-1)(r-1),\alpha} = \chi^2_{2,0.05} = 5.9915$.

Step 7. The decision is:

Because $(\chi_c^2 = 0.889) \le (\chi^2_{table} = 5.9915)$, we **accept** the null hypothesis that the concentration fields do not affect/determine/explain the love/hate of GSB420 → Equivalently, there is no difference in loving/hating GSB420 among concentration fields → Also, one can conclude that a statistically equal proportion of MBA students with different concentration fields love/hate GSB420 → Thus, one can conclude that about 75% of all students in each of the three concentration fields love GSB420 and 25% hate it.

D. Conditions and Assumptions for the Chi-Square Test

Despite the fact that it is a nonparametric test, the Chi-Square test is a very useful tool for us to know. However, it has certain conditions or assumptions under which its result will be valid. The main ones are as follows:

1. The data must be from an unbiased sample.

2. The data must in the form of raw frequency counts of two or more characteristics or outcomes that are independent, mutually exclusive, and exhaustive → i.e., any observation must fall into only one category or box in a contingency table.

3. To increase the validity of the test, both the observed frequency and the expected frequency in each cell have to be sufficiently large. Therefore, the number of observed data and expected data in each cell of the contingency table should be greater than 5.

Exercise Problems on Chapter 15. The Chi-Square Test

This set of exercise problems has 23 problems, worth a total of 25 points.

1. The Chi-square test used for analyzing a contingency table is a _____ test and often called a _____ test.

   a. one-tail; goodness-of-fit b. two-tail; goodness-of-fit
   c. parametric; goodness-of-fit d. one-tail; parametric
   e. two-tail; nonparametric

2. The Chi-square test is often used to analyze _____ data and thus, often called a _____ test.

   a. categorical; parametric b. numerical; parametric
   c. categorical; nonparametric d. numerical; nonparametric
   e. categorical; nonsignificance

3. The Chi-square test can be considered as a test of ______ for two _____ variables.

   a. independence; parametric b. dependence; parametric
   c. independence; continuous d. independence; categorical
   e. either (a) or (d) of the above

Given the following description, answer Questions 4 – 11.

In your manufacturing firm, you normally expect to see a total of 21 accident-days per 350-day period for each of the morning and the afternoon shifts. However, you have recently noticed that over the past 200 day-period, the morning-shift workers had 20 accident-days whereas the afternoon-shift workers had only 15. Alternatively, it can be tabulated as follows:

	Accident-days	No-Accident-days	Total Work Days
Morning Shift	20	180	200
Afternoon Shift	15	185	200
Total	35	365	400

4. Suppose that you wish to examine if the morning-shift workers are more careless than the afternoon-shift. Your null hypothesis can be stated as _____.

   a. The numbers of accidents and the work-shifts are not related.
   b. The numbers of accidents are independent from the work-shift difference.

c. The numbers of accidents are the same between the morning-shift and the afternoon-shift.
d. any of the above.
e. only (a) and (c) of the above.

5. What is the average expected accident-days for your firm?

a. 0.075 b. 0.1 c. 0.0875
d. 0.06 e. 0.175

6. The expected number of accident-days for the morning shift during the study period is _____.

a. 12 b. 15 c. 20
d. 35 e. unknown

7. The calculated chi-square value is _____.

a. 3.8 b. 0.9167 c. 37.9167
d. 6.4716 e. 0

8. The appropriate degree of freedom for the table Chi-square value is _____.

a. 1 b. 2 c. 3
d. 4 e. unknown

9. The critical (table) value of the Chi-square at a 5% significance level is _____ whereas that at a 1% significance level is _____.

a. 3.8415; 5.0239 b. 3.8415; 6.6349 c. 5.0239; 6.6349
d. 5.0239; 7.8779 e. 5.9915; 9.2103

10. If the calculated Chi-square is 3; and the critical (table) Chi-square at a 5% significance level is 2; the critical (table) Chi-square at a 1% significance level is 4, your conclusion is to _____ the null hypothesis at a 5% significance level and to _____ the null hypothesis at a 1% significance level.

a. reject; reject b. accept; reject c. reject; accept
d. accept; accept e. be indeterminate; be indeterminate .

11. If you are to reject the null hypothesis that the work-shift does not affect/determine/explain the difference in the number of accident-days, what are you really saying?

a. the work-shift is independent from the number of accident-days.
b. the work-shift does affect/determine/explain the difference in number of accident-days.
c. different work-shifts have different number of accident-days.
d. all work-shifts have statistically the same number of accident days.
e. only (b) and (c) of the above.

Given the following description, answer Questions 12 – 20.

The North Korean government, despite its many starving citizens, conducted a nuclear bomb explosion test on October 9, 2006. In light of this situation, the opinion of the international community is split in terms of whether to help starving North Koreans or not.

In order to identify if there is any difference in opinion among countries to help or not to help North Koreans, a survey was conducted in three countries and its results were tabulated as follows. (Note: This is a fictitious, made-up survey.)

Country	Help North Koreans	Don't Help North Koreans
U.S.A.	60	250
South Korea	100	20
Japan	40	30
Total	200	300

12. Which of the following represent(s) a null hypothesis to be tested?

a. There is no difference in opinion among the three countries for helping North Koreans.
b. Regardless of the nationality, the proportion of people who favor helping North Koreans is the same.
c. Nationality does not determine/explain the desire to help (or not to help) North Koreans.
d. any of the above.
e. A testable hypothesis can not be formulated in this case.

13. What is the average expected proportion of people in these three countries who favor helping North Koreans?

a. 0.667 b. 0.6 c. 0.4
d. 0.2 e. 0.1

14. The expected frequency in number of people in Korea who favor helping North Koreans is _____ and that in the USA is _____.

a. 124; 186 b. 124; 48 c. 186; 124

d. 48; 124 e. 100; 60

15. The calculated chi-square value is _____.

a. 157.5 b. 100 c. 58.2
d. 4.78 e. 500

16. The appropriate degree of freedom for the table Chi-square value is _____.

a. 1 b. 2 c. 3
d. 4 e. unknown

17. The critical (table) value of the Chi-square at a 5% significance level is _____ whereas that at a 1% significance level is _____.

a. 3.8415; 6.6349 b. 5.0239; 7.8778 c. 5.9915; 9.2103
d. 7.3778; 10.597 e. 7.8147; 11.3449

18. If the calculated Chi-square is 10 and the critical (table) Chi-square at a chosen significance level is 11, your conclusion is to _____.

a. reject the null hypothesis.
b. accept the null hypothesis.
c. fail to reject the null hypothesis.
d. only (b) and (c) of the above.
e. be indeterminate.

19. If the calculated Chi-square is 12 and the critical (table) Chi-square at a chosen significance level is 11, your conclusion is to _____.

a. reject the null hypothesis.
b. accept the null hypothesis.
c. fail to reject the null hypothesis.
d. only (b) and (c) of the above.
e. be indeterminate.

20. If you are to reject the null hypothesis that the nationality is independent from the difference in opinion to help or not to help North Koreans, what are you really saying?

a. the nationality does affect/determine/explain the difference in opinion.
b. different nations have different opinions.
c. all nations have the same opinion.
d. opinions have no role in nationalistic decisions.
e. only (a) and (b) of the above.

Given the following description, answer Questions 21 – 23.

Suppose the following survey results were found among DePaul students for their preferences for fast-food restaurants.

Restaurant\Sex	Male Students	Female Students
McDonald's	170	140
Wendy's	70	50
Chipotle	40	30
Burger King	50	50
Total	330	270

21. If you wish to know whether the preference for fast-food restaurants is influenced/determined by the sex of students, you would calculate Chi-square value as _____. Assume that the expected mix of male vs. female students to be 50%-50%. (2 points)

a. 2.90 b. 3.83 c. 6.234
d. 4.78 e. 7.66

22. The critical (table) value of the Chi-square at a 5% significance level is _____ whereas that at a 1% significance level is _____. (2 points)

a. 7.3778; 10.597 b. 9.3484; 12.8325 c. 5.9915; 9.2103
d. 7.8147; 11.3449 e. 12.5916; 16.8119

23. If the calculated Chi-square is 2 and the critical (table) Chi-square at a chosen significance level is 1.1, your conclusion is to _____. (1 point)

a. reject the null hypothesis.
b. accept the null hypothesis.
c. the preference for fast-food restaurants is influenced by student sex
d. only (a) and (c) of the above.
e. only (b) and (c) of the above.

Table A-1. (Relative) Binomial Probabilities

		P								
n	X	0.1	0.2	0.3	0.4	0.5	0.6	0.7	0.8	0.9
2	0	0.8100	0.6400	0.4900	0.3600	0.2500	0.1600	0.0900	0.0400	0.0100
	1	0.1800	0.3200	0.4200	0.4800	0.5000	0.4800	0.4200	0.3200	0.1800
	2	0.0100	0.0400	0.0900	0.1600	0.2500	0.3600	0.4900	0.6400	0.8100
3	0	0.7290	0.5120	0.3430	0.2160	0.1250	0.0640	0.0270	0.0080	0.0010
	1	0.2430	0.3840	0.4410	0.4320	0.3750	0.2880	0.1890	0.0960	0.0270
	2	0.0270	0.0960	0.1890	0.2880	0.3750	0.4320	0.4410	0.3840	0.2430
	3	0.0010	0.0080	0.0270	0.0640	0.1250	0.2160	0.3430	0.5120	0.7290
4	0	0.6561	0.4096	0.2401	0.1296	0.0625	0.0256	0.0081	0.0016	0.0001
	1	0.2916	0.4096	0.4116	0.3456	0.2500	0.1536	0.0756	0.0256	0.0036
	2	0.0486	0.1536	0.2646	0.3456	0.3750	0.3456	0.2646	0.1536	0.0486
	3	0.0036	0.0256	0.0756	0.1536	0.2500	0.3456	0.4116	0.4096	0.2916
	4	0.0001	0.0016	0.0081	0.0256	0.0625	0.1296	0.2401	0.4096	0.6561
5	0	0.5905	0.3277	0.1681	0.0778	0.0313	0.0102	0.0024	0.0003	0.0000
	1	0.3281	0.4096	0.3602	0.2592	0.1563	0.0768	0.0284	0.0064	0.0005
	2	0.0729	0.2048	0.3087	0.3456	0.3125	0.2304	0.1323	0.0512	0.0081
	3	0.0081	0.0512	0.1323	0.2304	0.3125	0.3456	0.3087	0.2048	0.0729
	4	0.0005	0.0064	0.0284	0.0768	0.1563	0.2592	0.3602	0.4096	0.3281
	5	0.0000	0.0003	0.0024	0.0102	0.0313	0.0778	0.1681	0.3277	0.5905
6	0	0.5314	0.2621	0.1176	0.0467	0.0156	0.0041	0.0007	0.0001	0.0000
	1	0.3543	0.3932	0.3025	0.1866	0.0938	0.0369	0.0102	0.0015	0.0001
	2	0.0984	0.2458	0.3241	0.3110	0.2344	0.1382	0.0595	0.0154	0.0012
	3	0.0146	0.0819	0.1852	0.2765	0.3125	0.2765	0.1852	0.0819	0.0146
	4	0.0012	0.0154	0.0595	0.1382	0.2344	0.3110	0.3241	0.2458	0.0984
	5	0.0001	0.0015	0.0102	0.0369	0.0938	0.1866	0.3025	0.3932	0.3543
	6	0.0000	0.0001	0.0007	0.0041	0.0156	0.0467	0.1176	0.2621	0.5314
7	0	0.4783	0.2097	0.0824	0.0280	0.0078	0.0016	0.0002	0.0000	0.0000
	1	0.3720	0.3670	0.2471	0.1306	0.0547	0.0172	0.0036	0.0004	0.0000
	2	0.1240	0.2753	0.3177	0.2613	0.1641	0.0774	0.0250	0.0043	0.0002
	3	0.0230	0.1147	0.2269	0.2903	0.2734	0.1935	0.0972	0.0287	0.0026
	4	0.0026	0.0287	0.0972	0.1935	0.2734	0.2903	0.2269	0.1147	0.0230
	5	0.0002	0.0043	0.0250	0.0774	0.1641	0.2613	0.3177	0.2753	0.1240
	6	0.0000	0.0004	0.0036	0.0172	0.0547	0.1306	0.2471	0.3670	0.3720
	7	0.0000	0.0000	0.0002	0.0016	0.0078	0.0280	0.0824	0.2097	0.4783
8	0	0.4305	0.1678	0.0576	0.0168	0.0039	0.0007	0.0001	0.0000	0.0000
	1	0.3826	0.3355	0.1977	0.0896	0.0313	0.0079	0.0012	0.0001	0.0000
	2	0.1488	0.2936	0.2965	0.2090	0.1094	0.0413	0.0100	0.0011	0.0000
	3	0.0331	0.1468	0.2541	0.2787	0.2188	0.1239	0.0467	0.0092	0.0004
	4	0.0046	0.0459	0.1361	0.2322	0.2734	0.2322	0.1361	0.0459	0.0046
	5	0.0004	0.0092	0.0467	0.1239	0.2188	0.2787	0.2541	0.1468	0.0331
	6	0.0000	0.0011	0.0100	0.0413	0.1094	0.2090	0.2965	0.2936	0.1488
	7	0.0000	0.0001	0.0012	0.0079	0.0313	0.0896	0.1977	0.3355	0.3826
	8	0.0000	0.0000	0.0001	0.0007	0.0039	0.0168	0.0576	0.1678	0.4305

Generated by the Excel Spreadsheet command of "=Binomdist(n, X, p, false)".

Table A-2. (Relative) Poisson Probabilities

	Lambda									
X	1	2	3	4	5	6	7	8	9	10
0	0.3679	0.1353	0.0498	0.0183	0.0067	0.0025	0.0009	0.0003	0.0001	0.0000
1	0.3679	0.2707	0.1494	0.0733	0.0337	0.0149	0.0064	0.0027	0.0011	0.0005
2	0.1839	0.2707	0.2240	0.1465	0.0842	0.0446	0.0223	0.0107	0.0050	0.0023
3	0.0613	0.1804	0.2240	0.1954	0.1404	0.0892	0.0521	0.0286	0.0150	0.0076
4	0.0153	0.0902	0.1680	0.1954	0.1755	0.1339	0.0912	0.0573	0.0337	0.0189
5	0.0031	0.0361	0.1008	0.1563	0.1755	0.1606	0.1277	0.0916	0.0607	0.0378
6	0.0005	0.0120	0.0504	0.1042	0.1462	0.1606	0.1490	0.1221	0.0911	0.0631
7	0.0001	0.0034	0.0216	0.0595	0.1044	0.1377	0.1490	0.1396	0.1171	0.0901
8	0.0000	0.0009	0.0081	0.0298	0.0653	0.1033	0.1304	0.1396	0.1318	0.1126
9	0.0000	0.0002	0.0027	0.0132	0.0363	0.0688	0.1014	0.1241	0.1318	0.1251
10	0.0000	0.0000	0.0008	0.0053	0.0181	0.0413	0.0710	0.0993	0.1186	0.1251
11	0.0000	0.0000	0.0002	0.0019	0.0082	0.0225	0.0452	0.0722	0.0970	0.1137
12	0.0000	0.0000	0.0001	0.0006	0.0034	0.0113	0.0263	0.0481	0.0728	0.0948
13	0.0000	0.0000	0.0000	0.0002	0.0013	0.0052	0.0142	0.0296	0.0504	0.0729
14	0.0000	0.0000	0.0000	0.0001	0.0005	0.0022	0.0071	0.0169	0.0324	0.0521
15	0.0000	0.0000	0.0000	0.0000	0.0002	0.0009	0.0033	0.0090	0.0194	0.0347
16	0.0000	0.0000	0.0000	0.0000	0.0000	0.0003	0.0014	0.0045	0.0109	0.0217
17	0.0000	0.0000	0.0000	0.0000	0.0000	0.0001	0.0006	0.0021	0.0058	0.0128
18	0.0000	0.0000	0.0000	0.0000	0.0000	0.0000	0.0002	0.0009	0.0029	0.0071
19	0.0000	0.0000	0.0000	0.0000	0.0000	0.0000	0.0001	0.0004	0.0014	0.0037
20	0.0000	0.0000	0.0000	0.0000	0.0000	0.0000	0.0000	0.0002	0.0006	0.0019
21	0.0000	0.0000	0.0000	0.0000	0.0000	0.0000	0.0000	0.0001	0.0003	0.0009
22	0.0000	0.0000	0.0000	0.0000	0.0000	0.0000	0.0000	0.0000	0.0001	0.0004
23	0.0000	0.0000	0.0000	0.0000	0.0000	0.0000	0.0000	0.0000	0.0000	0.0002
24	0.0000	0.0000	0.0000	0.0000	0.0000	0.0000	0.0000	0.0000	0.0000	0.0001
25	0.0000	0.0000	0.0000	0.0000	0.0000	0.0000	0.0000	0.0000	0.0000	0.0000

Generated by the Excel Spreadsheet command of "=Poisson(X, Lambda, false)".

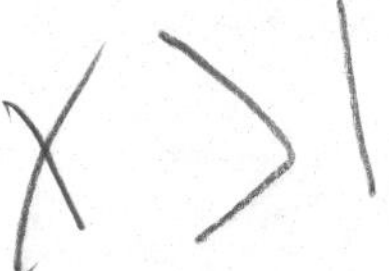

Table A-3a. Cumulative Standard Normal Probability for Positive (Upper-Tail) Z Values

Z	0.00	0.01	0.02	0.03	0.04	0.05	0.06	0.07	0.08	0.09
0.0	0.5000	0.4960	0.4920	0.4880	0.4840	0.4801	0.4761	0.4721	0.4681	0.4641
0.1	0.4602	0.4562	0.4522	0.4483	0.4443	0.4404	0.4364	0.4325	0.4286	0.4247
0.2	0.4207	0.4168	0.4129	0.4090	0.4052	0.4013	0.3974	0.3936	0.3897	0.3859
0.3	0.3821	0.3783	0.3745	0.3707	0.3669	0.3632	0.3594	0.3557	0.3520	0.3483
0.4	0.3446	0.3409	0.3372	0.3336	0.3300	0.3264	0.3228	0.3192	0.3156	0.3121
0.5	0.3085	0.3050	0.3015	0.2981	0.2946	0.2912	0.2877	0.2843	0.2810	0.2776
0.6	0.2743	0.2709	0.2676	0.2643	0.2611	0.2578	0.2546	0.2514	0.2483	0.2451
0.7	0.2420	0.2389	0.2358	0.2327	0.2296	0.2266	0.2236	0.2206	0.2177	0.2148
0.8	0.2119	0.2090	0.2061	0.2033	0.2005	0.1977	0.1949	0.1922	0.1894	0.1867
0.9	0.1841	0.1814	0.1788	0.1762	0.1736	0.1711	0.1685	0.1660	0.1635	0.1611
1.0	0.1587	0.1562	0.1539	0.1515	0.1492	0.1469	0.1446	0.1423	0.1401	0.1379
1.1	0.1357	0.1335	0.1314	0.1292	0.1271	0.1251	0.1230	0.1210	0.1190	0.1170
1.2	0.1151	0.1131	0.1112	0.1093	0.1075	0.1056	0.1038	0.1020	0.1003	0.0985
1.3	0.0968	0.0951	0.0934	0.0918	0.0901	0.0885	0.0869	0.0853	0.0838	0.0823
1.4	0.0808	0.0793	0.0778	0.0764	0.0749	0.0735	0.0721	0.0708	0.0694	0.0681
1.5	0.0668	0.0655	0.0643	0.0630	0.0618	0.0606	0.0594	0.0582	0.0571	0.0559
1.6	0.0548	0.0537	0.0526	0.0516	0.0505	0.0495	0.0485	0.0475	0.0465	0.0455
1.7	0.0446	0.0436	0.0427	0.0418	0.0409	0.0401	0.0392	0.0384	0.0375	0.0367
1.8	0.0359	0.0351	0.0344	0.0336	0.0329	0.0322	0.0314	0.0307	0.0301	0.0294
1.9	0.0287	0.0281	0.0274	0.0268	0.0262	0.0256	0.0250	0.0244	0.0239	0.0233
2.0	0.0228	0.0222	0.0217	0.0212	0.0207	0.0202	0.0197	0.0192	0.0188	0.0183
2.1	0.0179	0.0174	0.0170	0.0166	0.0162	0.0158	0.0154	0.0150	0.0146	0.0143
2.2	0.0139	0.0136	0.0132	0.0129	0.0125	0.0122	0.0119	0.0116	0.0113	0.0110
2.3	0.0107	0.0104	0.0102	0.0099	0.0096	0.0094	0.0091	0.0089	0.0087	0.0084
2.4	0.0082	0.0080	0.0078	0.0075	0.0073	0.0071	0.0069	0.0068	0.0066	0.0064
2.5	0.0062	0.0060	0.0059	0.0057	0.0055	0.0054	0.0052	0.0051	0.0049	0.0048
2.6	0.0047	0.0045	0.0044	0.0043	0.0041	0.0040	0.0039	0.0038	0.0037	0.0036
2.7	0.0035	0.0034	0.0033	0.0032	0.0031	0.0030	0.0029	0.0028	0.0027	0.0026
2.8	0.0026	0.0025	0.0024	0.0023	0.0023	0.0022	0.0021	0.0021	0.0020	0.0019
2.9	0.0019	0.0018	0.0018	0.0017	0.0016	0.0016	0.0015	0.0015	0.0014	0.0014
3.0	0.0013	0.0013	0.0013	0.0012	0.0012	0.0011	0.0011	0.0011	0.0010	0.0010
3.1	0.0010	0.0009	0.0009	0.0009	0.0008	0.0008	0.0008	0.0008	0.0007	0.0007
3.2	0.0007	0.0007	0.0006	0.0006	0.0006	0.0006	0.0006	0.0005	0.0005	0.0005
3.3	0.0005	0.0005	0.0005	0.0004	0.0004	0.0004	0.0004	0.0004	0.0004	0.0003
3.4	0.0003	0.0003	0.0003	0.0003	0.0003	0.0003	0.0003	0.0003	0.0003	0.0002
3.5	0.0002	0.0002	0.0002	0.0002	0.0002	0.0002	0.0002	0.0002	0.0002	0.0002
3.6	0.0002	0.0002	0.0001	0.0001	0.0001	0.0001	0.0001	0.0001	0.0001	0.0001
3.7	0.0001	0.0001	0.0001	0.0001	0.0001	0.0001	0.0001	0.0001	0.0001	0.0001
3.8	0.0001	0.0001	0.0001	0.0001	0.0001	0.0001	0.0001	0.0001	0.0001	0.0001
3.9	0.0000	0.0000	0.0000	0.0000	0.0000	0.0000	0.0000	0.0000	0.0000	0.0000
4.0	0.0000	0.0000	0.0000	0.0000	0.0000	0.0000	0.0000	0.0000	0.0000	0.0000

Generated by the Excel Spreadsheet command of "=Normsdist(Z)".

Table A-3b. Cumulative Standard Normal Probability for Negative (Lower-Tail) Z Values

Z	0.00	0.01	0.02	0.03	0.04	0.05	0.06	0.07	0.08	0.09
0.0	0.5000	0.4960	0.4920	0.4880	0.4840	0.4801	0.4761	0.4721	0.4681	0.4641
-0.1	0.4602	0.4562	0.4522	0.4483	0.4443	0.4404	0.4364	0.4325	0.4286	0.4247
-0.2	0.4207	0.4168	0.4129	0.4090	0.4052	0.4013	0.3974	0.3936	0.3897	0.3859
-0.3	0.3821	0.3783	0.3745	0.3707	0.3669	0.3632	0.3594	0.3557	0.3520	0.3483
-0.4	0.3446	0.3409	0.3372	0.3336	0.3300	0.3264	0.3228	0.3192	0.3156	0.3121
-0.5	0.3085	0.3050	0.3015	0.2981	0.2946	0.2912	0.2877	0.2843	0.2810	0.2776
-0.6	0.2743	0.2709	0.2676	0.2643	0.2611	0.2578	0.2546	0.2514	0.2483	0.2451
-0.7	0.2420	0.2389	0.2358	0.2327	0.2296	0.2266	0.2236	0.2206	0.2177	0.2148
-0.8	0.2119	0.2090	0.2061	0.2033	0.2005	0.1977	0.1949	0.1922	0.1894	0.1867
-0.9	0.1841	0.1814	0.1788	0.1762	0.1736	0.1711	0.1685	0.1660	0.1635	0.1611
-1.0	0.1587	0.1562	0.1539	0.1515	0.1492	0.1469	0.1446	0.1423	0.1401	0.1379
-1.1	0.1357	0.1335	0.1314	0.1292	0.1271	0.1251	0.1230	0.1210	0.1190	0.1170
-1.2	0.1151	0.1131	0.1112	0.1093	0.1075	0.1056	0.1038	0.1020	0.1003	0.0985
-1.3	0.0968	0.0951	0.0934	0.0918	0.0901	0.0885	0.0869	0.0853	0.0838	0.0823
-1.4	0.0808	0.0793	0.0778	0.0764	0.0749	0.0735	0.0721	0.0708	0.0694	0.0681
-1.5	0.0668	0.0655	0.0643	0.0630	0.0618	0.0606	0.0594	0.0582	0.0571	0.0559
-1.6	0.0548	0.0537	0.0526	0.0516	0.0505	0.0495	0.0485	0.0475	0.0465	0.0455
-1.7	0.0446	0.0436	0.0427	0.0418	0.0409	0.0401	0.0392	0.0384	0.0375	0.0367
-1.8	0.0359	0.0351	0.0344	0.0336	0.0329	0.0322	0.0314	0.0307	0.0301	0.0294
-1.9	0.0287	0.0281	0.0274	0.0268	0.0262	0.0256	0.0250	0.0244	0.0239	0.0233
-2.0	0.0228	0.0222	0.0217	0.0212	0.0207	0.0202	0.0197	0.0192	0.0188	0.0183
-2.1	0.0179	0.0174	0.0170	0.0166	0.0162	0.0158	0.0154	0.0150	0.0146	0.0143
-2.2	0.0139	0.0136	0.0132	0.0129	0.0125	0.0122	0.0119	0.0116	0.0113	0.0110
-2.3	0.0107	0.0104	0.0102	0.0099	0.0096	0.0094	0.0091	0.0089	0.0087	0.0084
-2.4	0.0082	0.0080	0.0078	0.0075	0.0073	0.0071	0.0069	0.0068	0.0066	0.0064
-2.5	0.0062	0.0060	0.0059	0.0057	0.0055	0.0054	0.0052	0.0051	0.0049	0.0048
-2.6	0.0047	0.0045	0.0044	0.0043	0.0041	0.0040	0.0039	0.0038	0.0037	0.0036
-2.7	0.0035	0.0034	0.0033	0.0032	0.0031	0.0030	0.0029	0.0028	0.0027	0.0026
-2.8	0.0026	0.0025	0.0024	0.0023	0.0023	0.0022	0.0021	0.0021	0.0020	0.0019
-2.9	0.0019	0.0018	0.0018	0.0017	0.0016	0.0016	0.0015	0.0015	0.0014	0.0014
-3.0	0.0013	0.0013	0.0013	0.0012	0.0012	0.0011	0.0011	0.0011	0.0010	0.0010
-3.1	0.0010	0.0009	0.0009	0.0009	0.0008	0.0008	0.0008	0.0008	0.0007	0.0007
-3.2	0.0007	0.0007	0.0006	0.0006	0.0006	0.0006	0.0006	0.0005	0.0005	0.0005
-3.3	0.0005	0.0005	0.0005	0.0004	0.0004	0.0004	0.0004	0.0004	0.0004	0.0003
-3.4	0.0003	0.0003	0.0003	0.0003	0.0003	0.0003	0.0003	0.0003	0.0003	0.0002
-3.5	0.0002	0.0002	0.0002	0.0002	0.0002	0.0002	0.0002	0.0002	0.0002	0.0002
-3.6	0.0002	0.0002	0.0001	0.0001	0.0001	0.0001	0.0001	0.0001	0.0001	0.0001
-3.7	0.0001	0.0001	0.0001	0.0001	0.0001	0.0001	0.0001	0.0001	0.0001	0.0001
-3.8	0.0001	0.0001	0.0001	0.0001	0.0001	0.0001	0.0001	0.0001	0.0001	0.0001
-3.9	0.0000	0.0000	0.0000	0.0000	0.0000	0.0000	0.0000	0.0000	0.0000	0.0000
-4.0	0.0000	0.0000	0.0000	0.0000	0.0000	0.0000	0.0000	0.0000	0.0000	0.0000

Generated by the Excel Spreadsheet command of "=Normsdist(|Z|)".

Table 4. Critical t-Values for an Upper-Tail Probability

Degrees of Freedom	Upper-Tail Probability							
	0.25	0.2	0.15	0.1	0.05	0.025	0.01	0.005
1	1.0000	1.3764	1.9626	3.0777	6.3138	12.7062	31.8205	63.6567
2	0.8165	1.0607	1.3862	1.8856	2.9200	4.3027	6.9646	9.9248
3	0.7649	0.9785	1.2498	1.6377	2.3534	3.1824	4.5407	5.8409
4	0.7407	0.9410	1.1896	1.5332	2.1318	2.7764	3.7469	4.6041
5	0.7267	0.9195	1.1558	1.4759	2.0150	2.5706	3.3649	4.0321
6	0.7176	0.9057	1.1342	1.4398	1.9432	2.4469	3.1427	3.7074
7	0.7111	0.8960	1.1192	1.4149	1.8946	2.3646	2.9980	3.4995
8	0.7064	0.8889	1.1081	1.3968	1.8595	2.3060	2.8965	3.3554
9	0.7027	0.8834	1.0997	1.3830	1.8331	2.2622	2.8214	3.2498
10	0.6998	0.8791	1.0931	1.3722	1.8125	2.2281	2.7638	3.1693
11	0.6974	0.8755	1.0877	1.3634	1.7959	2.2010	2.7181	3.1058
12	0.6955	0.8726	1.0832	1.3562	1.7823	2.1788	2.6810	3.0545
13	0.6938	0.8702	1.0795	1.3502	1.7709	2.1604	2.6503	3.0123
14	0.6924	0.8681	1.0763	1.3450	1.7613	2.1448	2.6245	2.9768
15	0.6912	0.8662	1.0735	1.3406	1.7531	2.1314	2.6025	2.9467
16	0.6901	0.8647	1.0711	1.3368	1.7459	2.1199	2.5835	2.9208
17	0.6892	0.8633	1.0690	1.3334	1.7396	2.1098	2.5669	2.8982
18	0.6884	0.8620	1.0672	1.3304	1.7341	2.1009	2.5524	2.8784
19	0.6876	0.8610	1.0655	1.3277	1.7291	2.0930	2.5395	2.8609
20	0.6870	0.8600	1.0640	1.3253	1.7247	2.0860	2.5280	2.8453
21	0.6864	0.8591	1.0627	1.3232	1.7207	2.0796	2.5176	2.8314
22	0.6858	0.8583	1.0614	1.3212	1.7171	2.0739	2.5083	2.8188
23	0.6853	0.8575	1.0603	1.3195	1.7139	2.0687	2.4999	2.8073
24	0.6848	0.8569	1.0593	1.3178	1.7109	2.0639	2.4922	2.7969
25	0.6844	0.8562	1.0584	1.3163	1.7081	2.0595	2.4851	2.7874
26	0.6840	0.8557	1.0575	1.3150	1.7056	2.0555	2.4786	2.7787
27	0.6837	0.8551	1.0567	1.3137	1.7033	2.0518	2.4727	2.7707
28	0.6834	0.8546	1.0560	1.3125	1.7011	2.0484	2.4671	2.7633
29	0.6830	0.8542	1.0553	1.3114	1.6991	2.0452	2.4620	2.7564
30	0.6828	0.8538	1.0547	1.3104	1.6973	2.0423	2.4573	2.7500
31	0.6825	0.8534	1.0541	1.3095	1.6955	2.0395	2.4528	2.7440
32	0.6822	0.8530	1.0535	1.3086	1.6939	2.0369	2.4487	2.7385
33	0.6820	0.8526	1.0530	1.3077	1.6924	2.0345	2.4448	2.7333
34	0.6818	0.8523	1.0525	1.3070	1.6909	2.0322	2.4411	2.7284
35	0.6816	0.8520	1.0520	1.3062	1.6896	2.0301	2.4377	2.7238
36	0.6814	0.8517	1.0516	1.3055	1.6883	2.0281	2.4345	2.7195
37	0.6812	0.8514	1.0512	1.3049	1.6871	2.0262	2.4314	2.7154
38	0.6810	0.8512	1.0508	1.3042	1.6860	2.0244	2.4286	2.7116
39	0.6808	0.8509	1.0504	1.3036	1.6849	2.0227	2.4258	2.7079
40	0.6807	0.8507	1.0500	1.3031	1.6839	2.0211	2.4233	2.7045
41	0.6805	0.8505	1.0497	1.3025	1.6829	2.0195	2.4208	2.7012
42	0.6804	0.8503	1.0494	1.3020	1.6820	2.0181	2.4185	2.6981
43	0.6802	0.8501	1.0491	1.3016	1.6811	2.0167	2.4163	2.6951
44	0.6801	0.8499	1.0488	1.3011	1.6802	2.0154	2.4141	2.6923
45	0.6800	0.8497	1.0485	1.3006	1.6794	2.0141	2.4121	2.6896
46	0.6799	0.8495	1.0483	1.3002	1.6787	2.0129	2.4102	2.6870
47	0.6797	0.8493	1.0480	1.2998	1.6779	2.0117	2.4083	2.6846
48	0.6796	0.8492	1.0478	1.2994	1.6772	2.0106	2.4066	2.6822
49	0.6795	0.8490	1.0475	1.2991	1.6766	2.0096	2.4049	2.6800
50	0.6794	0.8489	1.0473	1.2987	1.6759	2.0086	2.4033	2.6778
60	0.6786	0.8477	1.0455	1.2958	1.6706	2.0003	2.3901	2.6603
70	0.6780	0.8468	1.0442	1.2938	1.6669	1.9944	2.3808	2.6479
80	0.6776	0.8461	1.0432	1.2922	1.6641	1.9901	2.3739	2.6387
90	0.6772	0.8456	1.0424	1.2910	1.6620	1.9867	2.3685	2.6316
100	0.6770	0.8452	1.0418	1.2901	1.6602	1.9840	2.3642	2.6259
∞	0.6745	0.8417	1.0365	1.2816	1.6450	1.9602	2.3267	2.5763

Generated by the Excel Spreadsheet command of "=tinv(α, df)".

Table A-5a. Critical Values of F for a Upper-Tail Probability of 0.05 (=5%)

DF	Numerator Degrees of Freedom, n_1															
n_2	**1**	**2**	**3**	**4**	**5**	**6**	**7**	**8**	**9**	**10**	**12**	**15**	**20**	**30**	**40**	**50**
1	161	199	216	225	230	234	237	239	241	242	244	246	248	250	251	252
2	18.51	19.00	19.16	19.25	19.30	19.33	19.35	19.37	19.38	19.40	19.41	19.43	19.45	19.46	19.47	19.48
3	10.13	9.55	9.28	9.12	9.01	8.94	8.89	8.85	8.81	8.79	8.74	8.70	8.66	8.62	8.59	8.58
4	7.71	6.94	6.59	6.39	6.26	6.16	6.09	6.04	6.00	5.96	5.91	5.86	5.80	5.75	5.72	5.70
5	6.61	5.79	5.41	5.19	5.05	4.95	4.88	4.82	4.77	4.74	4.68	4.62	4.56	4.50	4.46	4.44
6	5.99	5.14	4.76	4.53	4.39	4.28	4.21	4.15	4.10	4.06	4.00	3.94	3.87	3.81	3.77	3.75
7	5.59	4.74	4.35	4.12	3.97	3.87	3.79	3.73	3.68	3.64	3.57	3.51	3.44	3.38	3.34	3.32
8	5.32	4.46	4.07	3.84	3.69	3.58	3.50	3.44	3.39	3.35	3.28	3.22	3.15	3.08	3.04	3.02
9	5.12	4.26	3.86	3.63	3.48	3.37	3.29	3.23	3.18	3.14	3.07	3.01	2.94	2.86	2.83	2.80
10	4.96	4.10	3.71	3.48	3.33	3.22	3.14	3.07	3.02	2.98	2.91	2.85	2.77	2.70	2.66	2.64
11	4.84	3.98	3.59	3.36	3.20	3.09	3.01	2.95	2.90	2.85	2.79	2.72	2.65	2.57	2.53	2.51
12	4.75	3.89	3.49	3.26	3.11	3.00	2.91	2.85	2.80	2.75	2.69	2.62	2.54	2.47	2.43	2.40
13	4.67	3.81	3.41	3.18	3.03	2.92	2.83	2.77	2.71	2.67	2.60	2.53	2.46	2.38	2.34	2.31
14	4.60	3.74	3.34	3.11	2.96	2.85	2.76	2.70	2.65	2.60	2.53	2.46	2.39	2.31	2.27	2.24
15	4.54	3.68	3.29	3.06	2.90	2.79	2.71	2.64	2.59	2.54	2.48	2.40	2.33	2.25	2.20	2.18
16	4.49	3.63	3.24	3.01	2.85	2.74	2.66	2.59	2.54	2.49	2.42	2.35	2.28	2.19	2.15	2.12
17	4.45	3.59	3.20	2.96	2.81	2.70	2.61	2.55	2.49	2.45	2.38	2.31	2.23	2.15	2.10	2.08
18	4.41	3.55	3.16	2.93	2.77	2.66	2.58	2.51	2.46	2.41	2.34	2.27	2.19	2.11	2.06	2.04
19	4.38	3.52	3.13	2.90	2.74	2.63	2.54	2.48	2.42	2.38	2.31	2.23	2.16	2.07	2.03	2.00
20	4.35	3.49	3.10	2.87	2.71	2.60	2.51	2.45	2.39	2.35	2.28	2.20	2.12	2.04	1.99	1.97
25	4.24	3.39	2.99	2.76	2.60	2.49	2.40	2.34	2.28	2.24	2.16	2.09	2.01	1.92	1.87	1.84
30	4.17	3.32	2.92	2.69	2.53	2.42	2.33	2.27	2.21	2.16	2.09	2.01	1.93	1.84	1.79	1.76
35	4.12	3.27	2.87	2.64	2.49	2.37	2.29	2.22	2.16	2.11	2.04	1.96	1.88	1.79	1.74	1.70
40	4.08	3.23	2.84	2.61	2.45	2.34	2.25	2.18	2.12	2.08	2.00	1.92	1.84	1.74	1.69	1.66
45	4.06	3.20	2.81	2.58	2.42	2.31	2.22	2.15	2.10	2.05	1.97	1.89	1.81	1.71	1.66	1.63
50	4.03	3.18	2.79	2.56	2.40	2.29	2.20	2.13	2.07	2.03	1.95	1.87	1.78	1.69	1.63	1.60

Generated by the Excel Spreadsheet command of "=Finv(α, n_1, n_2)".

Table A-5b. Critical Values of F for a Upper-Tail Probability of 0.025 (=2.5%)

DF	Numerator Degrees of Freedom, n_1														
n_2	1	2	3	4	5	6	7	8	9	10	12	15	20	30	40	50
1	648	799	864	900	922	937	948	957	963	969	977	985	993	1001	1006	1008
2	38.51	39.00	39.17	39.25	39.30	39.33	39.36	39.37	39.39	39.40	39.41	39.43	39.45	39.46	39.47	39.48
3	17.44	16.04	15.44	15.10	14.88	14.73	14.62	14.54	14.47	14.42	14.34	14.25	14.17	14.08	14.04	14.01
4	12.22	10.65	9.98	9.60	9.36	9.20	9.07	8.98	8.90	8.84	8.75	8.66	8.56	8.46	8.41	8.38
5	10.01	8.43	7.76	7.39	7.15	6.98	6.85	6.76	6.68	6.62	6.52	6.43	6.33	6.23	6.18	6.14
6	8.81	7.26	6.60	6.23	5.99	5.82	5.70	5.60	5.52	5.46	5.37	5.27	5.17	5.07	5.01	4.98
7	8.07	6.54	5.89	5.52	5.29	5.12	4.99	4.90	4.82	4.76	4.67	4.57	4.47	4.36	4.31	4.28
8	7.57	6.06	5.42	5.05	4.82	4.65	4.53	4.43	4.36	4.30	4.20	4.10	4.00	3.89	3.84	3.81
9	7.21	5.71	5.08	4.72	4.48	4.32	4.20	4.10	4.03	3.96	3.87	3.77	3.67	3.56	3.51	3.47
10	6.94	5.46	4.83	4.47	4.24	4.07	3.95	3.85	3.78	3.72	3.62	3.52	3.42	3.31	3.26	3.22
11	6.72	5.26	4.63	4.28	4.04	3.88	3.76	3.66	3.59	3.53	3.43	3.33	3.23	3.12	3.06	3.03
12	6.55	5.10	4.47	4.12	3.89	3.73	3.61	3.51	3.44	3.37	3.28	3.18	3.07	2.96	2.91	2.87
13	6.41	4.97	4.35	4.00	3.77	3.60	3.48	3.39	3.31	3.25	3.15	3.05	2.95	2.84	2.78	2.74
14	6.30	4.86	4.24	3.89	3.66	3.50	3.38	3.29	3.21	3.15	3.05	2.95	2.84	2.73	2.67	2.64
15	6.20	4.77	4.15	3.80	3.58	3.41	3.29	3.20	3.12	3.06	2.96	2.86	2.76	2.64	2.59	2.55
16	6.12	4.69	4.08	3.73	3.50	3.34	3.22	3.12	3.05	2.99	2.89	2.79	2.68	2.57	2.51	2.47
17	6.04	4.62	4.01	3.66	3.44	3.28	3.16	3.06	2.98	2.92	2.82	2.72	2.62	2.50	2.44	2.41
18	5.98	4.56	3.95	3.61	3.38	3.22	3.10	3.01	2.93	2.87	2.77	2.67	2.56	2.44	2.38	2.35
19	5.92	4.51	3.90	3.56	3.33	3.17	3.05	2.96	2.88	2.82	2.72	2.62	2.51	2.39	2.33	2.30
20	5.87	4.46	3.86	3.51	3.29	3.13	3.01	2.91	2.84	2.77	2.68	2.57	2.46	2.35	2.29	2.25
25	5.69	4.29	3.69	3.35	3.13	2.97	2.85	2.75	2.68	2.61	2.51	2.41	2.30	2.18	2.12	2.08
30	5.57	4.18	3.59	3.25	3.03	2.87	2.75	2.65	2.57	2.51	2.41	2.31	2.20	2.07	2.01	1.97
35	5.48	4.11	3.52	3.18	2.96	2.80	2.68	2.58	2.50	2.44	2.34	2.23	2.12	2.00	1.93	1.89
40	5.42	4.05	3.46	3.13	2.90	2.74	2.62	2.53	2.45	2.39	2.29	2.18	2.07	1.94	1.88	1.83
45	5.38	4.01	3.42	3.09	2.86	2.70	2.58	2.49	2.41	2.35	2.25	2.14	2.03	1.90	1.83	1.79
50	5.34	3.97	3.39	3.05	2.83	2.67	2.55	2.46	2.38	2.32	2.22	2.11	1.99	1.87	1.80	1.75

Generated by the Excel Spreadsheet command of "=Finv(α, n_1, n_2)".

Table A-5c. Critical Values of F for a Upper-Tail Probability of 0.01 (=1%)

DF	Numerator Degrees of Freedom, n_1														
n_2	1	2	3	4	5	6	7	8	9	10	12	15	20	30	40	50
1	4052	4999	5403	5625	5764	5859	5928	5981	6022	6056	6106	6157	6209	6261	6287	6303
2	98.50	99.00	99.17	99.25	99.30	99.33	99.36	99.37	99.39	99.40	99.42	99.43	99.45	99.47	99.47	99.48
3	34.12	30.82	29.46	28.71	28.24	27.91	27.67	27.49	27.35	27.23	27.05	26.87	26.69	26.50	26.41	26.35
4	21.20	18.00	16.69	15.98	15.52	15.21	14.98	14.80	14.66	14.55	14.37	14.20	14.02	13.84	13.75	13.69
5	16.26	13.27	12.06	11.39	10.97	10.67	10.46	10.29	10.16	10.05	9.89	9.72	9.55	9.38	9.29	9.24
6	13.75	10.92	9.78	9.15	8.75	8.47	8.26	8.10	7.98	7.87	7.72	7.56	7.40	7.23	7.14	7.09
7	12.25	9.55	8.45	7.85	7.46	7.19	6.99	6.84	6.72	6.62	6.47	6.31	6.16	5.99	5.91	5.86
8	11.26	8.65	7.59	7.01	6.63	6.37	6.18	6.03	5.91	5.81	5.67	5.52	5.36	5.20	5.12	5.07
9	10.56	8.02	6.99	6.42	6.06	5.80	5.61	5.47	5.35	5.26	5.11	4.96	4.81	4.65	4.57	4.52
10	10.04	7.56	6.55	5.99	5.64	5.39	5.20	5.06	4.94	4.85	4.71	4.56	4.41	4.25	4.17	4.12
11	9.65	7.21	6.22	5.67	5.32	5.07	4.89	4.74	4.63	4.54	4.40	4.25	4.10	3.94	3.86	3.81
12	9.33	6.93	5.95	5.41	5.06	4.82	4.64	4.50	4.39	4.30	4.16	4.01	3.86	3.70	3.62	3.57
13	9.07	6.70	5.74	5.21	4.86	4.62	4.44	4.30	4.19	4.10	3.96	3.82	3.66	3.51	3.43	3.38
14	8.86	6.51	5.56	5.04	4.69	4.46	4.28	4.14	4.03	3.94	3.80	3.66	3.51	3.35	3.27	3.22
15	8.68	6.36	5.42	4.89	4.56	4.32	4.14	4.00	3.89	3.80	3.67	3.52	3.37	3.21	3.13	3.08
16	8.53	6.23	5.29	4.77	4.44	4.20	4.03	3.89	3.78	3.69	3.55	3.41	3.26	3.10	3.02	2.97
17	8.40	6.11	5.18	4.67	4.34	4.10	3.93	3.79	3.68	3.59	3.46	3.31	3.16	3.00	2.92	2.87
18	8.29	6.01	5.09	4.58	4.25	4.01	3.84	3.71	3.60	3.51	3.37	3.23	3.08	2.92	2.84	2.78
19	8.18	5.93	5.01	4.50	4.17	3.94	3.77	3.63	3.52	3.43	3.30	3.15	3.00	2.84	2.76	2.71
20	8.10	5.85	4.94	4.43	4.10	3.87	3.70	3.56	3.46	3.37	3.23	3.09	2.94	2.78	2.69	2.64
25	7.77	5.57	4.68	4.18	3.85	3.63	3.46	3.32	3.22	3.13	2.99	2.85	2.70	2.54	2.45	2.40
30	7.56	5.39	4.51	4.02	3.70	3.47	3.30	3.17	3.07	2.98	2.84	2.70	2.55	2.39	2.30	2.25
35	7.42	5.27	4.40	3.91	3.59	3.37	3.20	3.07	2.96	2.88	2.74	2.60	2.44	2.28	2.19	2.14
40	7.31	5.18	4.31	3.83	3.51	3.29	3.12	2.99	2.89	2.80	2.66	2.52	2.37	2.20	2.11	2.06
45	7.23	5.11	4.25	3.77	3.45	3.23	3.07	2.94	2.83	2.74	2.61	2.46	2.31	2.14	2.05	2.00
50	7.17	5.06	4.20	3.72	3.41	3.19	3.02	2.89	2.78	2.70	2.56	2.42	2.27	2.10	2.01	1.95

Generated by the Excel Spreadsheet command of "=Finv(α, n_1, n_2)".

Table A-6. Critical Chi-Square Values for an Upper-Tail Significance Level

DF	Upper-Tail Significance Level							
	0.99	0.975	0.95	0.9	0.1	0.05	0.025	0.01
1	0.0002	0.0010	0.0039	0.0158	2.7055	3.8415	5.0239	6.6349
2	0.0201	0.0506	0.1026	0.2107	4.6052	5.9915	7.3778	9.2103
3	0.1148	0.2158	0.3518	0.5844	6.2514	7.8147	9.3484	11.3449
4	0.2971	0.4844	0.7107	1.0636	7.7794	9.4877	11.1433	13.2767
5	0.5543	0.8312	1.1455	1.6103	9.2364	11.0705	12.8325	15.0863
6	0.8721	1.2373	1.6354	2.2041	10.6446	12.5916	14.4494	16.8119
7	1.2390	1.6899	2.1673	2.8331	12.0170	14.0671	16.0128	18.4753
8	1.6465	2.1797	2.7326	3.4895	13.3616	15.5073	17.5345	20.0902
9	2.0879	2.7004	3.3251	4.1682	14.6837	16.9190	19.0228	21.6660
10	2.5582	3.2470	3.9403	4.8652	15.9872	18.3070	20.4832	23.2093
11	3.0535	3.8157	4.5748	5.5778	17.2750	19.6751	21.9200	24.7250
12	3.5706	4.4038	5.2260	6.3038	18.5493	21.0261	23.3367	26.2170
13	4.1069	5.0088	5.8919	7.0415	19.8119	22.3620	24.7356	27.6882
14	4.6604	5.6287	6.5706	7.7895	21.0641	23.6848	26.1189	29.1412
15	5.2293	6.2621	7.2609	8.5468	22.3071	24.9958	27.4884	30.5779
16	5.8122	6.9077	7.9616	9.3122	23.5418	26.2962	28.8454	31.9999
17	6.4078	7.5642	8.6718	10.0852	24.7690	27.5871	30.1910	33.4087
18	7.0149	8.2307	9.3905	10.8649	25.9894	28.8693	31.5264	34.8053
19	7.6327	8.9065	10.1170	11.6509	27.2036	30.1435	32.8523	36.1909
20	8.2604	9.5908	10.8508	12.4426	28.4120	31.4104	34.1696	37.5662
21	8.8972	10.2829	11.5913	13.2396	29.6151	32.6706	35.4789	38.9322
22	9.5425	10.9823	12.3380	14.0415	30.8133	33.9244	36.7807	40.2894
23	10.1957	11.6886	13.0905	14.8480	32.0069	35.1725	38.0756	41.6384
24	10.8564	12.4012	13.8484	15.6587	33.1962	36.4150	39.3641	42.9798
25	11.5240	13.1197	14.6114	16.4734	34.3816	37.6525	40.6465	44.3141
26	12.1981	13.8439	15.3792	17.2919	35.5632	38.8851	41.9232	45.6417
27	12.8785	14.5734	16.1514	18.1139	36.7412	40.1133	43.1945	46.9629
28	13.5647	15.3079	16.9279	18.9392	37.9159	41.3371	44.4608	48.2782
29	14.2565	16.0471	17.7084	19.7677	39.0875	42.5570	45.7223	49.5879
30	14.9535	16.7908	18.4927	20.5992	40.2560	43.7730	46.9792	50.8922
40	22.1643	24.4330	26.5093	29.0505	51.8051	55.7585	59.3417	63.6907
50	29.7067	32.3574	34.7643	37.6886	63.1671	67.5048	71.4202	76.1539
60	37.4849	40.4817	43.1880	46.4589	74.3970	79.0819	83.2977	88.3794
70	45.4417	48.7576	51.7393	55.3289	85.5270	90.5312	95.0232	100.4252
80	53.5401	57.1532	60.3915	64.2778	96.5782	101.8795	106.6286	112.3288
90	61.7541	65.6466	69.1260	73.2911	107.5650	113.1453	118.1359	124.1163
100	70.0649	74.2219	77.9295	82.3581	118.4980	124.3421	129.5612	135.8067

Generated by the Excel Spreadsheet command of "=Chiinv(α, df)".

Subject Index

CPSIA information can be obtained at www.ICGtesting.com
Printed in the USA
LVOW03s1510271214

420565LV00013B/364/P

9 781609 278724